COMPLEX ANALYSIS IN
BANACH SPACES

NORTH-HOLLAND MATHEMATICS STUDIES 120
Notas de Matemática (107)

NORTH-HOLLAND – AMSTERDAM • NEW YORK • OXFORD

COMPLEX ANALYSIS IN BANACH SPACES

Holomorphic Functions and Domains of Holomorphy in Finite and Infinite Dimensions

Jorge MUJICA

Universidade Estadual de Campinas
Campinas, Brazil

1986

NORTH-HOLLAND – AMSTERDAM • NEW YORK • OXFORD

ISBN: 0 444 87886 6

Publishers:

ELSEVIER SCIENCE PUBLISHERS B.V.
P.O. Box 1991
1000 BZ Amsterdam
The Netherlands

Sole distributors for the U.S.A. and Canada:

ELSEVIER SCIENCE PUBLISHING COMPANY, INC.
52 Vanderbilt Avenue
New York, N.Y. 10017
U.S.A.

Library of Congress Cataloging-in-Publication Data

Mujica, Jorge, 1946-
Complex analysis in Banach spaces.

(North-Holland mathematics studies ; 120) (Notas de matemática ; 107)
Bibliography: p.
Includes index.
1. Holomorphic functions. 2. Domains of holomorphy.
3. Banach spaces. I. Title. II. Series. III. Series:
Notas de matemática (Rio de Janeiro, Brazil) ; no. 107.
QA1.N86 no.107 [QA331] 510 s [515.9'8] 85-20922
ISBN 0-444-87886-6 (U.S.)

PRINTED IN THE NETHERLANDS

To my teacher,

Leopoldo Nachbin

FOREWORD

Problems arising from the study of holomorphic continuation and holomorphic approximation have been central in the development of complex analysis in finitely many variables, and constitute one of the most promising lines of current research in infinite dimensional complex analysis. This book is designed to present a unified view of these topics in both finite and infinite dimensions.

The contents of this book fall naturally into four parts. The first, comprising Chapters I through III, presents the basic properties of holomorphic mappings and domains of holomorphy in Banach spaces. The second part, comprising Chapters IV through VII, begins with the study of differentiable mappings, differential forms and the $\overline{\partial}$ operator in Banach spaces. Polynomially convex compact sets are investigated in detail, and some of the results obtained are applied to the study of Banach and Fréchet algebras.

The third part, comprising Chapters VIII through X, is devoted to the study of plurisubharmonic functions and pseudoconvex domains in Banach spaces. The identity of pseudoconvex domains and domains of holomorphy is established in the case of separable Banach spaces with the bounded approximation property. These results are extended to Riemann domains in the fourth part, in which envelopes of holomorphy are also studied in detail.

The present text evolved from a course taught at the Universidade Estadual de Campinas, Brazil, in 1982. It presupposes familiarity with the theory of Lebesgue integration, with the basic properties of holomorphic functions of a single variable, and with the basic principles of Banach and Hilbert spaces. Topics such as vector-valued integration, Schauder bases and the approximation property, are presented in detail in the book.

The presentation here has been affected by conversations and correspondence with several friends and colleagues, who were not always aware that they were speaking for posterity. In particular I would like to mention Richard Aron, Klaus-Dieter Bierstedt, Roberto Cignoli, Jean-François Colombeau, Seán Dineen, Dicesar Fernández, Klaus Floret, José M. Isidro, Mário Matos, Reinhold Meise and Martin Schottenloher.

One person has had more influence on this book than anybody else, long before this project was even conceived. Not only for accepting this text in his series Notas de Matemática, but mainly for having led me into research on this beautiful subject. I here acknowledge my greatest debt to Leopoldo Nachbin.

I am particularly grateful to my wife and children, Ana María, Ximena and Felipe, for their support and encouragement while I was writting the book, and for maintaining at home an atmosphere ideal for study and research.

Finally, I am very pleased to thank Miss Elda Mortari for her excellent typing of the manuscript.

Jorge Mujica

Campinas, June 1985

CONTENTS

CHAPTER I

POLYNOMIALS

1. MULTILINEAR MAPPINGS

This section is devoted to the study of multilinear mappings in Banach spaces. Besides their intrinsic interest, multilinear mappings will serve a twofold purpose. Whereas symmetric multilinear mappings will be helpful in the study of polynomials, alternating multilinear mappings will be used to introduce differential forms.

To begin with, we establish some notation. Throughout the whole book the letter $\mathbb{K}$ will stand either for the field $\mathbb{R}$ or all real numbers or for the field $\mathbb{C}$ of all complex numbers. The set of all strictly positive integers will be denoted by $\mathbb{N}$, whereas the set $\mathbb{N} \cup \{0\}$ will be denoted by $\mathbb{N}_o$. Unless stated otherwise, the letters E and F will always represent Banach spaces over the same field $\mathbb{K}$.

1.1. DEFINITION. For each $m \in \mathbb{N}$ we shall denote by $\mathcal{L}_a({}^mE;F)$ the vector space of all m-linear mappings $A : E^m \to F$, whereas we shall denote by $\mathcal{L}({}^mE;F)$ the subspace of all continuous members of $\mathcal{L}_a({}^mE;F)$. For each $A \in \mathcal{L}_a({}^mE;F)$ we define

$$\|A\| = sup\,\{\|A(x_1,\ldots,x_m)\| \;:\; x_j \in E,\; \max_j \|x_j\| \leq 1\}.$$

When $m = 1$ then as usual we shall write $\mathcal{L}_a({}^1E;F) = \mathcal{L}_a(E;F)$ and $\mathcal{L}({}^1E;F) = \mathcal{L}(E;F)$. When $F = \mathbb{K}$ then for short we shall write $\mathcal{L}_a({}^mE;\mathbb{K}) = \mathcal{L}_a({}^mE)$ and $\mathcal{L}({}^mE;\mathbb{K}) = \mathcal{L}({}^mE)$. Finally when $m = 1$ and $F = \mathbb{K}$ then as usual we shall write $\mathcal{L}_a(E) = E^*$ and $\mathcal{L}(E) = E'$.

1.2. PROPOSITION. *For each* $A \in \mathcal{L}_a(^mE;F)$ *the following conditions are equivalent:*

(a) A *is continuous.*

(b) A *is continuous at the origin.*

(c) $\|A\| < \infty$.

PROOF. The implication (a) ⇒ (b) is obvious.

(b) ⇒ (c): If (c) is not true then we can find a sequence of points $(x_1^k,\ldots,x_m^k)$ in E^m such that $\max_j \|x_j^k\| \leq 1$ and $\|A(x_1^k,\ldots,x_m^k)\| \geq k^m$ for every k. Hence

$$\max_j \left\| \frac{x_j^k}{k} \right\| \leq \frac{1}{k}$$

and

$$\left\| A\left(\frac{x_1^k}{k},\ldots,\frac{x_m^k}{k}\right)\right\| \geq 1$$

for every k, contradicting (b).

(c) ⇒ (a): Let $a = (a_1,\ldots,a_m) \in E^m$ and $x = (x_1,\ldots,x_m) \in E^m$ with $\max_j \|a_j\| \leq c$ and $\max_j \|x_j\| \leq c$. Then

$$\|A(x_1,\ldots,x_m) - A(a_1,\ldots,a_m)\|$$

$$= \left\| \sum_{j=1}^m [A(a_1,\ldots,a_{j-1},x_j,\ldots,x_m) - A(a_1,\ldots,a_j,x_{j+1},\ldots,x_m)] \right\|$$

$$\leq \sum_{j=1}^m \|A(a_1,\ldots,a_{j-1},x_j - a_j,a_{j+1},\ldots,a_m)\|$$

$$\leq \sum_{j=1}^m \|A\| c^{m-1} \|x_j - a_j\|$$

and (a) follows.

1.3. PROPOSITION. $\mathcal{L}(^mE;F)$ *is a Banach space under the norm* $A \to \|A\|$.

PROOF. One can readily see that the mapping $A \to \|A\|$ defines a norm on $\mathcal{L}(^mE;F)$. To establish completeness let (A_j) be a Cauchy sequence in $\mathcal{L}(^mE;F)$. Then for each $(x_1,\ldots,x_m)$ in E^m we have that

$$(1.1) \qquad \|A_j(x_1,\ldots,x_m) - A_k(x_1,\ldots,x_m)\| \leq \|A_j - A_k\| \, \|x_1\| \ldots \|x_m\|$$

and it follows that $(A_j(x_1,\ldots,x_m))$ is a Cauchy sequence in F. Since F is complete, the limit

$$(1.2) \qquad A(x_1,\ldots,x_m) = \lim A_j(x_1,\ldots,x_m)$$

exists. One can readily see that the mapping $A : E^m \to F$ thus defined is m-linear. Furthermore, since (A_j) is a Cauchy sequence in $\mathcal{L}(^mE;F)$ there is a constant $c > 0$ such that $\|A_j\| \leq c$ for every j. Then it follows from (1.2) that $\|A\| \leq c$ too, and A is therefore continuous, by Proposition 1.2. Finally, if follows readily from (1.1) that $\|A_j - A\| \to 0$ when $j \to \infty$.

1.4. PROPOSITION. *There is a canonical vector space isomorphism between* $\mathcal{L}_a(^{m+n}E;F)$ *and* $\mathcal{L}_a(^mE;\mathcal{L}_a(^nE;F))$. *This isomorphism induces an isometry between* $\mathcal{L}(^{m+n}E;F)$ *and* $\mathcal{L}(^mE;\mathcal{L}(^nE;F))$.

PROOF. One can readily verify that the mapping

$$A \in \mathcal{L}_a(^{m+n}E;F) \to \tilde{A} \in \mathcal{L}_a(^mE;\mathcal{L}_a(^nE;F))$$

defined by

$$\tilde{A}(x_1,\ldots,x_m)(y_1,\ldots,y_n) = A(x_1,\ldots,x_m,y_1,\ldots,y_n)$$

has the required properties.

For each $m \in \mathbb{N}$ we shall denote by S_m the group of all permutations of m elements. If $\sigma \in S_m$ then $(-1)^\sigma$ will denote the sign of the permutation σ.

1.5. DEFINITION. For each $m \in \mathbb{N}$ we shall denote by $\mathcal{L}_a^s({}^mE;F)$ the subspace of all $A \in \mathcal{L}_a({}^mE;F)$ which are *symmetric*, that is, such that

$$A(x_{\sigma(1)},\ldots,x_{\sigma(m)}) = A(x_1,\ldots,x_m)$$

for all $x_1,\ldots,x_m \in E$ and $\sigma \in S_m$. Likewise, we shall denote by $\mathcal{L}_a^a({}^mE;F)$ the subspace of all $A \in \mathcal{L}_a({}^mE;F)$ which are *alternating* or *antisymmetric*, that is, such that

$$A(x_{\sigma(1)},\ldots,x_{\sigma(m)}) = (-1)^{\sigma} A(x_1,\ldots,x_m)$$

for all $x_1,\ldots,x_m \in E$ and $\sigma \in S_m$. The spaces $\mathcal{L}^s({}^mE;F)$ and $\mathcal{L}^a({}^mE;F)$ are defined in the obvious way, that is

$$\mathcal{L}^s({}^mE;F) = \mathcal{L}_a^s({}^mE;F) \cap \mathcal{L}({}^mE;F)$$

and

$$\mathcal{L}^a({}^mE;F) = \mathcal{L}_a^a({}^mE;F) \cap \mathcal{L}({}^mE;F).$$

When $F = \mathbb{K}$ then we shall write $\mathcal{L}_a^s({}^mE;\mathbb{K}) = \mathcal{L}_a^s({}^mE)$, $\mathcal{L}_a^a({}^mE;\mathbb{K}) = \mathcal{L}_a^a({}^mE)$, etc.

1.6. PROPOSITION. *For each* $A \in \mathcal{L}_a({}^mE;F)$ *let* A^s *and* A^a *be defined by*

$$A^s(x_1,\ldots,x_m) = \frac{1}{m!} \sum_{\sigma \in S_m} A(x_{\sigma(1)},\ldots,x_{\sigma(m)})$$

and

$$A^a(x_1,\ldots,x_m) = \frac{1}{m!} \sum_{\sigma \in S_m} (-1)^{\sigma} A(x_{\sigma(1)},\ldots,x_{\sigma(m)}).$$

Then:

(a) *The mapping* $A \to A^s$ *is a projection from* $\mathcal{L}_a({}^mE;F)$ *onto* $\mathcal{L}_a^s({}^mE;F)$ *with* $\|A^s\| \le \|A\|$ *for every* $A \in \mathcal{L}_a({}^mE;F)$. *This mapping induces a continuous projection from* $\mathcal{L}({}^mE;F)$ *onto* $\mathcal{L}^s({}^mE;F)$.

(b) *The mapping* $A \to A^a$ *is a projection from* $\mathcal{L}_a({}^mE;F)$

onto $\mathcal{L}_a^a(^mE;F)$ *with* $\|A^a\| \leq \|A\|$ *for every* $A \in \mathcal{L}_a(^mE;F)$. *This mapping induces a continuous projection from* $\mathcal{L}(^mE;F)$ *onto* $\mathcal{L}^a(^mE;F)$.

The proof of this proposition is straightforward and is left as an exercise to the reader.

For convenience we also define, for $m = 0$, the spaces

$$\mathcal{L}_a(^0E;F) = \mathcal{L}_a^s(^0E;F) = \mathcal{L}_a^a(^0E;F)$$

$$= \mathcal{L}(^0E;F) = \mathcal{L}^s(^0E;F) = \mathcal{L}^a(^0E;F) = F$$

as Banach spaces.

For each $n \in \mathbb{N}$ and each multi-index $\alpha = (\alpha_1,\ldots,\alpha_n) \in \mathbb{N}_0^n$ we set

$$|\alpha| = \alpha_1 + \ldots + \alpha_n, \qquad \alpha! = \alpha_1! \ldots \alpha_n!$$

1.7. DEFINITION. Let $A \in \mathcal{L}_a(^mE;F)$. Then for each $(x_1, \ldots, x_n) \in E^n$ and each $\alpha = (\alpha_1,\ldots,\alpha_n) \in \mathbb{N}_0^n$ with $|\alpha| = m$ we define

$$Ax_1^{\alpha_1} \ldots x_n^{\alpha_n} = A(\underbrace{x_1, \ldots, x_1}_{\alpha_1}, \ldots, \underbrace{x_n, \ldots, x_n}_{\alpha_n})$$

if $m \geq 1$ and $Ax_1^{\alpha_1} \ldots x_n^{\alpha_n} = A$ if $m = 0$.

1.8. THEOREM. *Let* $A \in \mathcal{L}_a^s(^mE;F)$. *Then for all* $x_1,\ldots,x_n \in E$ *we have the Leibniz Formula*

$$A(x_1 + \ldots + x_n)^m = \Sigma \frac{m!}{\alpha!} Ax_1^{\alpha_1} \ldots x_n^{\alpha_n}$$

where the summation is taken over all multi-indices $\alpha = (\alpha_1,\ldots,\alpha_n) \in \mathbb{N}_0^n$ *such that* $|\alpha| = m$.

PROOF. By induction on m. The result is obvious for $m = 0$ and $m = 1$. Assuming the formula valid for a certain $m \geq 1$ one can readily establish it for $m+1$. Indeed, if $A \in \mathcal{L}_a^s({}^{m+1}E;F)$ then one can write

$$A(x_1 + \dots + x_n)^{m+1} = A(x_1 + \dots + x_n)(x_1 + \dots + x_n)^m$$

and apply the induction hypothesis to the mapping $A(x_1 + \dots + x_n)$, which belongs to $\mathcal{L}_a^s({}^mE;F)$. The details are left to the reader.

1.9. COROLLARY. *Let $A \in \mathcal{L}_a^s({}^mE;F)$. Then for all $x, y \in E$ we have the Newton Binomial Formula*

$$A(x+y)^m = \sum_{j=0}^{m} \binom{m}{j} Ax^{m-j}y^j .$$

1.10. THEOREM. *Let $A \in \mathcal{L}_a^s({}^mE;F)$. Then for all $x_o, \dots, x_m \in E$ we have the Polarization Formula*

$$A(x_1, \dots, x_m) = \frac{1}{m!\,2^m} \sum_{\varepsilon_j = \pm 1} \varepsilon_1 \dots \varepsilon_m A(x_o + \varepsilon_1 x_1 + \dots + \varepsilon_m x_m)^m .$$

PROOF. By the Leibniz Formula 1.8 we have that

$$A(x_o + \varepsilon_1 x_1 + \dots + \varepsilon_m x_m)^m = \sum \frac{m!}{\alpha_o! \dots \alpha_m!} \varepsilon_1^{\alpha_1} \dots \varepsilon_m^{\alpha_m} Ax_o^{\alpha_o} \dots x_m^{\alpha_m}$$

where the summation is taken ever all $\alpha_o, \dots, \alpha_m \in \mathbb{N}_o$ such that $\alpha_o + \dots + \alpha_m = m$. Hence

$$\sum_{\varepsilon_j = \pm 1} \varepsilon_1 \dots \varepsilon_m A(x_o + \varepsilon_1 x_1 + \dots + \varepsilon_m x_m)^m$$

$$= m! \sum_{\alpha_k} \frac{Ax_o^{\alpha_o} \dots x_m^{\alpha_m}}{\alpha_o! \dots \alpha_m!} \sum_{\varepsilon_j} \varepsilon_1^{\alpha_1 + 1} \dots \varepsilon_m^{\alpha_m + 1} .$$

Clearly

$$\sum_{\varepsilon_j = \pm 1} \varepsilon_1^{\alpha_1 + 1} \dots \varepsilon_m^{\alpha_m + 1} = 0$$

whenever $\alpha_i = 0$ for some i with $1 \le i \le m$. Since

$$\sum_{\varepsilon_j=\pm 1} \varepsilon_1^2 \ldots \varepsilon_m^2 = 2^m$$

the desired result follows.

One can generate multilinear forms out of linear forms in the following manner. Given $\varphi_1, \ldots, \varphi_m \in E^*$ one defines $A \in \mathcal{L}_a({}^mE;F)$ by

$$A(x_1, \ldots, x_m) = \varphi_1(x_1) \ldots \varphi_m(x_m)$$

for all $x_1, \ldots, x_m \in E$. This idea can be generalized as follows.

1.11. DEFINITION. Givem $A \in \mathcal{L}_a({}^mE)$ and $B \in \mathcal{L}_a({}^nE)$ their *tensor product* $A \otimes B \in \mathcal{L}_a({}^{m+n}E)$ is defined by

$$(A \otimes B)(x_1, \ldots, x_{m+n}) = A(x_1, \ldots, x_m)B(x_{m+1}, \ldots, x_{m+n})$$

for all $x_1, \ldots, x_{m+n} \in E$.

The following properties of the tensor product are clear:

(a) If A and B are continuous then $A \otimes B$ is continuous as well.

(b) The mapping $(A,B) \to A \otimes B$ is bilinear.

(c) $(A \otimes B) \otimes C = A \otimes (B \otimes C)$.

Clearly we can also define the tensor product $A \otimes B$ if one of the mappings A or B has values in a Banach space.

Until now we have studied $\mathbb{K}$-multilinear mappings without distinguishing whether $\mathbb{K}$ is $\mathbb{R}$ or $\mathbb{C}$. Now we study the relationship between these two notions.

Let E and F be complex Banach spaces and let $E_{\mathbb{R}}$ and $F_{\mathbb{R}}$ denote the underlying real Banach spaces. Then we have the following result, whose straightforward proof is left as an exercise to the reader.

1.12. PROPOSITION. *Let E and F be complex Banach spaces. Then each $A \in \mathcal{L}_a(E_{\mathbb{R}}, F_{\mathbb{R}})$ admits a unique decomposition of the form $A = A' + A''$, where A' is $\mathbb{C}$-linear and A'' is $\mathbb{C}$-antilinear. The mappings A' and A'' are given by the formulas*

$$A'(x) = \frac{1}{2}[A(x) - iA(ix)]$$

and

$$A''(x) = \frac{1}{2}[A(x) + iA(ix)]$$

for all $x \in E$. If A is continuous then A' and A'' are continuous as well.

To generalize this result to multilinear mappings we introduce the following definition.

1.13. DEFINITION. Let E and F be complex Banach spaces, and let $p, q \in \mathbb{N}_o$ with $p + q \geq 1$. Then we shall denote by $\mathcal{L}_a({}^{pq}E;F)$ the subspace of all $A \in \mathcal{L}_a({}^{p+q}E_{\mathbb{R}}, F_{\mathbb{R}})$ such that

$$A(\lambda x_1, \ldots, \lambda x_{p+q}) = \lambda^p \bar{\lambda}^q A(x_1, \ldots, x_{p+q})$$

for all $\lambda \in \mathbb{C}$ and $x_1, \ldots, x_{p+q} \in E$. We shall denote by $\mathcal{L}({}^{pq}E;F)$ the subspace of all continuous members of $\mathcal{L}_a({}^{pq}E;F)$.

As usual we shall write $\mathcal{L}_a({}^{pq}E;\mathbb{C}) = \mathcal{L}_a({}^{pq}E)$ and $\mathcal{L}({}^{pq}E;\mathbb{C}) = \mathcal{L}({}^{pq}E)$. For convenience we also define $\mathcal{L}_a({}^{oo}E;F) = \mathcal{L}({}^{oo}E;F) = F$.

1.14. EXAMPLE. Let E be a complex Banach space, and let $\varphi_1, \ldots, \varphi_{p+q} \in E^*$. Then the mapping

$$\varphi_1 \otimes \ldots \quad \varphi_p \otimes \bar{\varphi}_{p+1} \otimes \ldots \otimes \bar{\varphi}_{p+q},$$

where the bar means complex conjugate, belongs to $\mathcal{L}_a({}^{pq}E)$.

If E and F are complex Banach spaces then it is clear that $\mathcal{L}_a({}^mE;F) \subset \mathcal{L}_a({}^{mo}E;F)$ for every $m \in \mathbb{N}_o$. The opposite inclusion is far from obvious, but it is true, as the next theorem shows.

1.15. THEOREM. *Let* E *and* F *be complex Banach spaces. Then:*

(a) $\mathcal{L}_a({}^mE_{\mathbb{R}};F_{\mathbb{R}})$ *is the algebraic direct sum of the subspaces* $\mathcal{L}_a({}^{pq}E;F)$ *with* $p + q = m$. *Moreover,* $\mathcal{L}_a({}^{mo}E;F) = \mathcal{L}_a({}^mE;F)$.

(b) $\mathcal{L}({}^mE_{\mathbb{R}};F_{\mathbb{R}})$ *is the topological direct sum of the subspaces* $\mathcal{L}({}^{pq}E;F)$ *with* $p + q = m$. *Moreover,* $\mathcal{L}({}^{mo}E;F) = \mathcal{L}({}^mE;F)$.

PROOF. By induction on m. The theorem is obviously true for $m = 0$, and by Proposition 1.12 the theorem is also true for $m = 1$. Assuming the theorem true for a certain $m \geq 1$ we prove it for $m + 1$. By the induction hypothesis there are projections

$$v_k : \mathcal{L}_a({}^mE_{\mathbb{R}};F_{\mathbb{R}}) \to \mathcal{L}_a({}^{m-k,k}E;F)$$

such that $v_o + \ldots + v_m = identity$. Consider a mapping

$$A \in \mathcal{L}_a({}^{m+1}E_{\mathbb{R}};F_{\mathbb{R}}) = \mathcal{L}_a(E_{\mathbb{R}};\ \mathcal{L}_a({}^mE_{\mathbb{R}};F_{\mathbb{R}})).$$

Under this identification $Ax \in \mathcal{L}_a({}^mE_{\mathbb{R}};F_{\mathbb{R}})$ for each $x \in E$ and then we can write

$$Ax = \sum_{k=0}^{m} v_k(Ax).$$

Thus

$$v_k \circ A \in \mathcal{L}_a(E_{\mathbb{R}};\ \mathcal{L}_a({}^{m-k,k}E;F))$$

for each $k = 0,\ldots,m$. Now, by the case $m = 1$, for each $k = 0,\ldots,m$ there are projections

$$u_j^k : \mathcal{L}_a(E_{\mathbb{R}};\mathcal{L}_a({}^{m-k,k}E;F)) \to \mathcal{L}_a({}^{1-j,j}E;\mathcal{L}_a({}^{m-k,k}E;F))$$

with $j = 0,1$ such that $u_o^k + u_1^k = identity$. Hence

$$v_k \circ A = u_o^k(v_k \circ A) + u_1^k(v_k \circ A).$$

Thus

$$A = \sum_{k=o}^{m} v_k \circ A = \sum_{k=o}^{m} \sum_{j=o}^{1} u_j^k(v_k \circ A) = \sum_{q=o}^{m+1} w_q(A)$$

where

$$w_q(A) = \sum_{j+k=q} u_j^k(v_k \circ A).$$

Clearly

$$u_j^k(v_k \circ A) \in \mathcal{L}_a(E_{\mathbb{R}}; \mathcal{L}_a(^m E_{\mathbb{R}}; F_{\mathbb{R}})) = \mathcal{L}_a(^{m+1}E_{\mathbb{R}}; F_{\mathbb{R}}).$$

And since

$$u_j^k(v_k \circ A)(\lambda x_o)(\lambda x_1, \ldots, \lambda x_m)$$

$$= \lambda^{1-j}\,\bar{\lambda}^j u_j^k(v_k \circ A)(x_o)(\lambda x_1, \ldots, \lambda x_m)$$

$$= \lambda^{1-j}\,\bar{\lambda}^j\,\lambda^{m-k}\,\bar{\lambda}^k\, u_j^k(v_k \circ A)(x_o)(x_1, \ldots, x_m)$$

we see that $w_q(A) \in \mathcal{L}_a(^{m+1-q,q}E;F)$ for $q = 0,\ldots,m+1$. Thus we have found linear mappings

$$w_q : \mathcal{L}_a(^{m+1}E_{\mathbb{R}}; F_{\mathbb{R}}) \to \mathcal{L}_a(^{m+1-q,q}E;F)$$

such that $w_o + \ldots + w_m = identity$. Since it can be readily seen that

$$\mathcal{L}_a(^{m-j,j}E;F) \cap \mathcal{L}_a(^{m-k,k}E;F) = \{0\}$$

whenever $j \neq k$, we conclude that each w_q is a projection. Thus we have shown that $\mathcal{L}_a(^{m+1}E_{\mathbb{R}}; F_{\mathbb{R}})$ is the algebraic direct sum of the subspace $\mathcal{L}_a(^{pq}E;F)$ with $p + q = m + 1$. Now, since $w_o(A) = u_o^o(v_o \circ A)$ it follows from the induction hypothesis and from the case $m = 1$ that $w_o(A) \in \mathcal{L}_a(^{m+1}E;F)$. Whence it follows that

$$\mathcal{L}_a(^{m+1,o}E;F) = \mathcal{L}_a(^{m+1}E;F).$$

Thus (a) has been proved. But then an examination of the proof shows (b) as well.

EXERCISES

1.A. Let $A \in \mathcal{L}_a(^mE;F)$ be an m-linear mapping which is separately continuous in each variable. Using the Principle of Uniform Boundedness show that A is continuous.

1.B. Let (A_j) be a sequence in $\mathcal{L}_a(^mE;F)$ such that the limit $A(x) = \lim_j A_j(x)$ exists for every $x \in E^m$.

(a) Show that $A \in \mathcal{L}_a(^mE;F)$.

(b) If each A_j is symmetric (resp. alternating), show that A is symmetric (resp. alternating) as well.

(c) If each A_j is continuous, show that A is continuous as well.

1.C. Let E and F be Banach spaces over $\mathbb{K}$, with E finite dimensional. Let $(e_1, \ldots, e_n)$ be a basis for E and let $\xi_1, \ldots, \xi_n$ denote the corresponding coordinate functionals. Show that each $A \in \mathcal{L}_a(^mE;F)$ can be uniquely represented as a sum

$$A = \Sigma\, c_{j_1 \ldots j_m} \xi_{j_1} \otimes \ldots \otimes \xi_{j_m}$$

where $c_{j_1 \ldots j_m} \in F$ and where the summation is taken over all $j_1, \ldots, j_m$ varying from 1 to n. Conclude that $\mathcal{L}_a(^mE;F) = \mathcal{L}(^mE;F)$.

1.D. Let E and F be finite dimensional Banach spaces over $\mathbb{K}$. If E has dimension n and F has dimension p, show that $\mathcal{L}(^mE;F)$ has dimension $n^m p$.

1.E. Let $A \in \mathcal{L}_a^s(^mE;F)$ and let $x_o, \ldots, x_n \in E$. If $m < n$ show that

$$\sum_{\varepsilon_j = \pm 1} \varepsilon_1 \ldots \varepsilon_n A(x_o + \varepsilon_1 x_1 + \ldots + \varepsilon_n x_n)^m = 0.$$

2. POLYNOMIALS

This section is devoted to the study of polynomials in Banach spaces. Polynomials will be used to define power series, and these in turn will be used to define holomorphic mappings.

2.1. DEFINITION. A mapping $P : E \to F$ is said to be an *m-homogeneous polynomial* if there exists $A \in \mathcal{L}_a({}^mE;F)$ such that $P(x) = Ax^m$ for every $x \in E$. We shall denote by $P_a({}^mE;F)$ the vector space of all m-homogeneous polynomials from E into F. We shall represent by $P({}^mE;F)$ the subspace of all continuous members of $P_a({}^mE;F)$. For each $P \in P_a({}^mE;F)$ we shall set

$$\|P\| = \sup\{\|P(x)\| : x \in E, \|x\| \leq 1\}.$$

When $F = \mathbb{K}$ then for short we shall write $P_a({}^mE;\mathbb{K}) = P_a({}^mE)$ and $P({}^mE;\mathbb{K}) = P({}^mE)$.

2.2. THEOREM. *For each* $A \in \mathcal{L}_a({}^mE;F)$ *let* $\hat{A} \in P_a({}^mE;F)$ *be defined by* $\hat{A}(x) = Ax^m$ *for every* $x \in E$. *Then:*

(a) *The mapping* $A \to \hat{A}$ *induces a vector space isomorphism between* $\mathcal{L}_a^s({}^mE;F)$ *and* $P_a({}^mE;F)$.

(b) *We have the inequalities*

$$\|\hat{A}\| \leq \|A\| \leq \frac{m^m}{m!}\|\hat{A}\|$$

for every $A \in \mathcal{L}_a({}^mE;F)$.

PROOF. Given $P \in P_a({}^mE;F)$ we can find $A \in \mathcal{L}_a({}^mE;F)$ such that $P = \hat{A}$. But then

$$P = \hat{A} = (A^s)\hat{}$$

and $A^s \in \mathcal{L}_a^s({}^mE;F)$. If we apply the Polarization Formula 1.10 with $x_0 = 0$ to the mapping A^s then all the assertions follow at once.

2.3. COROLLARY. (a) *A polynomial* $P \in \mathcal{P}_a({}^mE;F)$ *is continuous if and only if* $\|P\| < \infty$.

(b) $\mathcal{P}({}^mE;F)$ *is a Banach space under the norm* $P \to \|P\|$.

(c) *The mapping* $A \to \hat{A}$ *induces a topological isomorphism between* $\mathcal{L}^s({}^mE;F)$ *and* $\mathcal{P}({}^mE;F)$.

2.4. PROPOSITION. *For each* $P \in \mathcal{P}_a({}^mE;F)$ *the following conditions are equivalent:*

(a) P *is continuous.*

(b) P *is bounded on every ball with finite radius.*

(c) P *is bounded on some open ball.*

PROOF. The implication (a) $\Rightarrow$ (b) follows from Corollary 2.3. The implication (b) $\Rightarrow$ (c) is obvious. And the implication (c) $\Rightarrow$ (a) follows from the following lemma.

2.5. LEMMA. *Let* $P \in \mathcal{P}_a({}^mE;F)$. *If* P *is bounded by* c *on an open ball* $B(a;r)$ *then* P *is bounded by* $cm^m/m!$ *on the ball* $B(0;r)$.

PROOF. There is $A \in \mathcal{L}_a^s({}^mE;F)$ such that $P = \hat{A}$. Then if suffices to apply the Polarization Formula 1.10 to A with $x_o = a$ and $x_1 = \ldots = x_m \in B(0;r/m)$.

Next we extend the Principle of Uniform Boundedness to homogeneous polynomials.

2.6. THEOREM. *A subset of* $\mathcal{P}({}^mE;F)$ *is norm bounded if and only if it is pointwise bounded.*

The proof of the theorem rests on the following lemma.

2.7. LEMMA. *Let* U *be an open subset of* E, *and let* (f_i) *be a family of continuous mappings from* U *into* F. *If the family* (f_i) *is pointwise bounded on* U *then there is an open set* $V \subset U$

where the family (f_i) *is uniformly bounded.*

PROOF. Set

$$A_n = \{x \in U : \|f_i(x)\| \leq n \text{ for every } i\}$$

for every $n \in \mathbb{N}$. Then $U = \bigcup_{n=1}^{\infty} A_n$ and each A_n is closed in U. Since U is a Baire space, some A_n has nonempty interior. Then the family (f_i) is uniformly bounded on the open set $V = \overset{\circ}{A}_n$.

PROOF OF THEOREM 2.6. To prove the nontrivial implication, let (P_i) be a subset of $P(^mE;F)$ which is pointwise bounded. By Lemma 2.7 the family (P_i) is uniformly bounded, by c say, on a ball $B(a;r)$. Then by Lemma 2.5 the family (P_i) is uniformly bounded by $cm^m/m!$ on the ball $B(0;r)$. The desired conclusion follows.

2.8. DEFINITION. A mapping $P : E \to F$ is said to be a *polynomial of degree at most* m if it can be represented as a sum

$$P = P_0 + P_1 + \ldots + P_m$$

where $P_j \in P_a(^jE;F)$ for $j = 0,\ldots,m$. We shall denote by $P_a(E;F)$ the vector space of all polynomials from E into F. We shall denote by $P(E;F)$ the subspace of all continuous members of $P_a(E;F)$.

When $F = \mathbb{K}$ then for short we shall write $P_a(E;\mathbb{K}) = P_a(E)$ and $P(E;\mathbb{K}) = P(E)$.

2.9. PROPOSITION. (a) $P_a(E;F)$ *is the algebraic direct sum of the subspaces* $P_a(^mE;F)$, *with* $m \in \mathbb{N}_0$.

(b) $P(E;F)$ *is the algebraic direct sum of the subspaces* $P(^mE;F)$, *with* $m \in \mathbb{N}_0$.

PROOF. (a) It suffices to show that if

$$P_0 + P_1 + \ldots + P_m = 0$$

with $P_j \in P_a({}^jE;F)$ for $j = 0,\dots,m$, then

$$P_o = P_1 = \dots = P_m = 0.$$

For each $x \in E$ and $\lambda \in \mathbb{K}$, $\lambda \neq 0$, we have that

$$0 = \sum_{j=0}^{m} P_j(\lambda x) = \sum_{j=0}^{m} \lambda^j P_j(x).$$

After dividing by λ^m and letting $|\lambda| \to \infty$ we get that $P_m = 0$. Proceeding inductively we get that $P_{m-1} = 0,\dots,P_o = 0$.

(b) If suffices to show that if the polynomial

$$P = P_o + P_1 + \dots + P_m$$

is continuous, where $P_j \in P_a({}^jE;F)$ for $j = 0,\dots,m$ then each P_j is continuous as well. We prove this by induction on m, the result being obvious for $m = 0$. If $m \geq 1$ then for all $x \in E$ and $\lambda \in \mathbb{K}$ we have that

$$P(\lambda x) - \lambda^m P(x) = \sum_{j=0}^{m-1} (\lambda^j - \lambda^m) P_j(x).$$

Choose $\lambda \in \mathbb{K}$ such that $\lambda^j - \lambda^m \neq 0$ for $j = 0,\dots,m-1$. Since the polynomial $x \to P(\lambda x) - \lambda^m P(x)$ is continuous and has degree at most $m - 1$, the induction hypothesis implies that each P_j with $j = 0,\dots,m-1$, is continuous. Then it follows that P_m is continuous as well, and the proof is complete.

EXERCISES

2.A. Let (P_j) be a sequence in $P_a({}^mE;F)$ such that the limit $P(x) = \lim P_j(x)$ exists for every $x \in E$.

(a) Show that $P \in P_a({}^mE;F)$.

(b) Show that if each P_j is continuous then P is continuous as well and (P_j) converges to P uniformly on compact subsets of E.

2.B. Let $P : E \to F$ be a mapping such that $P|M \in \mathcal{P}_a(^m M;F)$ for each subspace M of E of dimension $\leq m+1$. Show that $P \in \mathcal{P}_a(^m E;F)$.

2.C. If $P \in \mathcal{P}_a(E;F)$ satisfies $P(\lambda x) = \lambda^m P(x)$ for all $\lambda \in \mathbb{K}$ and $x \in E$, show that $P \in \mathcal{P}_a(^m E;F)$.

2.D. Let E, F, G, H be Banach spaces over $\mathbb{K}$. Given $S \in \mathcal{L}_a(E;F)$, $P \in \mathcal{P}_a(^m F;G)$ and $T \in \mathcal{L}_a(G;H)$, show that $P \circ S \in \mathcal{P}_a(^m E;G)$ and $T \circ P \in \mathcal{P}_a(^m F;H)$.

2.E. Given a sequence (a_m) of positive numbers, let (b_m) be defined by $b_m = \sqrt[m]{a_1 a_2 \ldots a_m}$. Show the following:

(a) If (a_m) converges to a, then (b_m) also converges to a.

(b) If (a_m) is increasing (resp. decreasing), then (b_m) is increasing (resp. decreasing) as well.

2.F. Apply 2.E. to the sequence $(1 + 1/m)^m$ to show that the sequence $(m / \sqrt[m]{m!})$ is increasing and converges to e.

2.G. Show that $\|A\| \leq e^m \|\hat{A}\|$ for every $A \in \mathcal{L}(^m E;F)$.

2.H. Let $\varphi \in E'$ with $\|\varphi\| = 1$ and let $P \in \mathcal{P}(^m E)$ be defined by $P = \varphi^m$.

(a) Show that $\|P\| = 1$.

(b) Show that $P = \hat{A}$ where $A \in \mathcal{L}^s(^m E)$ is given by $A = \varphi \otimes \ldots \otimes \varphi$.

(c) Show that $\|A\| = 1$.

2.I. Let (ξ_n) denote the sequence of coordinate functionals on $E = \ell^1$. Let $P \in \mathcal{P}(^m \ell^1)$ be defined by $P = \xi_1 \xi_2 \ldots \xi_m$.

(a) Show that $\|P\| = 1/m^m$.

(b) Show that $P = \hat{A}$ where $A \in \mathcal{L}^s({}^m\ell^1)$ is given by

$$A(x_1,\ldots,x_m) = \frac{1}{m!} \sum_{\sigma \in S_m} \xi_1(x_{\sigma(1)}) \ldots \xi_m(x_{\sigma(m)}).$$

(c) Show that $\|A\| = 1/m!$

The preceding two exercises show that in the inequalities in Theorem 2.2 the two extreme situations can actually occur.

2.J. Let E and F be Banach spaces over $\mathbb{K}$, with E finite dimensional. Let $(e_1,\ldots,e_n)$ be a basis for E and let $\xi_1,\ldots,\xi_n$ denote the corresponding coordinate functionals. Show that each $P \in P({}^mE;F)$ can be uniquely represented as a sum

$$P = \Sigma\, c_\alpha \xi_1^{\alpha_1} \ldots \xi_n^{\alpha_n}$$

where the summation is taken over all multi-indices $\alpha = (\alpha_1,\ldots,\alpha_n) \in \mathbb{N}_o^n$ such that $|\alpha| = m$.

2.K. Let $P_f({}^mE;F)$ denote the subspace of all $P \in P({}^mE;F)$ which can be written in the form

$$P = \sum_{j=1}^{p} c_j \varphi_j^m$$

where $c_j \in F$ and $\varphi_j \in E'$. Let $P_f(E;F)$ denote the algebraic sum of the spaces $P_f({}^mE;F)$ with $m \in \mathbb{N}_o$. Each $P \in P_f(E;F)$ is said to be a *continuous polynomial of finite type*. When $F = \mathbb{K}$ then as usual we shall write $P_f({}^mE;\mathbb{K}) = P_f({}^mE)$ and $P_f(E;\mathbb{K}) = P_f(E)$.

(a) Show that $P(E;F) = P_f(E;F)$ if E is finite dimensional.

(b) Show that $P \circ T \in P_f(E;F)$ for each polynomial $P \in P(E;F)$ and each operator $T \in \mathcal{L}(E;E)$ which has finite rank.

2.L. Show that if $P \in P_a(E;F)$ is a polynomial of degree at most m then $P(a + \lambda b)$ is a polynomial in λ of degree at most m for all $a, b \in E$.

2.M. Let $P = P_o + P_1 + \ldots + P_m$ where $P_j \in P_a({}^jE;F)$ for $j = 0,\ldots,m$. Using Exercise 1.E show that the homogeneous polynomial P_m is given by the formula

$$P_m(x) = \frac{1}{m!\, 2^m} \sum_{\varepsilon_j = \pm 1} \varepsilon_1 \ldots \varepsilon_m P(x_o + \varepsilon_1 x + \ldots + \varepsilon_m x) .$$

for all $x_o, x \in E$.

2.N. Given $P \in P_a(E;F)$ show that the following conditions are equivalent:

(a) P is continuous.

(b) P is bounded on every ball with finite radius.

(c) P is bounded on some open ball.

2.O. Show that a polynomial $P \in P_a(E;F)$ is continuous if and only if the polynomial $\psi \circ P \in P_a(E)$ is continuous for every $\psi \in F'$.

3. POLYNOMIALS OF ONE VARIABLE

We have seen in Exercise 2.L that if $P \in P_a(E;F)$ is a polynomial of degree at most m then $P(a + \lambda b)$ is a polynomial in λ of degree at most m for all $a, b \in E$. In this section we show that the converse is also true, thus presenting an alternative description of polynomials in Banach spaces. Before proving the main result, we present some preparatory results on interpolation polynomials.

3.1. PROPOSITION. *Let $\lambda_o,\ldots,\lambda_m$ be $m + 1$ distinct points in $\mathbb{K}$ and let $b_o,\ldots,b_m$ be arbitrary points in F. Then there is a unique polynomial $L \in P(\mathbb{K}; F)$ of degree at most m such that*

(3.1) $$L(\lambda_j) = b_j \quad \text{for} \quad j = 0,\ldots,m.$$

The polynomial L is called a Lagrange interpolation polynomial.

PROOF. We are looking for a polynomial L of the form

$$L(\lambda) = \sum_{k=0}^{m} c_k \lambda^k$$

with $c_k \in F$, which satisfies (3.1). Hence we have to solve the system of equations

$$\sum_{k=0}^{m} c_k \lambda_j^k = b_j \quad \textit{for} \quad j = 0,\dots,m.$$

Since the determinant of the coefficients is a nonzero Vandermonde determinant, this system of equations has a unique solution.

3.2. COROLLARY. *Let* $\lambda_o,\dots,\lambda_m$ *be* $m + 1$ *distinct points in* $\mathbb{K}$. *For each* $k = 0,\dots,m$ *let* $L_k \in P(\mathbb{K})$ *be the unique polynomial of degree at most* m *such that* $L_k(\lambda_j) = \delta_{kj}$ *for* $j = 0,\dots,m$. *Then each polynomial* $P \in P(\mathbb{K};F)$ *of degree at most* m *can be represented in the form*

$$P(\lambda) = \sum_{k=0}^{m} P(\lambda_k)L_k(\lambda) \quad \textit{for all} \quad \lambda \in \mathbb{K}.$$

PROOF. If we define

$$Q(\lambda) = \sum_{k=0}^{m} P(\lambda_k)L_k(\lambda) \quad \text{for all} \quad \lambda \in \mathbb{K}.$$

then both P and Q are polynomials of degree at most m and $Q(\lambda_j) = P(\lambda_j)$ for $j = 0,\dots,m$. By Proposition 3.1 the polynomials P and Q are identical.

3.3. LEMMA. *If a mapping* $P : \mathbb{K}^n \to F$ *is a polynomial in each variable separately, then* P *is a polynomial.*

PROOF. By induction on n, the result being obvious for $n = 1$.

Assuming the result true for a certain n we prove it for $n+1$. Thus let $P : \mathbb{K}^{n+1} \to F$ be a mapping which is a polynomial in each variable separately. For each $m \in \mathbb{N}$ Let A_m denote the set of all $\mu \in \mathbb{K}$ such that $P(\lambda_1,\ldots,\lambda_n,\mu)$ is a polynomial of degree at most m in each of the variables $\lambda_1,\ldots,\lambda_n$. Then the sequence (A_m) is increasing and it follows from the induction hypothesis that $\mathbb{K} = \bigcup_{m=1}^{\infty} A_m$. Hence some A_m is an infinite set. Choose $m+1$ distinct points $\xi_0,\ldots,\xi_m$ in $\mathbb{K}$. For each $k = 0,\ldots,m$ let $L_k \in P(\mathbb{K})$ denote the unique polynomial of degree at most m such that

(3.2) $$L_k(\xi_j) = \delta_{kj} \quad \text{for} \quad j = 0,\ldots,m.$$

Define $Q : \mathbb{K}^{n+1} \to F$ by

(3.3) $$Q(\lambda,\mu) = \Sigma\, L_{k_1}(\lambda_1) \ldots L_{k_n}(\lambda_n)P(\xi_{k_1},\ldots,\xi_{k_n},\mu)$$

where the summation is taken over all $k_1,\ldots,k_n$ varying from 0 to m. Then certainly $Q \in P(\mathbb{K}^{n+1};F)$ and to complete the proof it suffices to show that $P(\lambda,\mu) = Q(\lambda,\mu)$ for all $\lambda \in \mathbb{K}^n$ and $\mu \in \mathbb{K}$.

From (3.2) and (3.3) it follows that

(3.4) $$Q(\xi_{j_1},\ldots,\xi_{j_n},\mu) = P(\xi_{j_1},\ldots,\xi_{j_n},\mu)$$

for all $\mu \in \mathbb{K}$ and all $j_1,\ldots,j_n$ varying from 0 to m. If $\mu \in A_m$ then both $P(\lambda,\mu)$ and $Q(\lambda,\mu)$ are polynomials of degree at most m in each of the variables $\lambda_1,\ldots,\lambda_m$. Then it follows from (3.4) and Proposition 3.1 that $P(\lambda,\mu) = Q(\lambda,\mu)$ for all $\lambda \in \mathbb{K}^n$ and $\mu \in A_m$. Since A_m is an infinite set we conclude that $P(\lambda,\mu) = Q(\lambda,\mu)$ for all $\lambda \in \mathbb{K}^n$ and $\mu \in \mathbb{K}$. This completes the proof.

3.4. COROLLARY. *Let* $f : E \to F$ *be a mapping such that* $f(a+\lambda b)$ *is a polynomial in* λ *for all* $a, b \in E$. *Then* $f|M \in P(M;F)$

for each finite dimensional subspace M *of* E.

PROOF. Let $(e_1,\dots,e_n)$ be a basis for M. Then by hypothesis $f(\lambda_1 e_1 + \dots + \lambda_n e_n)$ is a polynomial in each of the variables $\lambda_1,\dots,\lambda_n$ separately. Then it follows from Lemma 3.3 that $f \mid M \in P(M;F)$.

3.5. LEMMA. *Let* $f : E \to F$ *be a mapping such that* $f(a + \lambda b)$ *is a polynomial in* λ *for all* $a, b \in E$. *Then there is a sequence of homogeneous polynomials* $P_k \in P_a({}^kE;F)$ *such that*

$$f(x) = \sum_{k=0}^{\infty} P_k(x) \quad \text{for every} \quad x \in E,$$

where for each $x \in E$ *we have that* $P_k(x) = 0$ *for all but finitely many indices.*

PROOF. By hypothesis for each $x \in E$ there is a sequence $(P_k(x))$ in F such that

$$(3.5) \qquad f(\lambda x) = \sum_{k=0}^{\infty} \lambda^k P_k(x) \quad \text{for every} \quad \lambda \in \mathbb{K},$$

where $P_k(x) = 0$ for all but finitely many indices. To prove the lemma it suffices to show that $P_k \in P_a({}^kE;F)$ for each k.

In view of Exercise 2.B it will be sufficient to show that $P_k|M \in P({}^kM;F)$ for each $k \in \mathbb{N}_0$ and each finite dimensional subspace M of E. Now, by Corollary 3.4, $f|M \in P(M;F)$. Hence there are homogeneous polynomials $Q_0, \dots, Q_m$, with $Q_k \in P({}^kM;F)$ for $k = 0,\dots,m$, such that

$$f(x) = \sum_{k=0}^{m} Q_k(x) \quad \text{for all} \quad x \in M$$

and therefore

$$(3.6) \quad f(\lambda x) = \sum_{k=0}^{m} \lambda^k Q_k(x) \quad \text{for all} \quad x \in M \quad \text{and} \quad \lambda \in \mathbb{K}.$$

From (3.5) and (3.6) we conclude that

$$P_k(x) = Q_k(x) \quad \text{for all} \quad x \in M \quad \text{and} \quad k \leq m$$

$$P_k(x) = 0 \quad \text{for all} \quad x \in M \quad \text{and} \quad k > m.$$

Thus $P_k|M \in P({}^kM;F)$ for every $k \in \mathbb{N}_0$ and the proof is complete.

3.6. **THEOREM.** *A mapping* $P : E \to F$ *is a polynomial of degree at most* m *if and only if* $P(a + \lambda b)$ *is a polynomial in* λ *of degree at most* m *for all* $a, b \in E$.

PROOF. The proof of the "only if" part was left to the reader as Exercise 2.L. The "if" part follows at once from the proof of Lemma 3.5. Indeed, it suffices to observe that in (3.5) we have that $P_k(x) = 0$ for every $k > m$.

Let $f : E \to F$ be a mapping such that $f(a + \lambda b)$ is a polynomial in λ for all $a, b \in E$. Then f need not be a polynomial in general. However, we have the following theorem.

3.7. **THEOREM.** *A continuous mapping* $P : E \to F$ *is a polynomial if and only if* $P(a + \lambda b)$ *is a polynomial in* λ *for all* $a, b \in E$.

To prove this theorem we need the following lemma.

3.8. **LEMMA.** *Let* $f : E \to F$ *be a mapping such that* $f(a + \lambda b)$ *is a polynomial in* λ *for all* $a, b \in E$. *If* f *is identically zero on a nonvoid open set, then* f *is identically zero on* E.

PROOF. Assume f is identically zero on a nonvoid open set $U \subset E$. Fix $a \in U$, $x \in E$ and $\psi \in F'$. Then $\psi \circ f[a + \lambda(x - a)]$ is a $\mathbb{K}$-valued polynomial in λ which is identically zero on a neighborhood of $\lambda = 0$. Hence $\psi \circ f[a + \lambda(x - a)] = 0$ for every $\lambda \in \mathbb{K}$, and in particular $\psi \circ f(x) = 0$. Since F' separates the points of F we conclude that $f(x) = 0$.

PROOF OF THEOREM 3.7. To prove the nontrivial implication assume $P(a + \lambda b)$ is a polynomial in λ for all $a, b \in E$. Then by Lemma 3.5 there is a sequence of homogeneous polynomials $P_k \in P_a({}^kE;F)$ such that

$$P(x) = \sum_{k=0}^{\infty} P_k(x) \quad \text{for every} \quad x \in E,$$

where for each $x \in E$ we have that $P_k(x) = 0$ for all but finitely many indices. We want to show the existence of an $m \in \mathbb{N}$ such that $P_k = 0$ for every $k > m$.

First we show that each P_k is continuous. We proceed by induction on k, the assertion being obvious for $k = 0$. Let $k > 0$ and assume P_j is continuous for every $j < k$. Then for all $x \in E$ and $\lambda \in \mathbb{K}$, $\lambda \neq 0$, we have that

$$P(\lambda x) = \sum_{j<k} \lambda^j P_j(x) + \lambda^k P_k(x) + \sum_{j>k} \lambda^j P_j(x)$$

and it follows that

$$P_k(x) = \lim_{\lambda \to 0} \lambda^{-k} [\, P(\lambda x) - \sum_{j<k} \lambda^j P_j(x)] . \tag{3.7}$$

Let (λ_n) be a sequence of nonzero scalars which converges to zero. For each $n \in \mathbb{N}$ let $Q_n : E \to F$ be defined by

$$Q_n(x) = \lambda_n^{-k} [\, P(\lambda_n x) - \sum_{j<k} \lambda_n^j P_j(x)] .$$

By the induction hypothesis each Q_n is continuous. By (3.7) the sequence (Q_n) converges pointwise to P_k, and is in particular pointwise bounded on E. Then by Lemma 2.7 the sequence (Q_n) is uniformly bounded on a ball $B(a,r)$. Hence P_k is bounded on $B(a,r)$ as well, and by Proposition 2.4 we may conclude that P_k is continuous, as asserted.

To complete the proof of the theorem we consider the sets

$$A_m = \{x \in E : P_k(x) = 0 \quad \text{for all} \quad k > m\}.$$

Then $E = \bigcup_{m=1}^{\infty} A_m$ and each A_m is closed in E. Since E is a Baire space, some A_m has nonempty interior. Then it follows from Lemma 3.8 that P_k is identically zero on E for every $k > m$. The proof is now complete.

Theorems 3.6 and 3.7 reduce the study of polynomials to the case where the domain space is one dimensional. Next we present results that reduce the study of polynomials to the case where the range space is one dimensional.

3.9. THEOREM. *A mapping* $P : E \to F$ *is a polynomial of degree at most* m *if and only if* $\psi \circ P : E \to \mathbb{K}$ *is a polynomial of degree at most* m *for every* $\psi \in F'$.

The nontrivial implication in Theorem 3.9 follows at once from Theorem 3.6 and the following lemma.

3.10. LEMMA. *A mapping* $P : \mathbb{K} \to F$ *is a polynomial of degree at most* m *if* $\psi \circ P : \mathbb{K} \to \mathbb{K}$ *is a polynomial of degree at most* m *for every* $\psi \in F'$.

PROOF. Let $\lambda_0, \ldots, \lambda_m$ be $m + 1$ distinct points in $\mathbb{K}$. For each $k = 0, \ldots, m$ let $L_k \in P(\mathbb{K})$ be the unique polynomial of degree at most m such that $L_k(\lambda_j) = \delta_{kj}$ for $j = 0, \ldots, m$. If $\psi \in F'$ then by Corollary 3.2 the polynomial $\psi \circ P$ can be represented in the form

$$\psi \circ P(\lambda) = \sum_{k=0}^{m} \psi \circ P(\lambda_k) L_k(\lambda) = \psi \left[\sum_{k=0}^{m} P(\lambda_k) L_k(\lambda) \right]$$

for every $\lambda \in \mathbb{K}$. Since F' separates the points of F we conclude that

$$P(\lambda) = \sum_{k=0}^{m} P(\lambda_k) L_k(\lambda) \quad \text{for every} \quad \lambda \in \mathbb{K}.$$

Thus P is a polynomial of degree at most m, as asserted.

We complete this section with the following continuous version of Theorem 3.9.

3.11. THEOREM. *A mapping* $P : E \to F$ *is a continuous polynomial if and only if* $\psi \circ P : E \to \mathbb{K}$ *is a continuous polynomial for every* $\psi \in F'$.

To prove this theorem we need two auxiliary lemmas.

3.12. LEMMA. *Let $(P_n) \subset P(\mathbb{K}; F)$ be a sequence of polynomials of degree at most m which converges pointwise to a mapping $P : \mathbb{K} \to F$. Then P is also a polynomial of degree at most m.*

PROOF. Let $\lambda_0, \ldots, \lambda_m$ be $m + 1$ distinct points in $\mathbb{K}$. For each $k = 0, \ldots, m$ let $L_k \in P(\mathbb{K})$ be the unique polynomial of degree at most m such that $L_k(\lambda_j) = \delta_{kj}$ for $j = 0, \ldots, m$. Then by Corollary 3.2 each P_n can be represented in the form

$$P_n(\lambda) = \sum_{k=0}^{m} P_n(\lambda_k) L_k(\lambda) \quad \text{for every} \quad \lambda \in \mathbb{K}.$$

By letting $n \to \infty$ we get that

$$P(\lambda) = \sum_{k=0}^{m} P(\lambda_k) L_k(\lambda) \quad \text{for every} \quad \lambda \in \mathbb{K}.$$

Hence P is also a polynomial of degree at most m, as asserted.

Now we can improve Lemma 3.10 as follows.

3.13. LEMMA. *A mapping $P : \mathbb{K} \to F$ is a polynomial if $\psi \circ P : \mathbb{K} \to \mathbb{K}$ is a polynomial for every $\psi \in F'$.*

PROOF. For each $m \in \mathbb{N}$ let B_m denote the set of all $\psi \in F'$ such that $\psi \circ P$ is a polynomial of degree at most m. By hypothesis $F' = \bigcup_{m=1}^{\infty} B_m$.

We claim that each B_m is closed in F'. Indeed let (ψ_n) be a sequence in B_m which converges to some $\psi \in F'$. Then the sequence $(\psi_n \circ P)$ converges pointwise to $\psi \circ P$, and it follows from Lemma 3.12 that $\psi \in B_m$ too. Thus each B_m is closed in F', as asserted.

Since F' is a Baire space, some B_m has nonvoid interior. But since clearly B_m is a vector subspace of F' we conclude that $B_m = F'$. Thus $\psi \circ P$ is a polynomial of degree at most m for every $\psi \in F'$ and then an application of Lemma 3.10 completes the proof.

PROOF OF THEOREM 3.11. To prove the nontrivial implication let $P : E \to F$ be a mapping such that $\psi \circ P \in \mathcal{P}(E)$ for every $\psi \in F'$. Then $\psi \circ P(a + \lambda b)$ is a polynomial in λ for all $\psi \in F'$ and $a, b \in E$. Then, by Lemma 3.13, $P(a + \lambda b)$ is a polynomial in λ for all $a, b \in E$. Then, by Lemma 3.5, there is a sequence of homogeneous polynomials $P_k \in \mathcal{P}_a({}^kE;F)$ such that

$$P(x) = \sum_{k=0}^{\infty} P_k(x) \quad \text{for every} \quad x \in E,$$

where for each $x \in E$ we have that $P_k(x) = 0$ for all but finitely many indices. Hence we get that

$$(3.8) \qquad \psi \circ P(x) = \sum_{k=0}^{\infty} \psi \circ P_k(x) \quad \text{for all} \quad x \in E \quad \text{and} \quad \psi \in F'.$$

By hypothesis $\psi \circ P \in \mathcal{P}(E)$ for each $\psi \in F'$. Then it follows from (3.8) that $\psi \circ P_k \in \mathcal{P}({}^kE)$ for every $\psi \in F'$. By Exercise 2.O we may conclude that $P_k \in \mathcal{P}({}^kE;F)$ for every $k \in \mathbb{N}_o$. After knowing that each P_k is continuous, we may proceed as in the proof of Theorem 3.7, and show the existence of an $m \in \mathbb{N}$ such that $P_k = 0$ for every $k > m$. This completes the proof.

EXERCISES

3.A. Let $\lambda_o, \ldots, \lambda_m$ be $m + 1$ distinct points in $\mathbb{K}$ and let $b_o, \ldots, b_m$ be arbitrary points in F. Let $L \in \mathcal{P}(\mathbb{K};F)$ be the unique polynomial of degree at most m such that $L(\lambda_j) = b_j$ for $j = 0, \ldots, m$. Show that L is given by the formula

$$L(\lambda) = \sum_{k=0}^{m} b_k \frac{\prod_{j \neq k} (\lambda - \lambda_j)}{\prod_{j \neq k} (\lambda_k - \lambda_j)}.$$

This is the *Lagrange interpolation formula*.

3.B. Given any infinite dimensional Banach space E, find a function $f : E \to \mathbb{K}$ such that $f(a + \lambda b)$ is a polynomial in λ

for all $a, b \in E$, but f is not a polynomial. Of course the function f most be discontinuous.

3.C. Let $(P_n) \subset P_a(E;F)$ be a sequence of polynomials of degree at most m which converges pointwise to a mapping $P : E \to F$.

(a) Show that P is a polynomial of degree at most m.

(b) Show that if each P_n is continuous then P is continuous as well.

4. POWER SERIES

In this section we introduce power series in Banach spaces in terms of homogeneous polynomials. We establish the Cauchy-Hadamard Formula for the radius of convergence.

4.1. DEFINITION. A *power series* from E into F around the point $a \in E$ is a series of mappings of the form $\sum_{m=0}^{\infty} P_m(x-a)$, where $P_m \in P_a(^mE;F)$ for every $m \in \mathbb{N}_0$.

Note that the power series $\sum_{m=0}^{\infty} P_m(x-a)$ can also be written in the form $\sum_{m=0}^{\infty} A_m(x-a)^m$, where $A_m \in \mathcal{L}_a^s(^mE;F)$, $\hat{A}_m = P_m$.

4.2. DEFINITION. The *radius of convergence*, or more precisely, the *radius of uniform convergence* of the power series $\sum_{m=0}^{\infty} P_m(x-a)$, is the supremum of all $r \geq 0$ such that the series converges uniformly on the ball $\overline{B}(a,r)$.

4.3. THEOREM. *Let R be the radius of convergence of the power series $\sum_{m=0}^{\infty} P_m(x-a)$. Then:*

(a) *R is given by the Cauchy-Hadamard Formula*

$$1/R = \limsup_{m\to\infty} \|P_m\|^{1/m}.$$

(b) *The series converges absolutely and uniformly on* $\overline{B}(a;r)$ *whenever* $0 \le r < R$.

PROOF. First we show the inequality

$$(4.1) \qquad 1/R \ge \limsup \|P_m\|^{1/m}.$$

Indeed, suppose $R > 0$ and let $0 \le r < R$. Then the series converges uniformly on $\overline{B}(a;r)$. Set

$$f(x) = \sum_{m=0}^{\infty} P_m(x-a) \quad \text{for} \quad x \in \overline{B}(a;r).$$

Choose $m_o \in \mathbb{N}$ such that

$$\left\| \sum_{j=0}^{m} P_j(x-a) - f(x) \right\| \le 1$$

for all $m \ge m_o$ and $x \in \overline{B}(a;r)$. Hence $\|P_m(t)\| \le 2$ for all $m > m_o$ and $t \in \overline{B}(0,r)$. Whence $\|P_m\| \le 2r^{-m}$ for all $m > m_o$, and therefore

$$\limsup \|P_m\|^{1/m} \le 1/r.$$

Letting $r \to R$ we obtain (4.1).

Next we show the opposite inequality:

$$(4.2) \qquad 1/R \le \limsup \|P_m\|^{1/m}.$$

Indeed, set $L = \limsup \|P_m\|^{1/m}$, assume $L < \infty$, and take r with $0 \le r < 1/L$. Choose s with $r < s < 1/L$ and choose $m_o \in \mathbb{N}$ such that $\|P_m\|^{1/m} < 1/s$ for all $m \ge m_o$. Hence

$$\|P_m(x-a)\| \le \left(\frac{r}{s}\right)^m \quad \text{for all} \quad x \in \overline{B}(a;r).$$

Thus the series $\sum_{m=0}^{\infty} P_m(x-a)$ converges absolutely and uniformly on the ball $\overline{B}(a;r)$. Hence $R \ge r$, and letting $r \to 1/L$

we obtain (4.2).

Thus we have shown (a). But in the course of proving (a) we have also shown (b).

4.4. PROPOSITION. *Let* $\sum_{m=0}^{\infty} P_m(x - a)$ *be a power series from* E *into* F. *If there is* $r > 0$ *such that* $\sum_{m=0}^{\infty} P_m(x - a) = 0$ *for all* $x \in B(a;r)$, *then* $P_m = 0$ *for every* $m \in \mathbb{N}_o$.

This proposition is an immediate consequence of the following lemma.

4.5. LEMMA. *Let* $(c_m)_{m=0}^{\infty}$ *be a sequence in* F. *If there is* $r > 0$ *such that* $\sum_{m=0}^{\infty} c_m \lambda^m = 0$ *for all* $\lambda \in \mathbb{K}$ *with* $|\lambda| \leq r$ *then* $c_m = 0$ *for every* $m \in \mathbb{N}_o$.

PROOF. By induction on m. Letting $\lambda = 0$ we get that $c_o = 0$. Assuming $c_o = \ldots = c_m = 0$ we show that $c_{m+1} = 0$ too. Indeed, since $\sum_{m=0}^{\infty} c_m r^m$ converges there is a constant C such that $\| c_m \| r^m \leq C$ for every m. Since $c_o = \ldots = c_m = 0$ it follows that

$$c_{m+1} = - \sum_{m+2}^{\infty} c_j \lambda^{j-m-1} \quad \text{for} \quad 0 < |\lambda| \leq r.$$

Then for $0 < |\lambda| \leq r/2$ we get that

$$\| c_{m+1} \| \leq |\lambda| \sum_{m+2}^{\infty} C r^{-j} \left(\frac{r}{2}\right)^{j-m-2} = 2|\lambda| C r^{-m-2}.$$

Letting $\lambda \to 0$ we get that $c_{m+1} = 0$ and the lemma has been proved.

Let E and F be Banach spaces over $\mathbb{K}$, with E n dimensional. We have seen in Exercise 2.J that each $P \in P(^mE;F)$ admits a unique representation of the form

$$P = \sum_{|\alpha|=m} c_\alpha \xi_1^{\alpha_1} \ldots \xi_n^{\alpha_n}$$

where $c_\alpha \in F$ and where $\xi_1, \dots, \xi_n$ denote the coordinate functionals associated with a given basis. Thus each power series $\sum_{m=0}^{\infty} P_m(x-a)$ from E into F can be written, at least formally, as a multiple power series

$$\sum_\alpha c_\alpha (\xi_1(x-a))^{\alpha_1} \dots (\xi_n(x-a))^{\alpha_n}$$

where the summation is taken over all multi-indices $\alpha = (\alpha_1, \dots, \alpha_n) \in \mathbb{N}_o^n$. If the multiple series converges absolutely and uniformly on a certain set X then it is clear that the power series $\sum_{m=0}^{\infty} P_m(x-a)$ also converges absolutely and uniformly on X and they coincide there. Conversely, we have the following proposition.

4.6. PROPOSITION. *Let* $\sum_{m=0}^{\infty} P_m(x) = \sum_{m=0}^{\infty} A_m x^m$ *be a power series from* E *into* F *with radius of convergence* $R > 0$. *Given* $e_1, \dots, e_n \in E$ *with* $\|e_1\| = \dots = \|e_n\| = 1$, *set*

$$c_\alpha = \frac{m!}{\alpha!} A_m e_1^{\alpha_1} \dots e_n^{\alpha_n}$$

for each $\alpha = (\alpha_1, \dots, \alpha_n) \in \mathbb{N}_o^n$ *with* $|\alpha| = m$. *Then we have that*

$$\sum_{m=0}^{\infty} P_m(\xi_1 e_1 + \dots + \xi_n e_n) = \sum_\alpha c_\alpha \xi_1^{\alpha_1} \dots \xi_n^{\alpha_n}$$

whenever $|\xi_1| + \dots + |\xi_n| < R/e$. *Both series converge absolutely and uniformly for* $|\xi_1| + \dots + |\xi_n| \leq r$ *whenever* $0 \leq r < R/e$.

PROOF. By the Leibniz Formula 1.8 we have that

$$P_m(\xi_1 e_1 + \dots + \xi_n e_n) = \sum_{|\alpha|=m} \frac{m!}{\alpha!} \xi_1^{\alpha_1} \dots \xi_n^{\alpha_n} A_m e_1^{\alpha_1} \dots e_n^{\alpha_n}.$$

On the other hand, using Exercise 2.G and applying the classical Leibniz formula, we get that

$$\sum_{m=0}^{\infty} \sum_{|\alpha|=m} \frac{m!}{\alpha!} |\xi_1^{\alpha_1} \ldots \xi_n^{\alpha_n}| \, \| A_m e_1^{\alpha_1} \ldots e_n^{\alpha_n} \|$$

$$\leq \sum_{m=0}^{\infty} \sum_{|\alpha|=m} \frac{m!}{\alpha!} |\xi_1|^{\alpha_1} \ldots |\xi_n|^{\alpha_n} e^m \|P_m\|$$

$$\leq \sum_{m=0}^{\infty} \|P_m\| \, e^m (|\xi_1| + \ldots + |\xi_n|)^m.$$

If $0 \leq r < R/e$ then the last written series converges uniformly for $|\xi_1| + \ldots + |\xi_n| \leq r$. The desired result follows.

The following lemma extends Lemma 4.5 to multiple power series.

4.7. LEMMA. *Let* $c_\alpha \in F$ *for each multi-index* $\alpha = (\alpha_1, \ldots, \alpha_n) \in \mathbb{N}_0^n$. *If there is* $r > 0$ *such that the series* $\sum_\alpha c_\alpha \lambda_1^{\alpha_1} \ldots \lambda_n^{\alpha_n}$ *converges absolutely to zero whenever* $|\lambda_1| \leq r, \ldots, |\lambda_n| < r$, *then* $c_\alpha = 0$ *for every* α.

PROOF. Repeated applications of Lemma 4.5 lead to the result.

EXERCISES

4.A. Let $\sum_{m=0}^{\infty} A_m(x-a)^m$ be a power series from E into F.

(a) Show that the series has a positive radius of convergence if and only if $\limsup \|A_m\|^{1/m} < \infty$.

(b) Show that the series has an infinite radius of convergence if and only if $\lim \|A_m\|^{1/m} = 0$.

4.B. Give an example of a power series $\sum_{m=0}^{\infty} P_m(x-a) = \sum_{m=0}^{\infty} A_m(x-a)^m$ such that $\limsup \|P_m\| \neq \limsup \|A_m\|$.

4.C. Let $\sum_\alpha c_\alpha (x_1 - a_1)^{\alpha_1} \ldots (x_n - a_n)^{\alpha_n}$ be a multiple power series from $\mathbb{K}^n$ into F. Let $b \in \mathbb{K}^n$ such that

$$\sup_{\alpha} \|c_\alpha\| \, |b_1 - a_1|^{\alpha_1} \cdots |b_n - a_n|^{\alpha_n} < \infty.$$

Show that the series converges absolutely and uniformly for $|x_1 - a_1| \leq r_1, \ldots, |x_n - a_n| \leq r_n$ whenever $0 \leq r_j < |b_j - a_j|$ for $j = 1, \ldots, n$. This is *Abel's lemma*.

4.D. Let (ξ_m) denote the sequence of coordinate functionals on $E = \ell^p$ where $1 \leq p < \infty$. Show that the power series $\sum_{m=0}^{\infty} (\xi_m(x))^m$ converges absolutely for every $x \in E$, but its radius of convergence equals one. Thus there is no analogue of Abel's lemma for power series in Banach spaces.

NOTES AND COMMENTS

Most of the results in Chapter I have been known for a long time and can already be found in the book of E. Hille and R. Phillips [1]. In Sections 1, 2 and 4 our presentation follows essentially the books of L. Nachbin [1], [2]. In Section 3 we have mostly followed an article of J. Bochnak and J. Siciak [1], which actually deals with spaces more general than Banach spaces. The least known result in Section 1 is perhaps Theorem 1.15, which I learned from R. Aron one Monday afternoon over coffee, at University College Dublin.

For additional results on the subject matter in this chapter see the books of T. Franzoni and E. Vesentini [1], S. Dineen [5] and J. F. Colombeau [1], the last two books being concerned more generally with locally convex spaces.

CHAPTER II

HOLOMORPHIC MAPPINGS

5. HOLOMORPHIC MAPPINGS

In this section we introduce holomorphic mappings in Banach spaces in terms of power series expansions. We derive several properties of these mappings in terms of the corresponding properties of holomorphic functions of one complex variable.

Throughout this chapter all Banach spaces considered will be complex. In particular, the letters E and F will always represent complex Banach spaces.

5.1. DEFINITION. Let U be an open subset of E. A mapping $f : U \to F$ is said to be *holomorphic* or *analytic* if for each $a \in U$ there exist a ball $B(a;r) \subset U$ and a sequence of polynomials $P_m \in \mathcal{P}(^mE;F)$ such that

$$f(x) = \sum_{m=0}^{\infty} P_m(x - a)$$

uniformly for $x \in B(a;r)$. We shall denote by $\mathcal{H}(U;F)$ the vector space of all holomorphic mapping from U into F. When $F = \mathbb{C}$ then we shall write $\mathcal{H}(U;\mathbb{C}) = \mathcal{H}(U)$.

5.2. REMARK. In view of Proposition 4.4 the sequence (P_m) which appears in Definition 5.1 is uniquely determined by f and a and we shall write $P_m = P^m f(a)$ for every $m \in \mathbb{N}_0$. The series $\sum_{m=0}^{\infty} P^m f(a)(x - a)$ is called the *Taylor series* of f at a. We shall denote by $A^m f(a)$ the unique member of $\mathcal{L}^s(^mE;F)$ such that $(A^m f(a))\hat{} = P^m f(a)$.

5.3. EXAMPLE. $\mathcal{P}(E;F) \subset \mathcal{H}(E;F)$.

PROOF. If suffices to show that $P \in \mathcal{H}(E;F)$ for each $P \in \mathcal{P}(^mE;F)$. Let $P = \hat{A}$, where $A \in \mathcal{L}^s(^mE;F)$. Given $a, x \in E$, by the Newton Binomial Formula 1.9 we have that

$$P(x) = Ax^m = \sum_{j=0}^{m} \binom{m}{j} A a^{m-j}(x-a)^j.$$

Thus P is holomorphic on E and

$$P^j P(a)(t) = \binom{m}{j} A a^{m-j} t^j \quad \text{if} \quad j \le m$$

$$P^j P(a)(t) = 0 \quad \text{if} \quad j > m.$$

5.4. EXAMPLE. Let $\sum_{m=0}^{\infty} P_m(x)$ be a power series from E into F with an infinite radius of convergence and with each P_m continuous. If we define

$$f(x) = \sum_{m=0}^{\infty} P_m(x) \quad \text{for each} \quad x \in E,$$

then $f \in \mathcal{H}(E;F)$.

PROOF. Set $P_m = \hat{A}_m$, with $A_m \in \mathcal{L}^s(^mE;F)$, for every $m \in \mathbb{N}_o$. We claim that

$$(5.1) \qquad \sum_{j=0}^{\infty} \sum_{m=j}^{\infty} \binom{m}{j} \|A_m\| \|a\|^{m-j} r^j < \infty$$

for each $a \in E$ and $r > 0$. Indeed, we have that

$$\sum_{j=0}^{\infty} \sum_{m=j}^{\infty} \binom{m}{j} \|A_m\| \|a\|^{m-j} r^j = \sum_{m=0}^{\infty} \sum_{j=0}^{m} \binom{m}{j} \|A_m\| \|a\|^{m-j} r^j$$

$$= \sum_{m=0}^{\infty} \|A_m\| (\|a\| + r)^m$$

and the last written series converges, since by Exercise 4.A we have that $\lim \|A_m\|^{1/m} = 0$.

From (5.1) we get on one hand that

$$\sum_{m=j}^{\infty} \binom{m}{j} \|A_m\| \; \|a\|^{m-j} < \infty$$

and hence the series $\sum_{m=j}^{\infty} \binom{m}{j} A_m a^{m-j}$ defines an element $Q_j \in P({}^jE;F)$ for each $j \in \mathbb{N}_0$. On the other hand if follows from (5.1) that

$$\begin{aligned} f(x) &= \sum_{m=0}^{\infty} P_m(x) \\ &= \sum_{m=0}^{\infty} \sum_{j=0}^{m} \binom{m}{j} A_m a^{m-j}(x-a)^j \\ &= \sum_{j=0}^{\infty} \sum_{m=j}^{\infty} \binom{m}{j} A_m a^{m-j}(x-a)^j \\ &= \sum_{j=0}^{\infty} Q_j(x-a) \end{aligned}$$

uniformly for $x \in \overline{B}(a;r)$. Thus $f \in \mathcal{H}(E;F)$.

5.5. **EXAMPLE.** Let (φ_m) be a sequence in E' which converges pointwise to zero. If we define

$$f(x) = \sum_{m=0}^{\infty} (\varphi_m(x))^m \quad \text{for every} \quad x \in E,$$

then $f \in \mathcal{H}(E)$.

PROOF. By the Principle of Uniform Boundedness these is a constant $c > 0$ such that $\|\varphi_m\| \leq c$ for every $m \in \mathbb{N}_0$. We claim that

$$(5.2) \qquad \sum_{j=0}^{\infty} \sum_{m=j}^{\infty} \binom{m}{j} |\varphi_m(a)|^{m-j} \|\varphi_m\|^j r^j < \infty$$

for each $a \in E$ and each r with $0 \leq r < 1/c$. Indeed, we have that

$$\sum_{j=0}^{\infty}\sum_{m=j}^{\infty}\binom{m}{j}|\varphi_m(a)|^{m-j}\|\varphi_m\|^j r^j = \sum_{m=0}^{\infty}\sum_{j=0}^{m}\binom{m}{j}|\varphi_m(a)|^{m-j}\|\varphi_m\|^j r^j$$

$$= \sum_{m=0}^{\infty}(|\varphi_m(a)| + \|\varphi_m\| r)^m$$

$$\leq \sum_{m=0}^{\infty}(|\varphi_m(a)| + cr)^m$$

and the last written series converges since $cr < 1$ and $\varphi_m(a) \to 0$.

From (5.2) we set on one hand that

$$\sum_{m=j}^{\infty}\binom{m}{j}|\varphi_m(a)|^{m-j}\|\varphi_m\|^j < \infty$$

and hence the series $\sum_{m=j}^{\infty}\binom{m}{j}(\varphi_m(a))^{m-j}\varphi_m^j$ defines an element $Q_j \in P({}^jE)$ for each $j \in \mathbb{N}_0$. On the other hand if follows from (5.2) that

$$f(x) = \sum_{m=0}^{\infty}[\varphi_m(a) + \varphi_m(x-a)]^m$$

$$= \sum_{m=0}^{\infty}\sum_{j=0}^{m}\binom{m}{j}(\varphi_m(a))^{m-j}(\varphi_m(x-a))^j$$

$$= \sum_{j=0}^{\infty}\sum_{m=j}^{\infty}\binom{m}{j}(\varphi_m(a))^{m-j}(\varphi_m(x-a))^j$$

$$= \sum_{j=0}^{\infty}Q_j(x-a)$$

uniformly for $x \in \overline{B}(a;r)$. Thus $f \in \mathcal{H}(E;F)$.

Many properties of holomorphic mappings in Banach spaces can be derived from the corresponding properties of holomorphic functions of one complex variable with the aid of the following simple result, whose straightforward proof is left as an exercise to the reader.

5.6. LEMMA. *Let U be an open subset of E, and let $f \in \mathcal{H}(U;F)$.*

Then:

(a) f *is continuous.*

(b) f *is locally bounded, that is,* f *is bounded on a suitable neighborhood of each point of* U.

(c) *For each* $a \in U$, $b \in E$ *and* $\psi \in F'$ *the function* $\lambda \to \psi \circ f(a + \lambda b)$ *is holomorphic on the open set* $\{\lambda \in \mathbb{C} : a + \lambda b \in U\}$.

To begin with, we extend the *Identity Principle.*

5.7. PROPOSITION. *Let* U *be a connected open subset of* E, *and let* $f \in \mathcal{H}(U;F)$. *If* f *is identically zero on a nonvoid open set* $V \subset U$ *then* f *is identically zero on all of* U.

PROOF. (a) First assume U convex. Let $a \in V$, let $x \in U$ and let

$$\Lambda = \{\lambda \in \mathbb{C} : a + \lambda(x - a) \in U\}.$$

Since U is convex the open set Λ is convex as well, and in particular connected. For each $\psi \in F'$ the function

$$g(\lambda) = \psi \circ f\,[a + \lambda(x - a)]$$

is holomorphic on Λ and is identically zero on an open disc $\Delta(0;\varepsilon)$. Then g is identically zero on Λ by the Identity Principle for holomorphic functions of one complex variable. In particular $\psi \circ f(x) = g(1) = 0$, and since F' separates the points of F we conclude that $f(x) = 0$.

(b) In the general case, let A denote the set of all points $a \in U$, such that f is identically zero on a neighborhood of a. Then A is obviously open, and to complete the proof it suffices to show that A is closed in U. Let (a_n) be a sequence in A which converges to a point $b \in U$. Choose $r > 0$ such that $B(b;r) \subset U$ and choose n such that $a_n \in B(b;r)$. Then if follows from (a) that f is identically zero on $B(b;r)$. Hence $b \in A$ and the proof is complete.

Next we extend the *Open Mapping Principle.*

5.8. PROPOSITION. *Let U be a connected open subset of E, and let $f \in \mathcal{H}(U)$. If f is not constant on U then $f(V)$ is an open subset of $\mathbb{C}$ for each open subset V of U.*

PROOF. Clearly if suffices to show that $f(V)$ is an open subset of $\mathbb{C}$ for each convex open subset V of U. Let V be a convex open subset of U and let $x \in V$. By the Identity Principle 5.7 the function f is not constant on V and hence there is a point $y \in V$ such that $f(x) \neq f(y)$. Since V is convex, the open set

$$\Lambda = \{\lambda \in \mathbb{C} : x + \lambda(y - x) \in V\}$$

is convex as well. The function

$$g(\lambda) = f[x + \lambda(y - x)]$$

is holomorphic on Λ and $g(0) = f(x) \neq f(y) = g(1)$. By the Open Mapping Principle for holomorphic functions of one complex variable the set $g(\Lambda)$ is open in $\mathbb{C}$. Since

$$f(x) = g(0) \in g(\Lambda) \subset f(V),$$

we conclude that $f(V)$ is also open in $\mathbb{C}$.

As an immediate consequence we obtain the *Maximum Principle.*

5.9. PROPOSITION. *Let U be a connected open subset of E, and let $f \in \mathcal{H}(U)$. If there exists $a \in U$ such that $|f(x)| \leq |f(a)|$ for every $x \in U$ then f is constant on U.*

PROOF. Assume f is not constant on U. Then by the Open Mapping Principle 5.8 the set $f(U)$ is open in $\mathbb{C}$, and hence contains an open disc $\Delta(f(a);r)$. But this is impossible, since by hypothesis $|f(x)| \leq |f(a)|$ for every $x \in U$.

To end this section we generalize *Liouville's Theorem.*

5.10. PROPOSITION. *If a mapping* $f \in \mathcal{H}(E;F)$ *is bounded on* E *then it is constant on* E.

PROOF. Let $x \in E$ and $\psi \in F'$. Then the function $g(\lambda) = \psi \circ f(\lambda x)$ is holomorphic on $\mathbb{C}$ and bounded there. By the classical Liouville's theorem, g is constant, and in particular $\psi \circ f(x) = \psi \circ f(0)$. Since F' separates the points of F we conclude that $f(x) = f(0)$ and the proof is complete.

EXERCISES

5.A. Let E, F, G, H be Banach spaces, and let V be an open subset of F. Given $S \in \mathcal{L}(E;F)$, $f \in \mathcal{H}(V;G)$ and $T \in \mathcal{L}(G;H)$, show that $f \circ S \in \mathcal{H}(S^{-1}(V);G)$ and $T \circ f \in \mathcal{H}(V;H)$.

5.B. Let U be an open subset of E, and let $a \in E$. For each $f \in \mathcal{H}(U;F)$ let $f_a : U - a \to F$ be defined by $f_a(t) = f(a+t)$ for every $t \in U - a$.

(a) Show that $f_a \in \mathcal{H}(U-a;F)$ and $P^m f_a(t) = P^m f(a+t)$ for every $t \in U - a$ and $m \in \mathbb{N}_0$.

(b) Show that the mapping $f \to f_a$ is a vector space isomorphism between $\mathcal{H}(U;F)$ and $\mathcal{H}(U-a;F)$.

5.C. Let U be an open subset of E. Given two functions $f, g \in \mathcal{H}(U)$ show that $fg \in \mathcal{H}(U)$ and

$$P^m(fg)(x) = \sum_{j=0}^{m} P^{m-j} f(x) P^j g(x)$$

for all $m \in \mathbb{N}_0$ and $x \in U$.

5.D. Let $\sum_{m=0}^{\infty} P_m(x-a)$ be a power series from E into F, with radius of convergence $R > 0$ and with each P_m continuous. Let $f : B(a;R) \to F$ be defined by

$$f(x) = \sum_{m=0}^{\infty} P_m(x-a) \quad \text{for each} \quad x \in B(a;R).$$

Show that f is holomorphic on the ball $B(a;R/e)$. Can you show that f is holomorphic on the ball $B(a;R)$?

5.E. Let X be a topological space.

(a) Show that each continuous mapping $f : X \to F$ is locally bounded.

(b) If X is metrizable show that a mapping $f : X \to F$ is locally bounded if and only if f is bounded on each compact subset of X.

5.F. Let U be a connected open subset of E, and let $f \in \mathcal{H}(U;F)$. Suppose there are a nonvoid open subset V of U and a closed subspace N of F such that $f(V) \subset N$. Show that $f(U) \subset N$.

5.G. Let U be a connected open subset of E, and let $f \in \mathcal{H}(U;F)$. If there is a point $a \in U$ such that $\|f(x)\| \leq \|f(a)\|$ for all $x \in U$, show that $\|f\|$ is constant on U.

5.H. Let $F = \mathbb{C}^2$ with the norm of the supremum. Let $f : \mathbb{C} \to F$ be defined by $f(z) = (1,z)$ for every $z \in \mathbb{C}$.

(a) Show that $f \in P(\mathbb{C};F)$.

(b) Show that $\|f\|$ is constant on $\Delta(0;1)$.

(c) Show that f is not constant on $\Delta(0;1)$.

(d) Show that $\|f\|$ is not constant on $\mathbb{C}$.

6. VECTOR-VALUED INTEGRATION

We assume that the reader is familiar with the theory of Lebesgue measure and integration, and by this we mean integration of scalar-valued functions. However, throughout this book we shall often find desirable to integrate functions with values in a Banach space. With this in mind we present a few elementary facts regarding the Bochner integral. These few facts will be sufficient for our needs.

6.1. DEFINITION. Let (X, Σ, μ) be a finite measure space. A mapping $f : X \to F$ is said to be *simple* if there are disjoint sets $A_1, \ldots, A_k \in \Sigma$ and vectors $b_1, \ldots, b_k \in F$ such that

$$f(x) = \sum_{j=1}^{k} \chi_{A_j}(x) b_j \quad \text{for all} \quad x \in X.$$

Then for each $A \in \Sigma$ we define

$$\int_A f d\mu = \sum_{j=1}^{k} \mu(A \cap A_j) b_j .$$

The verification of the following lemma is straightforward, and is left as an exercise to the reader.

6.2. LEMMA. *Let (X, Σ, μ) be a finite measure space, and let $f : X \to F$ be a simple mapping. Then for each $A \in \Sigma$ and $\psi \in F'$ we have that:*

(a) $\psi(\int_A f d\mu) = \int_A \psi \circ f d\mu$

(b) $\| \int_A f d\mu \| \leq \int_A \| f \| d\mu$.

6.3. DEFINITION. Let (X, Σ, μ) be a finite measure space.

(a) A mapping $f : X \to F$ is said to be *measurable* if there exists a sequence of simple mappings $f_n : X \to F$ which converges to f almost everywhere.

(b) A measurable mapping $f : X \to F$ is said to be *Bochner integrable* if there exists a sequence of simple mappings $f_n : X \to F$ such that

$$\lim_{n \to \infty} \int_X \| f_n - f \| d\mu = 0.$$

In this case we define

$$\int_A f d\mu = \lim_{n \to \infty} \int_A f_n \, d\mu$$

for each $A \in \Sigma$.

Lemma 6.2 guarantees that the Bochner integral $\int_A f d\mu$ is well defined. Indeed, on one hand Lemma 6.2 implies that $(\int_A f_n d\mu)$ is a Cauchy sequence, and on the other hand Lemma 6.2 guarantees that the definition of $\int_A f d\mu$ is independent of the choice of the sequence (f_n). Finally, from Lemma 6.2 and the definition of the Bochner integral we can easily obtain the following proposition. The details are left to the reader.

6.4. PROPOSITION. *Let (X, Σ, μ) be a finite measure space, and let $f : X \to F$ be a Bochner integrable mapping. Then:*

(a) *The function $\psi \circ f : X \to \mathbb{C}$ is integrable and*

$$\psi\left(\int_A f d\mu\right) = \int_A \psi \circ f d\mu$$

for each $\psi \in F'$ and $A \in \Sigma$.

(b) *The function $\| f \| : X \to \mathbb{R}$ is integrable and*

$$\left\| \int_A f d\mu \right\| \leq \int_A \| f \| \, d\mu$$

for each $A \in \Sigma$.

6.5. PROPOSITION. *Let μ be a finite Borel measure on a compact Hausdorff space X. Then each continuous mapping $f : X \to F$ is Bochner integrable.*

PROOF. Clearly it suffices to show that f is the uniform limit of a sequence of simple functions. Let $n \in \mathbb{N}$ be given. Since f is continuous and X is compact we can find points $a_1, \ldots, a_k \in X$ such that

$$X = \bigcup_{j=1}^{k} f^{-1} [B(f(a_j); 1/n)].$$

For each $j = 1,\ldots,k$ set

$$U_j = f^{-1}[B(f(a_j);\ 1/n)]$$

$$A_j = U_j \setminus \bigcup_{i<j} U_i .$$

Then $A_1,\ldots,A_k$ are disjoint Borel sets which cover X. If we define

$$f_n(x) = \sum_{j=1}^{k} \chi_{A_j}(x) f(a_j) \quad \text{for every} \quad x \in X,$$

then f_n is a simple function and

$$\| f_n(x) - f(x) \| < 1/n \quad \text{for every} \quad x \in X.$$

6.6. COROLLARY. *Let* μ *be a finite Borel measure on a compact Hausdorff space* X. *Let* (f_n) *be a sequence of continuous mappings from* X *into* F *which converges uniformly on* X *to a mapping* f. *Then* f *is continuous and*

$$\int_A f\,d\mu = \lim_{n\to\infty} \int_A f_n\,d\mu$$

For each $A \in \Sigma$.

So far we have only considered Bochner integration with respect to finite positive measures, but the extension to real measures or to complex measures may proceed exactly as in the scalar case.

EXERCISES

6.A. This is *Egoroff's Theorem* for vector-valued mappings. Let (X, Σ, μ) be a finite measure space. Let (f_n) be a sequence of measurable mappings from X into F which converges almost everywhere to a mapping f. By replacing absolute values by norms at the appropiate places in the standard proof of the scalar Egoroff's theorem, show that for each $\varepsilon > 0$ there exists a set $A \in \Sigma$ with $\mu(X \setminus A) \leq \varepsilon$ and such that (f_n) converges to f uniformly on A.

6.B. Let (X, Σ, μ) be a finite measure space. Let (f_n) be a sequence of measurable mappings from X into F which converges almost everywhere to a mapping f.

(a) Using Egoroff's Theorem 6.A find a sequence of sets $A_n \in \Sigma$ and a sequence of simple mappings $g_n : X \to F$ such that $\mu(X \setminus A_n) \leq 2^{-n}$ and $\|g_n(x) - f(x)\| \leq 2^{-n}$ for every $x \in A_n$.

(b) Let $B_j = \bigcap_{k=j}^{\infty} A_k$ for each $j \in \mathbb{N}$. Show that $\mu(X \setminus B_j) \leq 2^{-j+1}$ for every j and show that (g_n) converges to f uniformly on each B_j.

(c) Let $B = \bigcup_{j=1}^{\infty} B_j$. Show that $\mu(X \setminus B) = 0$ and show that $(g_n(x))$ converges to $f(x)$ for every $x \in B$. In particular this shows that f is measurable.

6.C. Let (X, Σ, μ) be a finite measure space. Using Exercise 6.B show that a measurable mapping $f : X \to F$ is Bochner integrable if and only if $\int_X \|f\| d\mu < \infty$. This is *Bochner's characterization* of Bochner integrable mappings.

6.D. This is the *Dominated Convergence Theorem* for Bochner integrable mappings. Let (X, Σ, μ) be a finite measure space. Let (f_n) be a sequence of Bochner integrable mappings from X into F which converges almost everywhere to a mapping f. Suppose there exists an integrable function $g : X \to \mathbb{R}$ such that $\|f_n(x)\| \leq g(x)$ for every $n \in \mathbb{N}$ and almost every $x \in X$. Using Bochner's characterization 6.C and the scalar Dominated Convergence Theorem show that f is Bochner integrable, $\int_X \|f_n - f\| d\mu \to 0$ and $\int_A f_n d\mu \to \int_A f d\mu$ for each $A \in \Sigma$.

6.E. Let (X, Σ, μ) be a finite measure space. Let (f_n) be a sequence of Bochner integrable mappings from X into F which converges uniformly on X to a mapping f. Show that f is Bochner integrable, $\int_X \| f_n - f \| d\mu \to 0$ and $\int_A f_n d\mu \to \int_A f d\mu$ for each $A \in \Sigma$.

6.F. Let μ be a Borel probability measure on a compact Hausdorff space X, and let $f : X \to F$ be a continuous mapping.

(a) Given $\psi_1, \ldots, \psi_n \in (F_{\mathbb{R}})'$ let

$$\eta_j = \int_X \psi \circ f \, d\mu \quad \text{for} \quad j = 1, \ldots, n,$$

and let $T \in \mathcal{L}(F_{\mathbb{R}}; \mathbb{R}^n)$ be defined by

$$Ty = (\psi_1(y), \ldots, \psi_n(y)) \quad \text{for every} \quad y \in F.$$

Using the Hahn-Banach separation theorem show that

$$(\eta_1, \ldots, \eta_n) \in co[\, T \circ f(X)] = T[\, co(f(X))],$$

where $co(B)$ denotes the convex hull of the set B.

(b) Using a compactness argument show the existence of a point $y \in \overline{co}(f(X))$ such that

$$\psi(y) = \int_X \psi \circ f \, d\mu \quad \text{for every} \quad \psi \in (F_{\mathbb{R}})'.$$

In particular this shows that the Bochner integral $\int_X f d\mu$ lies in the closed, convex hull $\overline{co}(f(X))$ of $f(X)$.

7. THE CAUCHY INTEGRAL FORMULAS

After the intermission on vector-valued integration in the proceding section, we continue our study of holomorphic mappings. In this section we establish the Cauchy integral formulas and derive some of their consequences. In particular we study more closely the question of convergence of the Taylor series.

7.1. THEOREM. *Let U be an open subset of E, and let $f \in \mathcal{H}(U;F)$.*

Let $a \in U$, $t \in E$ *and* $r > 0$ *be such that* $a + \zeta t \in U$ *for all* $\zeta \in \bar{\Delta}(0;r)$. *Then for each* $\lambda \in \Delta(0;r)$ *we have the Cauchy Integral Formula*

$$f(a + \lambda t) = \frac{1}{2\pi i} \int_{|\zeta|=r} \frac{f(a + \zeta t)}{\zeta - \lambda} \, d\zeta.$$

PROOF. If $\psi \in F'$ then the function $g(\zeta) = \psi \circ f(a + \zeta t)$ is holomorphic on a neighborhood of the closed disc $\bar{\Delta}(0;r)$. By the Cauchy integral formula for holomorphic functions of one complex variable we have that

$$\psi \circ f(a + \lambda t) = g(\lambda) = \frac{1}{2\pi i} \int_{|\zeta|=r} \frac{g(\zeta)}{\zeta - \lambda} \, d\zeta$$

$$= \frac{1}{2\pi i} \int_{|\zeta|=r} \frac{\psi \circ f(a + \zeta t)}{\zeta - \lambda} \, d\zeta .$$

for each $\lambda \in \Delta(0;r)$. Since F' separates the points of F the desired conclusion follows.

7.2. COROLLARY. *Let* U *be an open subset of* E, *and let* $f \in \mathcal{H}(U;F)$. *Let* $a \in U$, $t \in E$ *and* $r > 0$ *be such that* $a + \zeta t \in U$ *for all* $\zeta \in \bar{\Delta}(0;r)$. *The for each* $\lambda \in \Delta(0;r)$ *we have a series expansion of the form*

$$f(a + \lambda t) = \sum_{m=0}^{\infty} c_m \lambda^m$$

where

$$c_m = \frac{1}{2\pi i} \int_{|\zeta|=r} \frac{f(a + \lambda t)}{\zeta^{m+1}} \, d\zeta .$$

This series converges absolutely and uniformly for $|\lambda| \leq s$ *provided* $0 \leq s < r$.

PROOF. For $|\lambda| < |\zeta| = r$ we can write

$$\frac{f(a+\zeta t)}{\zeta-\lambda} = \frac{f(a+\zeta t)/\zeta}{1-\lambda/\zeta} = \sum_{m=0}^{\infty} \lambda^m \frac{f(a+\zeta t)}{\zeta^{m+1}},$$

and since f is bounded on the set $\{a+\zeta t : |\zeta| = r\}$, the series converges absolutely and uniformly for $|\zeta| = r$ and $|\lambda| \leq s < r$. By Corollary 6.6 we can integrate this series term by term to obtain

$$\int_{|\zeta|=r} \frac{f(a+\zeta t)}{\zeta-\lambda}\, d\zeta = \sum_{m=0}^{\infty} \lambda^m \int_{|\zeta|=r} \frac{f(a+\zeta t)}{\zeta^{m+1}}\, d\zeta,$$

and this last series converges absolutely and uniformly for $|\lambda| \leq s$. An application of Theorem 7.1 completes the proof.

7.3. COROLLARY. *Let U be an open subset of E, and let $f \in \mathcal{H}(U;F)$. Let $a \in U$, $t \in E$ and $r > 0$ be such that $a+\zeta t \in U$ for all $\zeta \in \bar{\Delta}(0;r)$. Then for each $m \in \mathbb{N}_0$ we have the Cauchy Integral Formula*

$$P^m f(a)(t) = \frac{1}{2\pi i} \int_{|\zeta|=r} \frac{f(a+\zeta t)}{\zeta^{m+1}}\, d\zeta .$$

PROOF. Since f is holomorphic we have a series expansion of the form

$$f(a+\lambda t) = \sum_{m=0}^{\infty} P^m f(a)(\lambda t) = \sum_{m=0}^{\infty} \lambda^m P^m f(a)(t),$$

for $|\lambda| \leq \varepsilon$, for a suitable $\varepsilon > 0$. After comparing this series expansion with the series expansion given by Corollary 7.2, an application of Lemma 4.5 completes the proof.

7.4. COROLLARY. *Let U be an open subset of E, and let $f \in \mathcal{H}(U;F)$. Let $a \in U$, $t \in E$ and $r > 0$ be such that $a+\zeta t \in U$ for all $\zeta \in \bar{\Delta}(0;r)$. The for each $m \in \mathbb{N}_0$ we have the Cauchy Inequality*

$$\|P^m f(a)(t)\| \leq r^{-m} \sup_{|\zeta|=r} \|f(a+\zeta t)\|$$

7.5. COROLLARY. *If* $P \in \mathcal{P}(^m E;F)$ *then for* $a, t \in E$ *we have the integral formula*

$$P(t) = \frac{1}{2\pi i} \int_{|\zeta|=1} \frac{P(a + \zeta t)}{\zeta^{m+1}} \, d\zeta$$

PROOF. By Example 5.3, $\hat{d}^m P(a)(t) = P(t)$. Then an application of Corollary 7.3 completes the proof.

7.6. COROLLARY. *Let* $P \in \mathcal{P}(^m E;F)$. *If* P *is bounded by* c *on an open ball* $B(a;r)$ *then* P *is also bounded by* c *on the ball* $B(0;r)$.

Thus Corollary 7.6 sharpens the conclusion of Lemma 2.5.

Before going on it is convenient to introduce some notation and terminology. A *polydisc* in $\mathbb{C}^n$ is a product of discs. The open polydisc with center $a = (a_1,\ldots,a_n)$ and polyradius $r = (r_1,\ldots,r_n)$ will be denoted by $\Delta^n(a;r)$. The corresponding closed polydisc will be denoted by $\bar{\Delta}^n(a;r)$. In other words.

$$\Delta^n(a;r) = \{z \in \mathbb{C}^n : |z_j - a_j| < r_j \quad \text{for} \quad j = 1,\ldots,n\},$$

$$\bar{\Delta}^n(a;r) = \{z \in \mathbb{C}^n : |z_j - a_j| \leq r_j \quad \text{for} \quad j = 1,\ldots,n\}.$$

If $a = 0 = (0,\ldots,0)$ and $r = 1 = (1,\ldots,1)$ then we shall simply write $\Delta^n(0;1) = \Delta^n$ and $\bar{\Delta}^n(0;1) = \bar{\Delta}^n$. The set

$$\{z \in \mathbb{C}^n : |z_j - a_j| = r_j \quad \text{for} \quad j = 1,\ldots,n\}$$

is contained in the boundary $\partial\Delta^n(a;r)$ of $\Delta^n(a;r)$, and will be denoted by $\partial_0\Delta^n(a;r)$. It is called the *distinguished boundary* of $\Delta^n(a;r)$.

7.7. THEOREM. *Let* U *be an open subset of* E, *and let* $f \in \mathcal{H}(U;F)$. *Let* $a \in U$, $t_1,\ldots,t_n \in E$ *and* $r_1,\ldots,r_n > 0$ *be such that* $a + \zeta_1 t_1 + \ldots + \zeta_n t_n \in U$ *for all* $\zeta \in \bar{\Delta}^n(0;r)$. *Then for each* $\lambda \in \Delta^n(0;r)$ *we have the Cauchy Integral Formula*

$$f(a + \lambda_1 t_1 + \ldots + \lambda_n t_n)$$

$$= \frac{1}{(2\pi i)^n} \int_{\partial_o \Delta^n(0;r)} \frac{f(a + \zeta_1 t_1 + \ldots + \zeta_n t_n)}{(\zeta_1 - \lambda_1) \ldots (\zeta_n - \lambda_n)} d\zeta_1 \ldots d\zeta_n .$$

PROOF. Since the polydisc $\overline{\Delta}^n(0;r)$ is compact, we can find $R_1 > r_1, \ldots, R_n > r_n$ such that $a + \zeta_1 t_1 + \ldots + \zeta_n t_n \in U$ for all $\zeta \in \Delta^n(0;R)$. If $\psi \in F'$ then the function

$$g(\zeta_1, \ldots, \zeta_n) = \psi \circ f(a + \zeta_1 t_1 + \ldots + \zeta_n t_n) \qquad (\zeta \in \Delta^n(0;R))$$

in separately holomorphic in each of the variables $\zeta_1, \ldots, \zeta_n$ when the other variables are held fixed. Then repeated applications of the Cauchy integral formula for holomorphic functions of one complex variable lead to the formula

$$\psi \circ f(a + \lambda_1 t_1 + \ldots + \lambda_n t_n)$$

$$= \frac{1}{(2\pi i)^n} \int_{|\zeta_1|=r_1} \frac{d\zeta_1}{\zeta_1 - \lambda_1} \int_{|\zeta_2|=r_2} \frac{d\zeta_2}{\zeta_2 - \lambda_2} \int_{|\zeta_n|=r_n} \frac{\psi \circ f(a+\zeta_1 t_1 + \ldots + \zeta_n t_n)}{\zeta_n - \lambda_n} d\zeta_n$$

for every $\lambda \in \Delta^n(0;r)$. Since the function

$$(\zeta_1, \ldots, \zeta_n) \to \frac{\psi \circ f(a + \zeta_1 t_1 + \ldots + \zeta_n t_n)}{(\zeta_1 - \lambda_1) \ldots (\zeta_n - \lambda_n)}$$

is continuous on the compact set $\partial_o \Delta^n(0;r)$, Fubini's Theorem allows us to replace the iterated integral by a multiple integral. And since F' separates points, the desired conclusion follows.

7.8. COROLLARY. *Let U be an open subset of E, and let $f \in \mathcal{H}(U;F)$. Let $a \in U$, $t_1, \ldots, t_n \in E$ and $r_1, \ldots, r_n > 0$ be such that $a + \zeta_1 t_1 + \ldots + \zeta_n t_n \in U$ for all $\zeta \in \overline{\Delta}^n(0;r)$. Then for each $\lambda \in \Delta^n(0;r)$ we have a series expansion of the form*

$$f(a + \lambda_1 t_1 + \ldots + \lambda_n t_n) = \sum_\alpha c_\alpha \lambda_1^{\alpha_1} \ldots \lambda_n^{\alpha_n}$$

where

$$c_\alpha = \frac{1}{(2\pi i)^n} \int_{\partial_o \Delta^n(0;r)} \frac{f(a + \zeta_1 t_1 + \ldots + \zeta_n t_n)}{\zeta_1^{\alpha_1+1} \ldots \zeta_n^{\alpha_n+1}} d\zeta_1 \ldots d\zeta_n .$$

This multiple series converges absolutely and uniformly for $\lambda \in \overline{\Delta}^n(0;s)$ *provided* $0 \leq s_j < r_j$ *for every* j.

PROOF. The proof is similar to that of Corollary 7.2. If $|\lambda_j| < |\zeta_j| = r_j$ for $j = 1,\ldots,n$ then we can write

$$\frac{f(a + \lambda_1 t_1 + \ldots + \lambda_n t_n)}{(\zeta_1 - \lambda_1) \ldots (\zeta_n - \lambda_n)}$$

$$= \sum_\alpha \lambda_1^{\alpha_1} \ldots \lambda_n^{\alpha_n} \frac{f(a + \zeta_1 t_1 + \ldots + \zeta_n t_n)}{\zeta_1^{\alpha_1+1} \ldots \zeta_n^{\alpha_n+1}}$$

and this multiple series converges absolutely and uniformly for $|\zeta_j| = r_j$ and $|\lambda_j| \leq s_j < r_j$. After integrating this series term by term, the desired conclusion follows from Theorem 7.7.

7.9. COROLLARY. *Let* U *be an open subset of* E, *and let* $f \in \mathcal{H}(U;F)$. *Let* $a \in U$, $t_1,\ldots,t_n \in E$ *and* $r_1,\ldots,r_n > 0$ *be such that* $a + \zeta_1 t_1 + \ldots + \zeta_n t_n \in U$ *for all* $\zeta \in \overline{\Delta}^n(0;r)$. *Then for each* $m \in \mathbb{N}_o$ *and each multi-index* $\alpha \in \mathbb{N}_o^n$ *with* $|\alpha| = m$ *we have the Cauchy Integral Formula*

$$A^m f(a) t_1^{\alpha_1} \ldots t_n^{\alpha_n} = \frac{\alpha!}{m!(2\pi i)^n} \int_{\partial_o \Delta^n(o;r)} \frac{f(a + \zeta_1 t_1 + \ldots + \zeta_n t_n)}{\zeta_1^{\alpha_1+1} \ldots \zeta_n^{\alpha_n+1}} d\zeta_1 \ldots d\zeta_n .$$

PROOF. By Proposition 4.6 we have a series expansion of the form

$$f(a + \lambda_1 t_1 + \ldots + \lambda_n t_n) = \sum_{m=0}^{\infty} P^m f(a)(\lambda_1 t_1 + \ldots + \lambda_n t_n)$$

$$= \sum_{\alpha} c_\alpha \lambda_1^{\alpha_1} \ldots \lambda_n^{\alpha_n}$$

where

$$c_\alpha = \frac{m!}{\alpha!} A^m f(a) t_1^{\alpha_1} \ldots t_n^{\alpha_n}$$

for each $\alpha \in \mathbb{N}_0^n$ with $|\alpha| = m$. This multiple series converges absolutely and uniformly on a suitable polydisc $\bar{\Delta}^n(0;\varepsilon)$. After comparing this series expansion with the series expansion given by Corollary 7.8, an application of Lemma 4.7 completes the proof.

7.10. COROLLARY. *Let* $A \in \mathcal{L}^s({}^mE;F)$ *and let* $P = \hat{A} \in \mathcal{P}({}^mE;F)$. *Then for all* $a, t_1, \ldots, t_n \in E$ *and all* $\alpha \in \mathbb{N}_0^n$ *with* $|\alpha| = m$ *we have the polarization formula*

$$At_1^{\alpha_1} \ldots t_n^{\alpha_n} = \frac{\alpha!}{m!(2\pi i)^n} \int_{|\zeta_j|=1} \frac{P(a + \zeta_1 t_1 + \ldots + \zeta_n t_n)}{\zeta_1^{\alpha_1+1} \ldots \zeta_n^{\alpha_n+1}} d\zeta_1 \ldots d\zeta_n .$$

PROOF. Apply Corollary 7.9 with $f = P$.

We recall that a set A in E containing the origin is said to be *balanced* if $\zeta x \in A$ for each $x \in A$ and each ζ in the closed unit disc $\bar{\Delta}$. If $a \in A$ then A is said to be *a-balanced* if the set $A - a$ is balanced.

7.11. THEOREM. *Let* U *be an a-balanced open subset of* E, *and let* $f \in \mathcal{H}(U;F)$. *Then the Taylor series of* f *at* a *converges to* f *uniformly on a suitable neighborhood of each compact subset of* U.

PROOF. Let K be a compact subset of U. Then the set

$$A = \{a + \zeta(x - a) : x \in K, \zeta \in \bar{\Delta}\}$$

is contained in U, and f is bounded on A. Using a compactness argument we can find $r > 1$ and a neighborhood V of K in U such that the set

$$B = \{a + \zeta(x - a) : x \in V, \ \zeta \in \overline{\Delta}(0;r)\}$$

is also contained in U, and f is bounded on B as well. Hence we can write

$$\frac{f[a + \zeta(x - a)]}{\zeta - 1} = \sum_{m=0}^{\infty} \frac{f[a + \zeta(x - a)]}{\zeta^{m+1}}$$

and this series converges absolutely and uniformly for $x \in V$ and $|\zeta| = r$. After integrating over the circle $|\zeta| = r$ and applying the Cauchy Integral Formulas 7.1 and 7.3 we conclude that

$$f(x) = \sum_{m=0}^{\infty} P^m f(a)(x - a)$$

and this series converges absolutely and uniformly for $x \in V$.

7.12. DEFINITION. Let U be an open subset of E, let $f \in \mathcal{H}(U;F)$ and let $a \in U$.

(a) The *radius of boundedness* of f at a is the supremum of all $r > 0$ such that $\overline{B}(a;r) \subset U$ and f is bounded on $\overline{B}(a;r)$. The radius of boundedness of f at a will be denoted by $r_b f(a)$.

(b) The radius of convergence of the Taylor series of f at a will be denoted by $r_c f(a)$. For short $r_c f(a)$ will be referred to as the *radius of convergence* of f at a.

(c) The distance from a to the boundary of U will be denoted by $d_U(a)$. When $U = E$ then for convenience we define $d_E(a) = \infty$.

7.13 THEOREM. *Let U be an open subset of E, let $f \in \mathcal{H}(U;F)$*

and let $a \in U$. *Then*

$$r_b f(a) = \min \{r_c f(a), d_U(a)\}.$$

PROOF. We observe at the outset that

$$d_U(a) = \sup \{r \geq 0 : \overline{B}(a;r) \subset U\},$$

and hence $r_b f(a) \leq d_U(a)$. Thus to show the inequality

$$(7.1) \qquad r_b f(a) \leq \min \{r_c f(a), d_U(a)\}$$

it suffices to show that $r_b f(a) \leq r_c f(a)$. Let $0 \leq r < r_b f(a)$. Then $\overline{B}(a;r) \subset U$ and f is bounded, by c say, on $\overline{B}(a;r)$. It follows from the Cauchy Inequalities 7.4 that $\| P^m f(a) \| \leq c r^{-m}$ for every $m \in \mathbb{N}$ and an application of the Cauchy-Hadamard Formula 4.3 shows that $r_c f(a) \geq r$. Letting $r \to r_b f(a)$ we get that $r_b f(a) \leq r_c f(a)$ and (7.1) follows.

Next we show that opposite inequality:

$$(7.2) \qquad r_b f(a) \geq \min \{r_c f(a), d_U(a)\}.$$

Let $0 \leq r < s < \min \{r_c f(a), d_U(a)\}$. Since $s < d_U(a)$ it follows that $B(a;s) \subset U$ and then Theorem 7.11 implies that

$$(7.3) \qquad f(x) = \sum_{m=0}^{\infty} P^m f(a)(x - a)$$

for every $x \in B(a;s)$. On the other hand, it follows from the Cauchy-Hadamard Formula that

$$\lim \sup \| P^m f(a) \|^{1/m} = \frac{1}{r_c f(a)} < \frac{1}{s}$$

and hence there exists $c > 1$ such that

$$(7.4) \qquad \| P^m f(a) \| < \frac{c}{s^m}$$

for every $m \in \mathbb{N}_0$. Then from (7.3) and (7.4) it follows that

$$\| f(x) \| \leq \sum_{m=0}^{\infty} c(\frac{r}{s})^m$$

for every $x \in \overline{B}(a;r)$. Hence $r_b f(a) \geq r$ and (7.2) follows.

7.14. REMARK. Let U be an open subset of E, and let $f \in \mathcal{H}(U;F)$. If E is finite dimensional then each closed ball with finite radius is compact, and whence it follows that $r_b f(a) = d_U(a)$ and $r_c f(a) \geq d_U(a)$ for every $a \in U$. In sharp contrast with the finite dimensional situation we have the following result.

7.15. PROPOSITION. *Suppose there exists a sequence* (φ_m) *in* E' *such that* $\| \varphi_m \| = 1$ *for every* m *and* $\lim \varphi_m(x) = 0$ *for every* $x \in E$. *Then there exists a function* $f \in \mathcal{H}(E)$ *whose radius of boundedness at the origin equals one.*

PROOF. By Example 5.5 the function

$$f(x) = \sum_{m=0}^{\infty} (\varphi_m(x))^m$$

is holomorphic on all of E. It follows from the Cauchy-Hadamard Formula that $r_c f(0) = 1$. An application of Theorem 7.13 completes the proof.

7.16. EXAMPLE. Let $E = c_o$ or ℓ^p $(1 \leq p < \infty)$ and let $(\xi_m) \subset E'$ denote the sequence of coordinate functionals. Then it is clear that $\| \xi_m \| = 1$ for every m and $\xi_m(x) \to 0$ for every $x \in E$. Thus the spaces c_o and ℓ^p $(1 \leq p < \infty)$ satisfy the hypothesis in Proposition 7.15.

7.17. THEOREM. *Let* U *be an open subset of* E, *and let* $f \in \mathcal{H}(U;F)$. *Then* $P^m f \in \mathcal{H}(U;\mathcal{P}(^mE;F))$ *and*

$$P^j(P^m f)(a) = P^m(P^{m+j} f(a))$$

for all $m, j \in \mathbb{N}_o$ *and* $a \in U$.

PROOF. (a) First we assume $0 \in U$. Choose $r > 0$ such that $\overline{B}(0;3r) \subset U$ and f is bounded on $\overline{B}(0;3r)$. By Theorem 7.13 the Taylor series of f at the origin converges to f uniformly on $\overline{B}(0;2r)$. Then given $\varepsilon > 0$ we can find $k_o \in \mathbb{N}$ such that

$$\| f(t) - \sum_{j=0}^{k} P^j f(0)(t) \| \leq \varepsilon r^m$$

for $\| t \| \leq 2r$ and $k \geq k_o$. By applying the Cauchy Inequalities 7.4 we get that

$$\| P^m f(t) - \sum_{j=0}^{k} P^m [P^j f(0)](t) \| \leq \varepsilon$$

for $\| t \| \leq r$ and $k \geq k_o$. Thus

$$P^m f(t) = \sum_{j=0}^{\infty} P^m [P^j f(0)](t)$$

and this series converges uniformly for $\| t \| \leq r$. Since $P^m[P^j f(0)] = 0$ whenever $m > j$ we can even write

$$P^m f(t) = \sum_{j=0}^{\infty} P^m [P^{m+j} f(0)](t).$$

(b) In the general case take an arbitrary point $a \in U$. If we define $f_a : U - a \to F$ by $f_a(t) = f(a + t)$ for every $t \in U - a$ then by Exercise 5.B we have that $f_a \in \mathcal{H}(U - a;F)$ and $P^m f_a(t) = P^m f(a + t)$ for every $t \in U - a$ and $m \in \mathbb{N}_o$. Moreover, if f is bounded on $\overline{B}(a;3r)$ then f_a is bounded on $\overline{B}(0;3r)$. Hence using (a) we get that

$$\begin{aligned} P^m f(a + t) &= P^m f_a(t) \\ &= \sum_{j=0}^{\infty} P^m [P^{m+j} f_a(0)](t) \\ &= \sum_{j=0}^{\infty} P^m [P^{m+j} f(a)](t) \end{aligned}$$

and the last written series converges uniformly for $\|t\| \leq r$. Since $P^m[P^{m+j}f(a)] \in \mathcal{P}({}^jE;\mathcal{P}({}^mE;F))$ the proof is complete.

7.18. COROLLARY. *Let U be an open subset of E, and let $f \in \mathcal{H}(U;F)$. If for each $m \in \mathbb{N}_o$ and $t \in E$ we define $P_t^m f : U \to F$ by*

$$P_t^m f(x) = P^m f(x)(t) \quad \text{for} \quad x \in U,$$

then $P_t^m f \in \mathcal{H}(U;F)$.

To end this section we generalize the classical *Schwarz' Lemma* as follows.

7.19. THEOREM. *Let $U = B(a;r) \subset E$ and let $f \in \mathcal{H}(U;F)$. Suppose that $\|f(x)\| \leq c$ for every $x \in B(a;r)$ and suppose there exists $m \in \mathbb{N}$ such that $P^j f(a) = 0$ for every $j < m$. Then*

$$\|f(x)\| \leq c\left(\frac{\|x - a\|}{r}\right)^m \quad \text{for every} \quad x \in B(a;r).$$

PROOF. Fix $x \in B(a;r)$ with $x \neq a$, and fix $\psi \in F'$ with $\|\psi\| = 1$. Let g be the function of one complex variable defined by

$$g(\lambda) = \lambda^{-m}\, \psi \circ f[a + \lambda(x - a)] \quad \text{for} \quad 0 < |\lambda| < \frac{r}{\|x - a\|},$$

$$g(0) = \psi \circ P^m f(a)(x - a).$$

By Theorem 7.11 the Taylor series of f at a converges pointwise to f on the ball $B(a;r)$. Whence the function g can be written as the sum of the power series

$$g(\lambda) = \sum_{j=m}^{\infty} \lambda^{j-m} \psi \circ P^j f(a)(x - a)$$

on the disc $\Delta(0;r/\|x - a\|)$. In particular g is holomorphic on that disc. Take s with $\|x - a\| < s < r$. Since $\|f\| \leq c$ on $B(a;r)$ it follows that

$$|g(\lambda)| \leq c\left(\frac{\|x - a\|}{s}\right)^m$$

for $|\lambda| = s/\|x - a\|$, and therefore for $|\lambda| \leq s/\|x - a\|$ by the classical Maximum Principle. By applying this inequality with $\lambda = 1$ we get that

$$|\psi \circ f(x)| \leq c\left(\frac{\|x - a\|}{s}\right)^m.$$

After letting $s \to r$, an application of the Hahn-Banach Theorem completes the proof.

EXERCISES

7.A. Let $P \in \mathcal{P}(E;F)$ be a continuous polynomial of degree at most m. Show that

$$\int_{|\zeta|=r} \frac{P(a + \zeta t)}{\zeta^{k+1}} = 0 \quad .$$

for all $a, t \in E$, $r > 0$ and $k > m$.

7.B. Let $f \in \mathcal{H}(E;F)$. Suppose there is an integer $m \in \mathbb{N}_0$ and a constant $c > 0$ such that $\|f(x)\| \leq c\|x\|^m$ for all $x \in E$. Show that f is a polynomial of degree at most m.

7.C. Let U be an open subset of E, and let $f \in \mathcal{H}(U;F)$. Suppose there is a closed subspace N of F such that $f(x) \in N$ for every $x \in U$. Show that $f \in \mathcal{H}(U;N)$.

7.D. Show that if U is a proper open subset of E then

$$|d_U(x) - d_U(y)| \leq \|x - y\|$$

for all $x, y \in U$.

7.E. Let U be an open subset of E, and let $f \in \mathcal{H}(U;F)$. Show that either $r_b f(x) = \infty$ for every $x \in U$ (in this case $U = E$), or else $r_b f(x) < \infty$ for every $x \in U$, and in the latter case

$$|r_b f(x) - r_b f(y)| \leq \|x - y\|$$

for all $x, y \in U$.

7.F. (a) Let E be a separable Banach space. Using Cantor's diagonal process show that each bounded sequence in E' has a $\sigma(E',E)$-convergent subsequence.

(b) Show that each infinite dimensional, separable Banach space satisfies the hypothesis in Proposition 7.15.

7.G. (a) Let E be a reflexive Banach space. By considering a suitable separable, closed subspace of E show that each bounded sequence in E has a $\sigma(E,E')$-convergent subsequence.

(b) Show that each infinite dimensional, reflexive Banach space satisfies the hypothesis in Proposition 7.15

8. G-HOLOMORPHIC MAPPINGS

In this section we show that a mapping is holomorphic if it is continuous and its restriction to each complex line is holomorphic. This is a useful characterization, and in many situations this is the easiest way to check that a given mapping is holomorphic.

8.1. DEFINITION. Let U be an open subset of E. A mapping $f : U \to F$ is said to be *G-holomorphic* or *G-analytic* (G for Goursat) if for all $a \in U$ and $b \in E$ the mapping $\lambda \to f(a + \lambda b)$ is holomorphic on the open set $\{\lambda \in \mathbb{C} : a + \lambda b \in U\}$. We shall denote by $\mathcal{H}_G(U;F)$ the vector space of all G-holomorphic mappings from U into F. If $F = \mathbb{C}$ then we shall write $\mathcal{H}_G(U;\mathbb{C}) = \mathcal{H}_G(U)$.

8.2. EXAMPLE. $P_a(E;F) \subset \mathcal{H}_G(E;F)$.

PROOF. $P(a + \lambda b)$ is a polynomial in λ for all $a, b \in E$.

8.3. REMARKS. (a) The Identity Principle, the Open Mapping Principle, the Maximum Principle and Liouville's Theorem, all of them established in Section 5 for holomorphic mappings, are actually true for G-holomorphic mappings. A glance at the corresponding proofs shows this at once.

(b) An examination of the corresponding proofs shows that Theorem 7.1 and Corollary 7.2 are still valid for G-holomorphic mappings.

(c) Finally, an examination of the corresponding proofs shows that Theorem 7.7 and Corollary 7.8 are still valid for those G-holomorphic mappings whose restrictions to finite dimensional subspaces are continuous.

8.4. PROPOSITION. *Let U be an open subset of E, and let $f \in \mathcal{H}_G(U;F)$. For each $a \in U$ and $m \in \mathbb{N}_0$ let $P^m f(a) : E \to F$ be defined by*

$$P^m f(a)(t) = \frac{1}{2\pi i} \int_{|\zeta|=r} \frac{f(a + \zeta t)}{\zeta^{m+1}} \, d\zeta$$

where $r > 0$ is chosen so that $a + \zeta t \in U$ for all $\zeta \in \bar{\Delta}(0;r)$. Then:

(a) *The definition of $P^m f(a)(t)$ is independent from the choice of r.*

(b) *The mapping $P^m f(a)$ is m-homogeneous, that is*

$$P^m f(a)(\mu t) = \mu^m P^m f(a)(t)$$

for all $t \in E$ and $\mu \in \mathbb{C}$.

(c) *If U is a-balanced then for each $x \in U$ we have the series expansion*

$$f(x) = \sum_{m=0}^{\infty} P^m f(a)(x - a).$$

PROOF. (a) Let $t \in E$ be given and let $0 < s < r$ be such that $a + \zeta t \in U$ for all $\zeta \in \overline{\Delta}(0;r)$. Then by Remark 8.3(b) we have that

$$\sum_{m=0}^{\infty} \lambda^m \int_{|\zeta|=r} \frac{f(a+\zeta t)}{\zeta^{m+1}}\, d\zeta = 2\pi i\, f(a+\lambda t) = \sum_{m=0}^{\infty} \lambda^m \int_{|\zeta|=s} \frac{f(a+\zeta t)}{\zeta^{m+1}}\, d\zeta$$

for every $\lambda \in \Delta(0;s)$. By Lemma 4.5 we conclude that

$$\int_{|\zeta|=r} \frac{f(a+\zeta t)}{\zeta^{m+1}}\, d\zeta = \int_{|\zeta|=s} \frac{f(a+\zeta t)}{\zeta^{m+1}}\, d\zeta$$

for every $m \in \mathbb{N}_o$.

(b) Let $t \in E$ and $\mu \in \mathbb{C}$ be given. If $r > 0$ is sufficiently small then again by Remark 8.3(b) we have that

$$\sum_{m=0}^{\infty} \lambda^m \int_{|\zeta|=r} \frac{f(a+\zeta\mu t)}{\zeta^{m+1}}\, d\zeta = 2\pi i\, f(a+\lambda\mu t) = \sum_{m=0}^{\infty} \lambda^m \mu^m \int_{|\zeta|=r} \frac{f(a+\zeta t)}{\zeta^{m+1}}\, d\zeta$$

for every $\lambda \in \Delta(0:r)$. Then another application of Lemma 4.5 yields the desired conclusion.

(c) Given $x \in U$ we have that $a + \zeta(x - a) \in U$ for all $\zeta \in \overline{\Delta}$. By a compactness argument we can find $r > 1$ such that $a + \zeta(x - a) \in U$ for all $\zeta \in \overline{\Delta}(0;r)$. Then by Remark 8.3(b) we have that

$$f[a + \lambda(x - a)] = \sum_{m=0}^{\infty} \lambda^m P^m f(a)(x - a)$$

for each $\lambda \in \Delta(0;r)$. Letting $\lambda = 1$ we get the desired conclusion.

8.5. REMARK. Under the setting of Proposition 8.4 it is true that $P^m f(a) \in P_a(^mE;F)$ for all $a \in U$ and $m \in \mathbb{N}_o$, but a proof of this fact, without additional hypotheses on f, will have to wait until Section 36, for it rests on a deep theorem of Hartogs on separate analyticity.

8.6. PROPOSITION. *Let U be an open subset of E, and let $f \in \mathcal{H}_G(U;F)$. Then f is continuous if and only if f is locally bounded.*

PROOF. To begin with we remark that Schwarz' Lemma 7.19 is valid for G-holomorphic mappings. Indeed, the same proof applies provided we use Proposition 8.4(c) instead of Theorem 7.11.

Now, let $f : U \to F$ be G-holomorphic and locally bounded. Given $a \in U$ we choose $r > 0$ and $c > 0$ such that $\|f(x)\| \le c$ for all $x \in B(a;r)$. By applying Schwarz' Lemma to the mapping $f(x) - f(a)$ we get that

$$\|f(x) - f(a)\| \le 2c \frac{\|x - a\|}{r}$$

for all $x \in B(a;r)$, proving that f is continuous at the point a. Since the reverse implication is clear, the proof is complete.

Now we can establish the characterization of holomorphic mappings announced at the beginning of this section.

8.7. THEOREM. *Let U be an open subset of E. Then for each mapping $f : U \to F$ the following conditions are equivalent:*

(a) *f is holomorphic.*

(b) *f is continuous and G-holomorphic.*

(c) *f is continuous and $f|U \cap M$ is holomorphic for each finite dimensional subspace M of E.*

PROOF. The implication (a) $\Rightarrow$ (b) is clear.

(b) $\Rightarrow$ (c): Let $f : U \to F$ be G-holomorphic and continuous, and let M be a finite dimensional subspace of E. Let $a \in U \cap M$ and let $(e_1,\ldots,e_n)$ be a basis for M. Then by Remark 8.3(c) we have a series expansion of the form

$$f(a + \lambda_1 e_1 + \ldots + \lambda_n e_n) = \sum_\alpha c_\alpha \lambda_1^{\alpha_1} \ldots \lambda_n^{\alpha_n}$$

where this multiple series converges absolutely and uniformly on a suitable polydisc $\Delta^n(0;r)$. If for each $m \in \mathbb{N}_o$ we define $P_m \in P(^mM;F)$ by

$$P_m(\lambda_1 e_1 + \ldots + \lambda_n e_n) = \sum_{|\alpha|=m} c_\alpha \lambda_1^{\alpha_1} \ldots \lambda_n^{\alpha_n}$$

then we have a power series expansion

$$f(a + \lambda_1 e_1 + \ldots + \lambda_n e_n) = \sum_{m=0}^{\infty} P_m(\lambda_1 e_1 + \ldots + \lambda_n e_n)$$

with uniform convergence on $\Delta^n(0;r)$. This shows (c).

(c) ⇒ (a): Let $B(a;r) \subset U$. If M is a finite dimensional subspace of E containing a then by hypothesis $f \mid U \cap M$ is holomorphic and then by Theorem 7.11 there is a power series $\sum_{m=0}^{\infty} P_m^M(x - a)$ from M into F such that

$$f(x) = \sum_{m=0}^{\infty} P_m^M(x - a)$$

for every $x \in M \cap B(a;r)$. If M and N are two finite dimensional subspaces of E containing a then it follows from the uniqueness of the Taylor series expansion that $P_m^M(t) = P_m^N(t)$ for all $t \in M \cap N$ and all $m \in \mathbb{N}_o$. Let $P_m : E \to F$ be defined by $P_m(t) = P_m^M(t)$ if M is any finite dimensional subspace of E containing a and t. Then $P_m \in P_a(^mE;F)$ by Exercise 2.B, and

$$f(x) = \sum_{m=0}^{\infty} P_m(x - a)$$

for every $x \in B(a,r)$. Now since f is continuous we can find a ball $\overline{B}(a;s) \subset B(a;r)$ and a constant $c > 0$ such that $\|f(x)\| \leq c$ for every $x \in \overline{B}(a;s)$. Given $t \in E$ with $\|t\| \leq 1$ let M be any finite dimensional subspace of E containing a and t. Then by the Cauchy Integral Formula 7.3 we get that

$$P_m(t) = P_m^M(t) = \frac{1}{2\pi i} \int_{|\zeta|=s} \frac{f(a + \zeta t)}{\zeta^{m+1}} \, d\zeta$$

and it follows that $\|P_m\| \leq cs^{-m}$. Hence each P_m is continuous and the power series $\sum_{m=0}^{\infty} P_m(x - a)$ has a radius of convergence greater than or equal to s. This show (a) and the theorem.

Let U be an open subset of $\mathbb{C}^n$. Then each G-holomorphic mapping $f : U \to F$ is *separately holomorphic*, that is, $f(\zeta_1,\ldots,\zeta_n)$ is holomorphic in each ζ_j when the other variables are held fixed. The following result on separately holomorphic mappings parallels Proposition 8.6.

8.8. LEMMA. *Let U be an open subset of $\mathbb{C}^n$, and let $f: U \to F$ be separately holomorphic. Then f is continuous if and only if f is locally bounded.*

PROOF. Let $f : U \to F$ be separately holomorphic and locally bounded. Given $a \in U$ choose $r > 0$ and $c > 0$ such that $\|f(\zeta)\| \leq c$ for every $\zeta \in \Delta^n(a;r)$. Then for each $\zeta \in \Delta^n(a;r)$ we can write

$$f(\zeta) - f(a)$$

$$= \sum_{j=1}^{n} [f(a_1,\ldots,a_{j-1},\zeta_j,\ldots,\zeta_n) - f(a_1,\ldots,a_j,\zeta_{j+1},\ldots,\zeta_n)].$$

If follows from the hypothesis that the difference

$$g_j(\zeta_j) = f(a_1,\ldots,a_{j-1},\zeta_j,\ldots,\zeta_n) - f(a_1,\ldots,a_j,\zeta_{j+1},\ldots,\zeta_n)$$

is a holomorphic function of ζ_j when the other variables are held fixed. Furthermore, $\|g_j(\zeta_j)\| \leq 2c$ for $|\zeta_j - a_j| < r$. Hence, by applying Schwarz' Lemma to each g_j we get that

$$\|f(\zeta) - f(a)\| \leq \sum_{j=1}^{n} 2c \frac{|\zeta_j - a_j|}{r}$$

for every $\zeta \in \Delta^n(a;r)$. This shows that f is continuous. Since the opposite implication is clear, the proof of the lemma is complete.

8.9. **LEMMA.** *Let U be an open subset of $\mathbb{C}^n$. Then a mapping $f : U \to F$ is holomorphic if and only if f is separately holomorphic and continuous.*

PROOF. To prove the nontrivial implication let $f : U \to F$ be separately holomorphic and continuous, and let $a \in U$. Then the proofs of Theorem 7.7 and Corollary 7.8 apply and hence we obtain a series expansion of the form

$$f(a + \lambda) = \sum_{\alpha} c_\alpha \lambda_1^{\alpha_1} \dots \lambda_n^{\alpha_n}$$

with absolute and uniform convergence on a suitable polydisc. Hence f is holomorphic.

Lemma 8.9 extends at once to arbitrary Banach spaces.

8.10. **PROPOSITION.** *Let $E_1,\dots,E_n$, F be Banach spaces, and let U be an open subset of $E_1 \times \dots \times E_n$. Then a mapping $f : U \to F$ is holomorphic if and only if f is separately holomorphic and continuous.*

PROOF. To prove the nontrivial implication, let $f : U \to F$ be separately holomorphic and continuous. Let $a = (a_1,\dots,a_n) \in U$ and $b = (b_1,\dots,b_n) \in E_1 \times \dots \times E_n$. Then the mapping

$$(\lambda_1,\dots,\lambda_n) \to f(a_1 + \lambda_1 b_1,\dots,a_n + \lambda_n b_n)$$

is separately holomorphic and continuous, and therefore holomorphic by Lemma 8.9. Whence the mapping

$$\lambda \to f(a_1 + \lambda b_1,\dots,a_n + \lambda b_n)$$

is holomorphic as well. Thus f is G-holomorphic and continuous, and therefore holomorphic by Theorem 8.7. The proof is complete.

In Section 36 we shall prove that the hypothesis of continuity in Lemma 8.9 and Proposition 8.10 is superfluous. This is a deep theorem of Hartogs.

By introducing the notion of G-holomorphic mapping we are reducing the study of holomorphic mappings to the case where the domain space is one dimensional. Now we introduce a notion that reduces the study of holomorphic mappings to the case where the range space is one dimensional.

8.11. DEFINITION. Let U be an open subset of E. A mapping $f : U \to F$ is said to be *weakly holomorphic* or *weakly analytic* if $\psi \circ f$ is holomorphic for every $\psi \in F'$. Likewise f is said to be *weakly G-holomorphic* or *weakly G-analytic* if $\psi \circ f$ is G-holomorphic for every $\psi \in F'$.

8.12. THEOREM. *Let U be an open subset of E, and let $f : U \to F$.*

(a) *f is G-holomorphic if and only if f is weakly G-holomorphic.*

(b) *f is holomorphic if and only if f is weakly holomorphic.*

Before proving this theorem we establish the following lemma.

8.13. LEMMA. *Let U be an open set in $\mathbb{C}$. Then a mapping $f : U \to F$ is holomorphic if and only if f is weakly holomorphic.*

PROOF. To prove the nontrivial implication let $f : U \to F$ be weakly holomorphic. First we shall prove that f is continuous. Given $\lambda_o \in U$ we choose $r > 0$ such that $\overline{\Delta}(\lambda_o; 2r) \subset U$. Let $\psi \in F'$ and let $\lambda \in \overline{\Delta}(\lambda_o; r)$, $\lambda \neq \lambda_o$. Using the Cauchy Integral Formula for holomorphic functions of one complex variable we can write

$$\psi \circ f(\lambda) - \psi \circ f(\lambda_o) = \frac{1}{2\pi i} \int_{|\zeta|=2r} \left(\frac{\psi \circ f(\zeta)}{\zeta - \lambda} - \frac{\psi \circ f(\zeta)}{\zeta - \lambda_o} \right) d\zeta$$

$$= \frac{\lambda - \lambda_o}{2\pi i} \int_{|\zeta|=2r} \frac{\psi \circ f(\zeta)}{(\zeta - \lambda)(\zeta - \lambda_o)} d\zeta$$

and it follows that

$$|\psi(\frac{f(\lambda) - f(\lambda_o)}{\lambda - \lambda_o})| \leq \frac{1}{2\pi} \cdot \frac{M}{r \cdot 2r} 4\pi r = \frac{M}{r}$$

where $M = sup\{|\psi \circ f(\zeta)| : |\zeta - \lambda_o| = 2r\}$. By the Principle of Uniform Boundedness there is a constant $c > 0$ such that

$$\| \frac{f(\lambda) - f(\lambda_o)}{\lambda - \lambda_o} \| \leq c$$

for every $\lambda \in \overline{\Delta}(\lambda_o;r)$ with $\lambda \neq \lambda_o$. This shows that f is continuous at λ_o.

Now we can prove that f is holomorphic, proceeding as in the proofs of Theorem 7.1 and Corollary 7.2. Indeed, if $\overline{\Delta}(\lambda_o;r) \subset U$ then first we get that

$$f(\lambda) = \frac{1}{2\pi i} \int_{|\zeta-\lambda_o|=r} \frac{f(\zeta)}{\zeta - \lambda} d\zeta$$

for each $\lambda \in \Delta(\lambda_o;r)$, and from this we get the series expansion

$$f(\lambda) = \frac{1}{2\pi i} \sum_{m=0}^{\infty} (\lambda - \lambda_o)^m \int_{|\zeta-\lambda_o|=r} \frac{f(\zeta)}{(\zeta - \lambda_o)^{m+1}} d\zeta$$

with uniform convergence on each polydisc $\overline{\Delta}(\lambda_o;s)$ with $0 \leq s < r$. this shows that f is holomorphic.

PROOF OF THEOREM 8.12. Part (a) is an immediate consequence of Lemma 8.13. To prove the nontrivial implication in (b) let $f : U \to F$ be weakly holomorphic. Then f is weakly G-holomorphic, and therefore G-holomorphic by (a). We claim that f is locally bounded. To show this let K be a compact subset of U. Since f is weakly holomorphic the set $f(K)$ is weakly bounded, and therefore norm bounded by the Principle of Uniform Boundedness. Thus f is bounded on each compact subset of U, and f is therefore locally bounded, by Exercise 5.E. Thus f is continuous by Proposition 8.6, and holomorphic by Theorem 8.7. The proof is complete.

EXERCISES

8.A. Let U be an open subset of E. Let $(f_n)_{n=1}^{\infty}$ be a sequence of holomorphic mappings from U into F which converges to a mapping $f : U \to F$ uniformly on each compact subset of U. Using Theorems 8.7 and 8.12 show that f is holomorphic.

8.B. Let $\sum_{m=0}^{\infty} P_m(x - a)$ be a power series from E into F with radius of convergence $R > 0$ and with each P_m continuous. Show that the sum $f(x) = \sum_{m=0}^{\infty} P_m(x - a)$ of the power series is holomorphic on the ball $B(a;R)$. This improves the conclusion in Exercise 5.D.

8.C. Let U be an open subset of E, and let $f \in \mathcal{H}(U;F)$.

(a) Using Theorem 7.13 and Exercises 7.E and 8.B show that

$$r_c f(x) \geq r_c f(a) - \| x - a \|$$

for each $a \in U$ and each $x \in B(a;d_U(a))$.

(b) Show that

$$r_c f(x) \leq r_c f(a) + \| x - a \|$$

for each $a \in U$ and each $x \in B(a; \frac{1}{2} d_U(a))$.

(c) If U is connected show that either $r_c f(x) = \infty$ for every $x \in U$, or else $r_c f(x) < \infty$ for every $x \in U$, and in the latter case

$$| r_c f(x) - r_c f(a) | \leq \| x - a \|$$

for each $a \in U$ and each $x \in B(a; \frac{1}{2} d_U(a))$.

8.D. Let U be an open subset of E. By adapting the proofs of Lemma 8.13 and Theorem 8.12 show that a mapping $f : U \to F'$ is holomorphic if and only if the function $x \in U \to f(x)(y) \in \mathbb{C}$ is holomorphic for each $y \in F$.

8.E. More generally, let E, F, G be Banach spaces, let U be an open subset of E, and let $f : U \to \mathcal{L}(F;G)$. Show that the following conditions are equivalent:

(a) f is holomorphic.

(b) The mapping $x \in U \to f(x)(y) \in G$ is holomorphic for each $y \in F$.

(c) The function $x \in U \to \eta[f(x)(y)] \in \mathbb{C}$ is holomorphic for each $y \in F$ and $\eta \in G'$.

Show that the same conclusion is true when the space $\mathcal{L}(F;G)$ is replaced by the space $P(^mF;G)$.

8.F. Let U be an open subset of E, and let $(f_n)_{n=1}^{\infty}$ be a sequence in $\mathcal{H}(U)$ which converges to zero uniformly on each compact subset of U. Show that the mapping $f : U \to c_o$ defined by $f(x) = (f_n(x))_{n=1}^{\infty}$ is holomorphic.

8.G. Let U be an open subset of E, let $1 \leq p < \infty$, and let $(f_n)_{n=1}^{\infty}$ be a sequence in $\mathcal{H}(U)$ such that $\sup_{x \in K} \sum_{n=1}^{\infty} |f_n(x)|^p < \infty$ for each compact subset K of U. Show that the mapping $f : U \to \ell^p$ defined by $f(x) = (f_n(x))_{n=1}^{\infty}$ is holomorphic.

8.H. Let U be an open subset of E, and let $(f_n)_{n=1}^{\infty}$ be a sequence of holomorphic functions from U into $\mathbb{C}$ which are uniformly bounded on each compact subset of U. Show that the mapping $f : U \to \ell^{\infty}$ defined by $f(x) = (f_n(x))_{n=1}^{\infty}$ is holomorphic.

8.I. Let $V \subset U$ be open subsets of E, with U connected. Let

$g \in \mathcal{H}(V;F)$ and suppose that for each $\psi \in F'$ there exists a function $f_\psi \in \mathcal{H}(U)$ such that $f_\psi \mid V = \psi \circ g$.

(a) Using Exercise 8.D. find a mapping $f \in \mathcal{H}(U;F'')$ such that $f \mid V = g$.

(b) Using Exercises 5.F and 7.C show that $f \in \mathcal{H}(U;F)$.

9. THE COMPACT-OPEN TOPOLOGY

Our aim in this section is the study of the completeness and compactness properties of collections of holomorphic mapping. With this in mind we introduce the compact-open topology, which is the most natural topology on the space of holomorphic mappings.

9.1. DEFINITION. Let $C(X;F)$ denote the vector space of all continuous mappings from a topological space X into a Banach space F. When $F = \mathbb{C}$ we shall write $C(X;\mathbb{C}) = C(X)$. The *compact-open topology* or *topology of compact convergence* is the locally convex topology τ_c on $C(X;F)$ which is generated by the seminorms of the form $f \to \sup_{x \in K} \| f(x) \|$, where K varies among all compact subsets of X.

9.2. DEFINITION. A topological space X is said to be a *k-space* if a set $A \subset X$ is open whenever $A \cap K$ is open in K for each compact subset K of X.

9.3. EXAMPLES. Every first countable space is a k-space. Every locally compact space is a k-space.

The verification of these examples, as well as the proof of the following lemma, are left as exercises to the reader.

9.4. LEMMA. *Let X be a k-space and let Y be an arbitrary topological space. Then a mapping $f : X \to Y$ is continuous if and only if $f \mid K$ is continuous for each compact subset K of X.*

9.5. PROPOSITION. *If X is a k-space then $(C(X;F), \tau_c)$ is complete* *for each Banach space* F.

PROOF. Let (f_i) be a Cauchy net in $(C(X;F),\tau_c)$. Then $(f_i(x))$ is a Cauchy net in F for each $x \in X$. If we define $f : X \to F$ by $f(x) = \lim f_i(x)$ then it is clear that (f_i) converges to f uniformly on each compact subset of X. Hence $f\,|\,K$ is continuous for each compact subset K of X, and since X is a k-space we conclude that f is continuous.

9.6. DEFINITION. A topological space X is *hemicompact* or *countable at infinity* if there exists a sequence $(K_n)_{n=1}^{\infty}$ of compact subsets of X such that each compact subset of X is contained in some K_n.

9.7. EXAMPLE. Each open set $U \subset \mathbb{C}^m$ is hemicompact. Indeed it suffices to consider the compact sets

$$K_n = \{x \in U : \|x\| \leq n \quad \text{and} \quad d_U(x) \geq 1/n\}.$$

The following result is clear.

9.8. PROPOSITION. *If* X *is a hemicompact space then* $(C(X;F),\tau_c)$ *is metrizable for each Banach space* F.

Let X be a topological space and let F be a Banach space. Then as usual F^X denotes the vector space of all mappings from X into F. The *topology of pointwise convergence* is the locally convex topology τ_p on F^X which is generated by the seminorms of the form $f \to \sup_{x \in A} \|f(x)\|$, where A varies among all finite subsets of X. The topology of pointwise convergence on F^X is nothing but the Tychonoff product topology.

9.9. DEFINITION. Let X be a topological space and let F be a Banach space.

(a) A family $\mathcal{F} \subset F^X$ is said to be *equicontinuous* if for each $a \in X$ and $\varepsilon > 0$ there is a neighborhood V of a in X

such that $\|f(x) - f(a)\| \leq \varepsilon$ for all $x \in V$ and $f \in \mathcal{F}$.

(b) A family $\mathcal{F} \subset F^X$ is said to be *locally bounded* if for each $a \in X$ there are a neighborhood V of a in X and a constant $c > 0$ such that $\|f(x)\| \leq c$ for all $x \in V$ and $f \in \mathcal{F}$.

The proof of the following lemma is left as an exercise to the reader.

9.10. LEMMA. *Let X be a topological space and let F be a Banach space. If a family $\mathcal{F} \subset F^X$ is equicontinuous (resp. locally bounded) then the closure $\overline{\mathcal{F}}$ of $\mathcal{F}$ for the topology of pointwise convergence is equicontinuous (resp. locally bounded) as well.*

9.11. PROPOSITION. *Let X be a topological space and let F be a Banach space. Then the topology of compact convergence and the topology of pointwise convergence induce the same topology on each equicontinuous subset of $C(X;F)$.*

PROOF. Let $\mathcal{F}$ be an equicontinuous subset of $C(X;F)$. We always have that $\tau_p \leq \tau_c$, and to show that these two topologies coincide on $\mathcal{F}$ let K be a compact subset of X and let $\varepsilon > 0$. Since $\mathcal{F}$ is equicontinuous each point $a \in K$ has a neighborhood V_a such that $\|f(x) - f(a)\| \leq \varepsilon$ for all $x \in V_a$ and $f \in \mathcal{F}$. Since K is compact there is a finite set $A \subset K$ such that $K \subset \cup \{V_a : a \in A\}$. Whence it follows that

$$\sup_{x \in K} \|f(x)\| \leq \sup_{x \in A} \|f(x)\| + \varepsilon$$

for all $f \in \mathcal{F}$. By applying this argument to the set $\mathcal{F} - \mathcal{F}$ (which is also equicontinuous) we can find a finite set $B \subset K$ such that

$$\sup_{x \in K} \|f(x) - g(x)\| \leq \sup_{x \in B} \|f(x) - g(x)\| + \varepsilon$$

for all $f, g \in \mathcal{F}$. It follows that

$$\{f \in F : \sup_{x \in K} \|f(x) - f_o(x)\| \leq 2\varepsilon\}$$

$$\supset \{f \in F : \sup_{x \in B} \|f(x) - f_o(x)\| \leq \varepsilon\}$$

for each $f_o \in F$ and the proof is complete.

Now it is easy to prove *Ascoli's Theorem.*

9.12. THEOREM. *Let X be a topological space. Then each equicontinuous, pointwise bounded subset of $C(X)$ is relatively compact in $C(X)$ for the compact-open topology.*

PROOF. Let F be an equicontinuous, pointwise bounded subset of $C(X)$, and let $\overline{F}$ denote the closure of F in $\mathbb{C}^X$. Then $\overline{F}$ is clearly pointwise bounded, and therefore compact in $\mathbb{C}^X$ by Tychonoff's product theorem. Now, the set $\overline{F}$ is equicontinuous by Lemma 9.10, and hence the product topology and the compact-open topology coincide on $\overline{F}$ by Proposition 9.11. Thus $\overline{F}$ is a compact subset of $(C(X), \tau_c)$ and the proof is complete.

After establishing some topological properties of the spaces of continuous mappings, we devote our attention to the spaces of holomorphic mappings.

9.13. PROPOSITION. *If U is an open subset of E then $\mathcal{H}(U;F)$ is a closed vector subspace of $(C(U;F), \tau_c)$. In particular $(\mathcal{H}(U;F), \tau_c)$ is complete.*

PROOF. The proposition is essentially a restatement of Exercise 8.A. Let (f_i) be a net in $\mathcal{H}(U;F)$ which converges to a mapping $f \in C(U;F)$ for the compact-open topology. Given $a \in U$, $b \in E$ and $\psi \in F'$ set $g_i(\lambda) = \psi \circ f_i(a + \lambda b)$ and $g(\lambda) = \psi \circ f(a + \lambda b)$ for every $\lambda \in \Lambda = \{\lambda \in \mathbb{C} : a + \lambda b \in U\}$. Then each g_i is holomorphic on Λ and the net (g_i) converges to g uniformly on each compact subset of Λ. By the well known theorem of Weierstrass for holomorphic functions of one complex variable, the function g is holomorphic on Λ. Then it follows from Theorems 8.7 and 8.12 that $f \in \mathcal{H}(U;F)$. The last assertion in the proposition follows from Proposition 9.5.

9.14. COROLLARY. *If* U *is an open subset of* $\mathbb{C}^n$ *then* $(\mathcal{H}(U;F), \tau_c)$ *is a Fréchet space.*

9.15. PROPOSITION. *Let* U *be an open subset of* E. *Then for each family* $\mathcal{F} \subset \mathcal{H}(U;F)$ *the following conditions are equivalent.*

(a) $\mathcal{F}$ *is bounded in* $(\mathcal{H}(U;F), \tau_c)$.

(b) $\mathcal{F}$ *is locally bounded.*

(c) $\mathcal{F}$ *is equicontinuous and pointwise bounded.*

PROOF. (a) ⇒ (b): If $\mathcal{F}$ is not locally bounded then we can find a point $a \in U$, a sequence $(f_n) \subset \mathcal{F}$ and a sequence $(a_n) \subset U$ such that $\| a_n - a \| < 1/n$ and $|f_n(a_n)| > n$ for every n. If we set

$$K = \{a_n : n \in \mathbb{N}\} \cup \{a\}$$

then K is a compact subset of U and the sequence (f_n) is unbounded on K. Hence $\mathcal{F}$ is not bounded in $(\mathcal{H}(U;F), \tau_c)$.

(b) ⇒ (a): Assume $\mathcal{F}$ is locally bounded, that is uniformly bounded on a suitable neighborhood of each point of U. Then clearly $\mathcal{F}$ is uniformly bounded on each compact subset of U, that is $\mathcal{F}$ is bounded in $(\mathcal{H}(U;F), \tau_c)$

(b) ⇒ (c): If $\mathcal{F}$ is locally bounded then $\mathcal{F}$ is obviously pointwise bounded. To show that $\mathcal{F}$ is equicontinuous let $a \in U$, $r > 0$ and $c > 0$ be such that $\overline{B}(a;r) \subset U$ and $\| f(x) \| \leq c$ for every $x \in \overline{B}(a;r)$ and $f \in \mathcal{F}$. Then it follows from the Cauchy inequalities that

$$\| f(x) - f(a) \| \leq \sum_{m=1}^{\infty} \| P^m f(a)(x - a) \|$$

$$\leq \sum_{m=1}^{\infty} c \left(\frac{\| x - a \|}{r} \right)^m$$

$$\leq \frac{c \, \| x - a \|}{r - \| x - a \|}$$

for every $x \in B(a;r)$ and $f \in F$. Hence F is equicontinuous.

(c) ⇒ (b): Let $a \in U$. Since F is pointwise bounded there is $c > 0$ such that $\|f(a)\| \leq c$ for every $f \in F$. Since F is equicontinuous there is a neighborhood V of a in U such that $\|f(x) - f(a)\| \leq 1$ for all $x \in V$ and $f \in F$. Then $\|f(x)\| \leq c + 1$ for all $x \in V$ and $f \in F$, completing the proof.

By combining Ascoli's Theorem 9.12 and Propositions 9.13 and 9.15 we obtain at once the following result, which extends the classical *Montel's Theorem*.

9.16. PROPOSITION. *Let U be an open subset of E. Then each bounded subset of $(\mathcal{H}(U), \tau_c)$ is relatively compact.*

EXERCISES

9.A. Show that each closed subspace of a k-space is a k-space. Show that each open subspace of a Hausdorff k-space is a k-space.

9.B. Let K be a compact subset of a Hausdorff space X. Let U_1 and U_2 be open subsets of X such that $K \subset U_1 \cup U_2$. Find compact sets $K_1 \subset U_1$ and $K_2 \subset U_2$ such that $K = K_1 \cup K_2$.

9.C. Let U_1 and U_2 be two open subsets of E. Show that the space $(\mathcal{H}(U_1 \cup U_2), \tau_c)$ can be canonically identified with a closed vector subspace of the product $(\mathcal{H}(U_1), \tau_c) \times (\mathcal{H}(U_2), \tau_c)$. Generalize this to an arbitrary family $(U_i)_{i \in I}$ of open subsets of E.

9.D. Let (x_i) be a net in a topological space X with the property that every subnet of (x_i) has a subnet which converges to a fixed point x. Show that (x_i) converges to x.

9.E. Let U be a connected open subset of E. Let (f_n) be a bounded sequence in $(\mathcal{H}(U), \tau_c)$ and suppose that the sequence $(f_n(x))$ converges in $\mathbb{C}$ for every point x in a nonvoid open set $V \subset U$.

(a) Using Montel's Theorem 9.16 and Exercise 9.D. show that the limit $f(x) = \lim f_n(x)$ exists for every $x \in U$.

(b) Show that (f_n) converges to f uniformly on each compact subset of U.

This result extends the classical *Vitali's Theorem*.

9.F. Let U be an open subset of a separable Banach space E. Using Cantor's diagonal process show that each bounded sequence in $(\mathcal{H}(U), \tau_c)$ has a convergent subsequence. This sharpens the conclusion in Montel's Theorem 9.16.

9.G. Show that if U is an open subset of E then $(P(^mE;F), \tau_c)$ is a complemented subspace of $(\mathcal{H}(U;F), \tau_c)$. More precisely, show that for each $a \in U$ the mapping $f \to P^m f(a)$ is a continuous projection from $(\mathcal{H}(U;F), \tau_c)$ onto $(P(^mE;F), \tau_c)$.

9.H. Show that if U is an open subset of E then the mapping

$$f \in (\mathcal{H}(U;F), \tau_c) \to P_t^m f \in (\mathcal{H}(U;F), \tau_c)$$

is continuous for each $m \in \mathbb{N}_o$ and $t \in E$.

9.I. Show that if U is an open subset of a finite dimensional Banach space E then the mapping

$$f \in (\mathcal{H}(U;F), \tau_c) \to P^m f \in (\mathcal{H}(U;P(^mE;F)), \tau_c)$$

is continuous for each $m \in \mathbb{N}_o$.

9.J. Show that if U is an open subset of an infinite dimensional Banach space E, and if $F \neq \{0\}$, then the mapping

$$f \in (\mathcal{H}(U;F), \tau_c) \to P^m f \in (\mathcal{H}(U;P(^mE;F)), \tau_c)$$

is not continuous for any $m \in \mathbb{N}$. Furthermore, show that if E

satisfies the hypothesis in Proposition 7.15 then for each $m \in \mathbb{N}$ one can even find a sequence (f_n) in $\mathcal{H}(U;F)$ such that (f_n) converges to zero in $(\mathcal{H}(U;F), \tau_c)$ but $(P^m f_n)$ does not converge to zero in $(\mathcal{H}(U;P(^m E;F)), \tau_c)$.

9.K. A locally convex space is said to be *barrelled* if each closed, convex, balanced, absorbing set is a neighborhood of zero.

(a) Using the Baire Category Theorem show that each Fréchet space is barrelled. Conclude that $(\mathcal{H}(U;F), \tau_c)$ is barrelled if U is an open subset of a finite dimensional Banach space E.

(b) Show that if U is an open subset of an infinite dimensional Banach space E, and if $F \neq \{0\}$, then for each $a \in U$ the set

$$B = \{f \in \mathcal{H}(U;F) : \| P^1 f(a) \| \leq 1\}$$

is a closed, convex, balanced, absorbing subset of $(\mathcal{H}(U;F), \tau_c)$, but is not a neighborhood of zero. Hence $(\mathcal{H}(U;F), \tau_c)$ is not barrelled.

NOTES AND COMMENTS

Most of the results in Chapter II have been known for a long time and can already be found in the book of E. Hille and R. Phillips [1]. Among the results that appeared within the last twenty years we mention Theorem 7.13, due to L. Nachbin [2], Proposition 7.15, due to S. Dineen [4], and Proposition 9.16, noticed by H. Alexander [1]. It was also H. Alexander [1] who showed that $(\mathcal{H}(U), \tau_c)$ is never barrelled when U is an open subset of an infinite dimensional Banach space, a result that was left to the reader as Exercise 9.K.

Proposition 7.15 has an interesting sequel, for it raised a natural question in the theory of Banach spaces. Indeed, we know from Exercises 7.F and 7.G that each separable, or reflexive, infinite dimensional Banach space satisfies the hypothesis in Proposition 7.15. It is then natural to ask whether every infinite dimensional Banach space satisfies the hypothesis in Proposition 7.15. This question was answered in the affirmative by B. Josefson [2], and independently by A. Nissenzweig [1]. This is a deep result and the interested reader is referred to the original papers of B. Josefson [2] and A. Nissenzweig [1], or to the recent book of J. Diestel [1], for a proof of this theorem.

Many of the results in Sections 5,7 and 8 can be found in the books of L. Nachbin [1], [2] and T. Franzoni and E. Vesentini [1]. Our brief presentation of the Bochner integral in Section 6 follows essentially the book of J. Diestel [1]. Our presentation of the compact-open topology in Section 9 is quite standard and can be found for instance in the book of S. Willard [1].

For the properties of holomorphic mappings between locally convex spaces the reader is referred to the books of M. Hervé [1], P. Noverraz [3], G. Coeuré [1], S. Dineen [5] and J. F. Colombeau [1].

CHAPTER III

DOMAINS OF HOLOMORPHY

10. DOMAINS OF HOLOMORPHY

In this section we introduce the notions of domain of holomorphy and domain of existence, and study their elementary properties.

We begin by presenting some examples to motivate the definitions. As in the preceding chapter all Banach spaces considered will be complex.

10.1. EXAMPLE. If U and V are two open sets in $\mathbb{C}$ with $U \subsetneqq V$ and V connected, then there is a function $f \in \mathcal{H}(U)$ which has no holomorphic extension to V.

PROOF. Since V is connected there is a point $a \in V \cap \partial U$. Then the function $f \in \mathcal{H}(U)$ defined by $f(z) = (z - a)^{-1}$ has no holomorphic extension to V.

For holomorphic functions of $n \geq 2$ variables the situation is entirely different, as the following example shows.

10.2. EXAMPLE. Let $D = \Delta^2(0;R)$ and let

$$H = \{z \in D : |z_1| > r_1 \quad \text{or} \quad |z_2| < r_2\}$$

where $0 < r_j < R_j \leq \infty$ for $j = 1,2$. Then each $f \in \mathcal{H}(H)$ has a unique extension $\tilde{f} \in \mathcal{H}(D)$. The pair (H,D) is called a *Hartogs figure* in $\mathbb{C}^2$.

PROOF. Choose ρ_1 with $r_1 < \rho_1 < R_1$. Given $f \in \mathcal{H}(H)$ define

$$g(z) = \frac{1}{2\pi i} \int_{|\zeta_1| = \rho_1} \frac{f(\zeta_1, z_2)}{\zeta_1 - z_1} d\zeta_1$$

for every z in the polydisc $D' = \Delta^2(D;R')$ where $R' = (\rho_1, R_2)$. After expanding $(\zeta_1 - z_1)^{-1}$ in powers of z_1, a term by term integration shows that g is a holomorphic function of z_1 for each z_2 fixed. On the other hand, by differentiation under the integral sign we see that g is a holomorphic function of z_2 for each z_1 fixed. Since g is clearly locally bounded, an application of Lemmas 8.8 and 8.9 shows that $g \in \mathcal{H}(D')$. Now, by the Cauchy Integral Formula for holomorphic functions of one variable, we have that $g(z) = f(z)$ for every $z \in \mathbb{C}^2$ with $|z_1| < \rho_1$ and $|z_2| < r_2$, and therefore for every $z \in D' \cap H$, since $D' \cap H$ is connected. Then the function $\tilde{f} \in \mathcal{H}(D)$ defined by $\tilde{f} = f$ on H and $\tilde{f} = g$ on D' is the required extension. The uniqueness of the extension is clear.

This example motivates the following definition.

10.3. DEFINITION. Let U be an open subset of E. An open subset V of E containing U is said to be a *holomorphic extension* or *holomorphic continuation* of U if each $f \in \mathcal{H}(U)$ has a unique extension $\tilde{f} \in \mathcal{H}(V)$.

We want to study those open sets U in E which are in some sense the largest common domains of definition for all the functions $f \in \mathcal{H}(U)$. These open sets will be called domains of holomorphy. How should we define domains of holomorphy? We might be inclined to define a domain of holomorphy as an open set in E which has no proper holomorphic continuation, but such a definition would turn out to be inadequate. Actually, such a definition would be adequate if we enlarged the class of objects under discussion by replacing open sets in Banach spaces by Riemann domains over Banach spaces. We shall indeed do this in Section 52, but for the time being we shall restrict our study to domains of holomorphy in Banach spaces, and in this case the definition is the following.

10.4. DEFINITION. An open set U in E is said to be a *domain of holomorphy* if there are no open sets V and W in E with the following properties:

(a) V is connected and not contained in U.

(b) $\phi \neq W \subset U \cap V$.

(c) For each $f \in \mathcal{H}(U)$ there exists $\tilde{f} \in \mathcal{H}(V)$ (necessarily unique) such that $\tilde{f} = f$ on W.

If U is a domain of holomorphy then clearly U has no proper holomorpic continuation, but the converse is not true in general.

10.5. PROPOSITION. *Let U be an open subset of E. Assume that for each sequence (a_j) in U which converges to a point $a \in \partial U$ there exists a function $f \in \mathcal{H}(U)$ which is unbounded on (a_j). Then U is a domain of holomorphy.*

PROOF. Suppose U is not a domain of holomorphy, and let V and W be two open sets satisfying the conditions in Definition 10.4. By the Identity Principle we may assume that W is a connected component of $U \cap V$. By Exercise 10.F there is a sequence (a_j) in W which converges to a point $a \in V \cap \partial U \cap \partial W$. By hypothesis there is a function $f \in \mathcal{H}(U)$ which is unbounded on (a_j). Then on one hand $\tilde{f}(a_j)$ converges to $\tilde{f}(a)$, and on the other hand $\tilde{f}(a_j) = f(a_j)$ is unbounded. This is impossible.

10.6. COROLLARY. *Every open set in $\mathbb{C}$ is a domain of holomorphy.*

PROOF. Let U be an open set in $\mathbb{C}$ and let (a_j) be a sequence in U which converges to a point $a \in \partial U$. Then the function $f(z) = (z - a)^{-1}$ is holomorphic on U and unbounded on (a_j).

10.7. COROLLARY. *Every convex open set in E is a domain of holomorphy.*

PROOF. Let U be a convex open set in E and let (a_j) be a sequence in U which converges to a point $a \in \partial U$. By the Hahn-Banach Theorem there exists $\varphi \in E'$ such that $Re\varphi(x) < Re\varphi(a)$ for every $x \in U$. Then the function $f(x) = [\varphi(x-a)]^{-1}$ is holomorphic on U and unbounded on (a_j).

If U is a domain of holomorphy then for each pair of open sets V and W satisfying the conditions (a) and (b) in Definition 10.4 there exists a function $f \in \mathcal{H}(U)$, which cannot be extended to V in the sense that there is no $\tilde{f} \in \mathcal{H}(V)$ such that $\tilde{f} = f$ on W. In general the function f depends on the open sets V and W. If we can take the same function f for all V and W then we shall say that U is a domain of existence. More precisely, we have the following definition.

10.8. DEFINITION. An open set U in E is said to be the *domain of existence* of a function $f \in \mathcal{H}(U)$ if there are no open sets V and W in E and no function $\tilde{f} \in \mathcal{H}(V)$ with the following properties:

(a) V is connected and not contained in U.

(b) $\phi \neq W \subset U \cap V$.

(c) $\tilde{f} = f$ on W.

Clearly every domain of existence is a domain of holomorphy. The next theorem shows that in $\mathbb{C}^n$ the two concepts coincide.

10.9. THEOREM. *Every domain of holomorphy in* $\mathbb{C}^n$ *is a domain of existence.*

We shall presently give an existencial proof of Theorem 10.9 based on the Baire Category Theorem. In the next section we shall give a constructive proof of a theorem of H. Cartan and P. Thullen, which implies Theorem 10.9. The key to the first proof of Theorem 10.9 is the following lemma.

10.10. LEMMA. *Let* U *be a domain of holomorphy in a separable Banach space* E. *Let* $\mathcal{F}$ *denote the set of all function* $f \in \mathcal{H}(U)$

such that U *is not the domain of existence of* f. *Then* F *is a set of the first category in* $(\mathcal{H}(U), \tau_c)$.

PROOF. For each pair of open sets V and W satisfying the conditions (a) and (b) in Definition 10.4 let $\mathcal{H}(U,V,W)$ denote the vector subspace of all $f \in \mathcal{H}(U)$ for which there exists $\tilde{f} \in \mathcal{H}(V)$ (necessarily unique) such that $\tilde{f} = f$ on W. Since U is a domain of holomorphy, $\mathcal{H}(U,V,W)$ is a proper vector subspace of $\mathcal{H}(U)$. For each $m \in \mathbb{N}$ let $\mathcal{H}_m(U,V,W)$ denote the set of all $f \in \mathcal{H}(U,V,W)$ such that $|\tilde{f}| \leq m$ on V. We claim that $\mathcal{H}_m(U,V,W)$ is a closed subset of $(\mathcal{H}(U), \tau_c)$. Indeed, let (f_i) be a net in $\mathcal{H}_m(U,V,W)$ which converges to some f in $(\mathcal{H}(U), \tau_c)$. Since $|\tilde{f}_i| \leq m$ on V for every i an application of Montel's Theorem 9.16 yields a subnet of $(\tilde{f}_i)$ which converges to a function g in $(\mathcal{H}(V), \tau_c)$. Whence it follows that $f \in \mathcal{H}_m(U,V,W)$ and $\tilde{f} = g$, and our claim has been proved. Since $\mathcal{H}(U,V,W)$ is a proper vector subspace of $\mathcal{H}(U)$ and has therefore empty interior in $(\mathcal{H}(U), \tau_c)$, we conclude that the smaller set $\mathcal{H}_m(U,V,W)$ is a closed, nowhere dense subset of $(\mathcal{H}(U), \tau_c)$.

Now, let D denote a countable dense subset of ∂U. Let $\mathcal{V}$ denote the collection of all open balls V whose centers belong to D and whose radii are rational. Let P denote the collection of all pairs (V,W) such that $V \in \mathcal{V}$ and W is a connected component of $U \cap V$. Clearly P is countable, and to complete the proof of the lemma we shall show that F is the union of the sets $\mathcal{H}_m(U,V,W)$ with $(V,W) \in P$ and $m \in \mathbb{N}$. Let $f \in F$. Then we can find open sets V and W in E and a function $\tilde{f} \in \mathcal{H}(V)$ satisfying the conditions (a), (b), (c) in Definition 10.8. Without loss of generality we may assume that W is a connected component of $U \cap V$. By Exercise 10.F there is a point $a \in V \cap \partial U \cap \partial W$. Choose $V' \in \mathcal{V}$ such that $a \in V' \subset V$ and $|\tilde{f}|$ is bounded, by m say, on V'. Since $a \in \partial W$ there is a point $b \in W \cap V'$. Let W' denote the connected component of $U \cap V'$ which contains b. Then $(V',W') \in P$ and $f \in \mathcal{H}_m(U,V',W')$, completing the proof.

PROOF OF THEOREM 10.9. Let U be a domain of holomorphy in $\mathbb{C}^n$.

Then $(\mathcal{H}(U), \tau_c)$ is a Fréchet space and, by Lemma 10.10, the set of all $f \in \mathcal{H}(U)$ such that U is the domain of existence of f, is of the second category in $(\mathcal{H}(U), \tau_c)$, and is in particular nonempty.

Theorem 10.9 does not generalize to arbitrary Banch spaces. Indeed, A. Hirschowitz [1] has given an example of a nonseparable Banach space whose open unit ball is not a domain of existence. But the following problem remains open.

10.11. PROBLEM. Let E be a separable Banach space. Is every domain of holomorphy in E a domain of existence ?

In Section 45 we shall present a partial positive solution to Problem 10.11.

EXERCISES

10.A. Let $D = \Delta^n(0;R)$ and let

$$H = \{z \in D : |z_1| > r_1 \quad \text{or} \quad |z_j| < r_j \quad \text{for} \quad j = 2,\ldots,n\}$$

where $n \geq 2$ and $0 < r_j < R_j \leq \infty$ for $j = 1,\ldots,n$. Show that D is a holomorphic continuation of H. The pair (H,D) is called a *Hartogs figure* in $\mathbb{C}^n$.

10.B. Let V be a connected open set in $\mathbb{C}^n$ with $n \geq 2$. Let $\bar{\Delta}^n(a;r)$ be a compact polydisc contained in V, and let $U = V \setminus \bar{\Delta}^n(a;r)$. Show that V is a holomorphic continuation of U.

10.C. Let U and V be two open subsets of E with $U \subset V$. Show that V is a holomorphic continuation of U if and only if each $f \in \mathcal{H}(U)$ has an extension $\tilde{f} \in \mathcal{H}(V)$, and each connected component of V contains points of U.

10.D. Let V be a holomorphic continuation of an open set U in E. Using Exercise 8.I show that if F is any Banach space then each $f \in \mathcal{H}(U;F)$ has a unique extension $\tilde{f} \in \mathcal{H}(V;F)$.

10.E. Show that an open set U in E is a domain of holomorphy (resp. a domain of existence) if and only if each connected component of U is a domain of holomorphy (resp. a domain of existence).

10.F. Let U and V be open subsets of E, with V connected and not contained in U. Let W be a connected component of $U \cap V$. Show the existence of a point $a \in V \cap \partial U \cap \partial W$.

10.G. Let (H,D) be a Hartogs figure in $\mathbb{C}^2$, and let $U = H \cup (D \setminus \bar{H})$. Show that U has no proper holomorphic continuation, but U is not a domain of holomorphy.

10.H. Let U_i be a domain of holomorphy in E for each $i \in I$. Show that the set $U = \operatorname{int} \bigcap_{i \in I} U_i$ is a domain of holomorphy as well.

10.I. Let $\varphi \in \mathcal{H}(E)$ and let A be an open set in $\mathbb{C}$. Show that the set $U = \varphi^{-1}(A)$ is a domain of holomorphy in E.

10.J. Given open sets $A_1, \ldots, A_n$ in $\mathbb{C}$ show that the product $U = A_1 \times \ldots \times A_n$ is a domain of holomorphy in $\mathbb{C}^n$.

10.K. Show that an open subset U of E is the domain of existence of a function $f \in \mathcal{H}(U)$ if and only if $r_c f(x) \leq d_U(x)$ for every $x \in U$.

10.L. Let U be an open subset of E, and let $f \in \mathcal{H}(U)$. Suppose that for each ball $B(a;r)$ with center $a \in \partial U$ and sufficiently small radius, the function f is unbounded on each connected component of $U \cap B(a;r)$. Show that U is the domain of existence of f.

11. HOLOMORPHICALLY CONVEX DOMAINS

In this section we introduce the notion of holomorphic convexity and establish a classical theorem of H. Cartan and P.

Thullen, which characterizes domains of holomorphy in terms of holomorphic convexity.

To motivate the definition of holomorphically convex domains we begin by presenting some properties of convex sets.

11.1. PROPOSITION. *Let* $\mathbb{C} \oplus E'$ *denote the vector space of all continuous affine forms on* E, *let* A *be a subset of* E *and let* $\hat{A}_{\mathbb{C} \oplus E'}$ *denote the set*

$$\hat{A}_{\mathbb{C} \oplus E'} = \{x \in E : |f(x)| \leq \sup_A |f| \quad \text{for all} \quad f \in \mathbb{C} \oplus E'\}.$$

Then:

(a) *The set* $\hat{A}_{\mathbb{C} \oplus E'}$ *is always convex and closed, and in particular contains the closed, convex hull* $\overline{co}(A)$ *of* A.

(b) *If* A *is bounded then* $\hat{A}_{\mathbb{C} \oplus E'} = \overline{co}(A)$.

(c) *If* A *is bounded (resp. compact) then* $\hat{A}_{\mathbb{C} \oplus E'}$ *is bounded (resp. compact) as well.*

PROOF. We shall prove that $\hat{A}_{\mathbb{C} \oplus E'} \subset \overline{co}(A)$ when A is bounded. All the other assertions are clear. Let $y \notin \overline{co}(A)$. By the Hahn-Banach Theorem there exist $\varphi \in E'$ and $\alpha \in \mathbb{R}$ such that $Re\varphi(x) < \alpha < Re\varphi(y)$ for all $x \in \overline{co}(A)$. Since $\varphi(\overline{co}(A))$ is bounded there is a disc $\overline{\Delta}(\zeta;r)$ such that $\varphi(\overline{co}(A)) \subset \overline{\Delta}(\zeta;r)$ and $\varphi(y) \notin \overline{\Delta}(\zeta;r)$. Let $f \in \mathbb{C} \oplus E'$ be defined by $f(x) = \varphi(x) - \zeta$ for every $x \in E$. Then $\sup_A |f| \leq \sup_{\overline{co}(A)} |f| \leq r < f(y)$, proving that $y \notin \hat{A}_{\mathbb{C} \oplus E'}$.

11.2. PROPOSITION. *For an open subset* U *of* E *the following conditions are equivalent:*

(a) U *is convex.*

(b) $\hat{K}_{\mathbb{C} \oplus E'} \subset U$ *for each compact set* $K \subset U$.

(c) $\hat{K}_{\mathbb{C} \oplus E'} \cap U$ *is compact for each compact set* $K \subset U$.

PROOF. (a) ⇒ (b): If K is a compact subset of U then there is a ball $V = B(0;r)$ such that $K + V \subset U$. Hence $\hat{K}_{\mathbb{C}\oplus E'} = \overline{co}(K) \subset co(K) + V = co(K + V) \subset U$.

(b) ⇒ (c): This is obvious since $\hat{K}_{\mathbb{C}\oplus E'}$ is compact for each compact subset K of E.

(c) ⇒ (a): Let $x, y \in U$ and set $K = \{x,y\}$. By Proposition 11.1 the line segment $[x,y]$ equals $\hat{K}_{\mathbb{C}\oplus E'}$. Hence we can write $[x,y] = A \cup B$, where $A = \hat{K}_{\mathbb{C}\oplus E'} \cap U$ and $B = \hat{K}_{\mathbb{C}\oplus E'} \setminus U$ are two disjoint compact sets. Since $[x,y]$ is connected we conclude that B must be empty. Thus $[x,y] \subset U$ and U is convex.

With this motivation in mind we introduce the following definition.

11.3. DEFINITION. Let U be an open subset of E.

(a) The $\mathcal{H}(U)$*-hull* of a set $A \subset U$ is defined by

$$\hat{A}_{\mathcal{H}(U)} = \{x \in U : |f(x)| \leq \sup_A |f| \quad \text{for all} \quad f \in \mathcal{H}(U)\}.$$

(b) The open set U is said to be *holomorphically convex* if $\hat{K}_{\mathcal{H}(U)}$ is compact for each compact set $K \subset U$.

Let U be an open subset of E. We shall set $d_U(A) = \inf_{x \in A} d_U(x)$ for each set $A \subset U$. If K is a compact subset of U then it is clear that $\hat{K}_{\mathcal{H}(U)}$ is contained in the compact set $\hat{K}_{\mathbb{C}\oplus E'}$. Since $\hat{K}_{\mathcal{H}(U)}$ is clearly closed in U we conclude that $\hat{K}_{\mathcal{H}(U)}$ is compact if and only if $d_U(\hat{K}_{\mathcal{H}(U)}) > 0$. Thus U is holomorphically convex if and only if $d_U(\hat{K}_{\mathcal{H}(U)}) > 0$ for each compact set $K \subset U$.

11.4. THEOREM. *For an open subset U of E consider the following conditions:*

(a) *U is a domain of existence.*

(b) *U is the union of an increasing sequence of open sets A_j such that $d_U((\hat{A}_j)_{\mathcal{H}(U)}) > 0$ for every j.*

(c) *For each sequence (a_j) in U which converges to a point $a \in \partial U$ there exists a function $f \in \mathcal{H}(U)$ which is unbounded on (a_j).*

(d) *U is a domain of holomorphy.*

(e) *$d_U(\hat{K}_{\mathcal{H}(U)}) = d_U(K)$ for each compact set $K \subset U$.*

(f) *U is holomorphically convex.*

Then the implications (a) $\Rightarrow$ (b) $\Rightarrow$ (c) $\Rightarrow$ (d) $\Rightarrow$ (e) $\Rightarrow$ (f) *are always true. If E is separable then* (a) $\Longleftrightarrow$ (b).

PROOF. (a) $\Rightarrow$ (b): Suppose U is the domain of existence of a function $f \in \mathcal{H}(U)$. Consider the following open sets:

$$B_j = \{x \in U : |f(x)| < j\}$$

and

$$A_j = \{x \in B_j : d_{B_j}(x) > 1/j\}.$$

Then $U = \bigcup_{j=1}^{\infty} A_j$ and $A_j \subset A_{j+1}$ for every j. Furthermore, the function f is bounded by j on the set $A_j + \overline{B}(0;1/j)$, and whence it follows that $|P^m f(x)(t)| \leq j$ for every $x \in A_j$ and $t \in \overline{B}(0;1/j)$. We claim that $r_c f(y) \geq 1/j$ for every $y \in (\hat{A}_j)_{\mathcal{H}(U)}$. To show this let $y \in (\hat{A}_j)_{\mathcal{H}(U)}$ and $t \in \overline{B}(0;1/j)$. Since $P^m_t f \in \mathcal{H}(U)$ it follows that $|P^m_t f(y)| \leq \sup_{A_j} |P^m_t f| \leq j$. Thus $\| P^m f(y)\| \leq j^{m+1}$ for every m and it follows from the Cauchy - Hadamard Formula that $r_c f(y) \geq 1/j$, as asserted. Hence the series $\sum_{m=0}^{\infty} P^m f(y)(t)$ defines a function f_y, holomorphic on the ball $B(y;1/j)$, and which coincides with f on a neighborhood of the point y. Since U is the domain of existence of f we conclude that $B(y;1/j) \subset U$. This shows that $d_U((\hat{A}_j)_{\mathcal{H}(U)}) \geq 1/j$ and (b) is satisfied.

(b) ⇒ (c): By hypothesis U is the union of an increasing sequence of open sets A_j such that $d_U((\hat{A}_j)_{\mathcal{H}(U)}) > 0$ for every j. Set $B_j = (\hat{A}_j)_{\mathcal{H}(U)}$ for every j and note that $(\hat{B}_j)_{\mathcal{H}(U)} = B_j$. Let (a_j) be a sequence in U which converges to a point in ∂U. After replacing (a_j) and (B_j) by suitable subsequences, if necessary, we may assume that $a_j \notin B_j$ and $a_j \in B_{j+1}$ for every j. Since $a_j \notin B_j = (\hat{B}_j)_{\mathcal{H}(U)}$ we can find a sequence (φ_j) in $\mathcal{H}(U)$ such that $\sup_{B_j} |\varphi_j| < 1 < |\varphi_j(a_j)|$ for every j. By taking sufficiently high powers of each φ_j we can inductively find a sequence (f_j) in $\mathcal{H}(U)$ such that $\sup_{B_j}|f_j| \leq 2^{-j}$ and

$$|f_j(a_j)| \geq j + 1 + |\sum_{i<j} f_i(a_j)|$$

for every j. Whence it follows that the series $\sum_{j=1}^{\infty} f_j$ converges uniformly on each B_i to a function $f \in \mathcal{H}(U)$ and $|f(a_j)| \geq j$ for every j. This shows (c).

(c) ⇒ (d): This is the content of Proposition 10.5.

(d) ⇒ (e): Let K be a compact subset of U and set $r = d_U(K)$. By modifying the proof of (a) ⇒ (b) we shall prove that for each $f \in \mathcal{H}(U)$ and $y \in \hat{K}_{\mathcal{H}(U)}$ the series $\sum_{m=0}^{\infty} P^m f(y)(t)$ defines a function f_y holomorphic on the ball $B(y;r)$. Since f_y will coincide with f on a neighborhood of y, and since U is by hypothesis a domain of holomorphy, we shall conclude that $B(y;r) \subset U$ and therefore $d_U(\hat{K}_{\mathcal{H}(U)}) = r$. Now, fix $f \in \mathcal{H}(U)$ and $y \in \hat{K}_{\mathcal{H}(U)}$. Given $t \in B(0;r)$ choose $\rho > 1$ such that $\rho t \in B(0;r)$. Then $K + \overline{\Delta}\rho t$ is a compact subset of U and we can find $\varepsilon > 0$ such that the set $B = K + \overline{\Delta}\rho t + B(0;\rho\varepsilon)$ is contained in U and f is bounded, by c say, on B. If $h \in B(0;\varepsilon)$ then it follows from the Cauchy Inequality 7.4 that

$$\begin{aligned} |P^m f(y)(t + h)| &\leq \sup_{x \in K} |P^m f(x)(t + h)| \\ &\leq \rho^{-m} \sup_B |f| = c\,\rho^{-m}. \end{aligned}$$

Thus for each $t \in B(0;r)$ there exists $\varepsilon > 0$ such that the series $\sum_{m=0}^{\infty} P^m f(y)(t+h)$ converges uniformly for $h \in B(0;\varepsilon)$. This shows that the series $\sum_{m=0}^{\infty} P^m f(y)(t)$ defines a holomorphic function f_y on the ball $B(y;r)$ and the proof of (d) $\Rightarrow$ (e) is complete.

Since the implication (e) $\Rightarrow$ (f) is obvious, it only remains to show that (b) $\Rightarrow$ (a) when E is separable. Let D be a countable dense subset of U. For each $x \in D$ let $B(x)$ denote the largest open ball centered at x and contained in U, that is $B(x) = B(x;d_U(x))$. By modifying the proof of (b) $\Rightarrow$ (c) we shall construct a function $f \in \mathcal{H}(U)$ which is unbounded on $B(x)$ for every $x \in D$. Now, let (x_j) be a sequence in D with the property that each point of D appears in the sequence (x_j) infinitely many times. By hypothesis U is the union of an increasing sequence of open sets A_j such that $d_U((\hat{A}_j)_{\mathcal{H}(U)}) > 0$ for every j. Set $B_j = (\hat{A}_j)_{\mathcal{H}(U)}$ for every j. Note that $B(x) \not\subset B_j$ for each $x \in D$ and $j \in \mathbb{N}$. Hence, after replacing (B_j) by a subsequence, if necessary, we can find a sequence (y_j) in U such that $y_j \in B(x_j)$, $y_j \notin B_j$ and $y_j \in B_{j+1}$ for every j. Then, proceeding as in the proof of (b) $\Rightarrow$ (c), we can construct a function $f \in \mathcal{H}(U)$ such that $|f(y_j)| \geq j$ for every j. We claim that f is unbounded on the ball $B(x)$ for every $x \in D$. Indeed, given $x \in D$ we can find a strictly increasing sequence (j_k) in $\mathbb{N}$ such that $x = x_{j_k}$ for every k. Hence $y_{j_k} \in B(x)$ for every k and f is unbounded on $B(x)$, as asserted. To complete the proof we shall prove that U is the domain of existence of f. Indeed, suppose there exist open sets V and W and a function $\tilde{f} \in \mathcal{H}(V)$ satisfying the conditions in Definition 10.8. Take a point $a \in V \cap \partial U \cap \partial W$ and consider any $r > 0$ such that $B(a;2r) \subset V$. Choose a point $x \in D \cap W \cap B(a;r)$. Then $d_U(x) < r$ and $B(x) \subset B(a;2r) \subset V$. Since $B(x)$ is connected and contained in $U \cap V$, we conclude that $B(x) \subset W$. Thus $\tilde{f} = f$ is unbounded on $B(x)$, and therefore on $B(a;2r)$. Since $r > 0$ can be taken arbitrarily small we conclude that $\tilde{f}$ is not locally bounded at a, a contradiction. Hence U is the domain of existence of f, and the proof of the theorem is complete.

Now it is easy to prove the *Cartan-Thullen Theorem:*

11.5. THEOREM. *For an open subset U of $\mathbb{C}^n$ the following conditions are equivalent:*

(a) *U is a domain of existence.*

(b) *U is a domain of holomorphy.*

(c) *U is holomorphically convex.*

PROOF. By Theorem 11.4 the implications (a) ⇒ (b) ⇒ (c) always hold. To show that (c) ⇒ (a) consider the compact sets

$$K_j = \{x \in U : \|x\| \leq j \quad \text{and} \quad d_U(x) \geq 1/j\}.$$

Then $U = \bigcup_{j=1}^{\infty} K_j$ and $K_j \subset \overset{\circ}{K}_{j+1}$ for every j. Since U is holomorphically convex the set $(\hat{K}_j)_{\mathcal{H}(U)}$ is compact for every j. If we set $A_j = \overset{\circ}{K}_j$ then $U = \bigcup_{j=1}^{\infty} A_j$, $A_j \subset A_{j+1}$ and $d_U((\hat{A}_j)_{\mathcal{H}(U)}) > 0$ for every j. By Theorem 11.4, U is a domain of existence.

Theorem 11.5 does not generalize to arbitrary Banach spaces. Indeed, B. Josefson [1] has given an example of a holomorphically convex open set in a nonseparable Banach space which is not a domain of holomorphy. But the following problem remains open.

11.6. PROBLEM. Let E be a separable Banach space. Is every holomorphically convex open set in E a domain of existence, or at least a domain of holomorphy ?

In Section 45 we shall present a partial positive solution to Problem 11.6. To complete this section we give two applications of Theorem 11.4.

11.7. PROPOSITION. *Let* V *be a domain of existence in* F, *let* $T \in \mathcal{L}(E;F)$ *and let* $U = T^{-1}(V)$. *Then:*

(a) U *is a domain of holomorphy.*

(b) *If* E *is separable then* U *is a domain of existence.*

PROOF. One can readily check that

$$\hat{A}_{\mathcal{H}(U)} \subset T^{-1}[(T(A))\hat{}_{\mathcal{H}(V)}] \tag{11.1}$$

for each set $A \subset U$. Now, since V is a domain of existence, Theorem 11.4 yields an increasing sequence of open sets B_j and a sequence of 0-neighborhoods V_j in F such that $V = \bigcup_{j=1}^{\infty} B_j$ and $(\hat{B}_j)_{\mathcal{H}(V)} + V_j \subset V$ for every j. Set $A_j = T^{-1}(B_j)$ and $U_j = T^{-1}(V_j)$ for every j. Then using (11.1) we get that $U = \bigcup_{j=1}^{\infty} A_j$ and $(\hat{A}_j)_{\mathcal{H}(U)} + U_j \subset U$ for every j. By Theorem 11.4 we may conclude that U is a domain of holomorphy if E is arbitrary, and that U is a domain of existence if E is separable.

Next we show that in the case of separable Banach spaces the conclusion of Corollary 10.7 can be improved as follows.

11.8. PROPOSITION. *Every convex open set in a separable Banach space is a domain of existence.*

PROOF. Let U be a convex open set in a separable Banach space E. Then U is the union of the increasing sequence of open sets A_j defined by

$$A_j = \{x \in U : \|x\| < j \quad \text{and} \quad d_U(x) > 1/j\}.$$

Using the identity

$$B(\alpha x + \beta y; r) = \alpha B(x;r) + \beta B(y;r)$$

for all $\alpha, \beta \geq 0$ with $\alpha + \beta = 1$, we can see that each A_j is convex. Then it follows from Proposition 11.1 that $(\hat{A}_j)_{\mathcal{H}(U)} \subset (\hat{A}_j)_{\mathbb{C} \oplus E'} = \overline{co}(A_j) = \overline{A_j}$. Hence $d_U((\hat{A}_j)_{\mathcal{H}(U)}) \geq 1/j$ for every j. Thus an application of Theorem 11.4 completes the proof.

EXERCISES

11.A. Show that an open subset U of E is holomorphically convex if and only if each connected component of U is holomorphically convex.

11.B. Let U_i be a holomorphically convex open set in E for each $i = 1, 2$. Show that the open set $U = U_1 \cap U_2$ is holomorphically convex as well.

11.C. Given a holomorphically convex open set U in E and a function $f \in \mathcal{H}(U)$ show that the open set $V = \{x \in U : |f(x)| < 1\}$ is holomorphically convex as well.

11.D. Let V be an open subset of F, let $T \in \mathcal{L}(E;F)$ and let $U = T^{-1}(V)$. Show that if V is holomorphically convex then U is holomorphically convex as well.

11.E. Let U_i be a holomorphically convex open subset of a Banach space E_i for $i = 1, 2$. Show that $U_1 \times U_2$ is a holomorphically convex open subset of $E_1 \times E_2$.

11.F. Let U_i be a domain of existence in a Banach space E_i for $i = 1, 2$. Show that $U_1 \times U_2$ is a domain of holomorphy in $E_1 \times E_2$.

11.G. If U is a convex open set in E show that $d_U(\hat{K}_{\mathbb{C} \oplus E'}) = d_U(K)$ for each compact set $K \subset U$.

11.H. Let U be a holomorphically convex open set in $\mathbb{C}^n$, and let (a_j) be a sequence in U such that each $f \in \mathcal{H}(U)$ is bounded on (a_j).

(a) Show that the set $B = \{f \in \mathcal{H}(U) : \sup|f(a_j)| \leq 1\}$ is a closed, convex, balanced, absorbing subset of $(\mathcal{H}(U), \tau_c)$.

(b) Using Exercise 9.K show the existence of a compact set K in U such that $a_j \in \hat{K}_{\mathcal{H}(U)}$ for every j.

(c) Using Proposition 10.5 conclude that U is a domain of holomorphy.

This exercise, together with Theorem 10.9, give an alternative proof of the Cartan-Thullen Theorem 11.5, based on the Fréchet space properties of $(\mathcal{H}(U), \tau_c)$.

12. BOUNDING SETS

In this section we introduce the notion of bounding set and study its connection with domains of holomorphy and domains of existence.

12.1. DEFINITION. Let U be an open subset of E. A set $B \subset U$ is said to be a *bounding* subset of U, or $\mathcal{H}(U)$*-bounding*, if each $f \in \mathcal{H}(U)$ is bounded on B.

12.2. EXAMPLES. Let U be an open subset of E. Then each relatively compact subset of U is $\mathcal{H}(U)$-bounding. Moreover, for each compact subset K of U the set $\hat{K}_{\mathcal{H}(U)}$ is $\mathcal{H}(U)$-bounding.

12.3. PROPOSITION. *Let U be an open subset of E, and let B be a bounding subset of U. Then, for each increasing sequence (A_j) of open subsets of U which cover U, there exists j such that $B \subset (\hat{A}_j)_{\mathcal{H}(U)}$.*

PROOF. Set $B_j = (\hat{A}_j)_{\mathcal{H}(U)}$ for every j. If $B \not\subset B_j$ for each j then, after replacing (B_j) by a suitable subsequence, if necessary, we can find a sequence (x_j) in B such that $x_j \notin B_j$ and $x_j \in B_{j+1}$ for every j. Then the proof of the implication

(b) ⇒ (c) in Theorem 11.4 yields a function $f \in \mathcal{H}(U)$ which is unbounded on (x_j), and therefore unbounded on B, a contradiction.

12.4. **COROLLARY.** *For an open subset U of E consider the following conditions:*

(a) *U is a domain of existence.*

(b) *$d_U(B) > 0$ for each bounding subset B of U.*

(c) *U is a domain of holomorphy.*

Then (a) ⇒ (b) ⇒ (c).

PROOF. Apply Theorem 11.4 and Proposition 12.3.

12.5. **THEOREM.** *Every bounding subset of a separable Banach space is relatively compact.*

PROOF. Let B be a bounding subset of a separable Banach space E. Let (a_j) be a dense sequence in E. Given $\varepsilon > 0$, let (A_j) be the increasing sequence of open sets defined by $A_j = \bigcup_{i=1}^{j} B(a_i;\varepsilon)$. Then $E = \bigcup_{j=1}^{\infty} A_j$ and by Proposition 12.3 there exists j such that $B \subset (\hat{A}_j)_{\mathcal{H}(E)}$. Then, by Proposition 11.1, $B \subset (\hat{A}_j)_{\mathbb{C} \oplus E'} = \overline{co}(A_j)$. Set $K_j = co\{a_1,\ldots,a_j\}$. Then $co(A_j) \subset K_j + B(0;\varepsilon)$ and $B \subset \overline{co}(A_j) \subset K_j + B(0;2\varepsilon)$. Since K_j is compact we conclude that B can be covered by finitely many balls of radius 3ε. Thus B is precompact and therefore relatively compact in E.

12.6. **COROLLARY.** *Let U be a domain of existence in a separable Banach space. Then each bounding subset of U is relatively compact in U.*

Theorem 12.5 does not generalize to arbitrary Banach spaces. Indeed, S. Dineen [2] has shown that the unit vectors $u_n = (0,\ldots,0,1,0,\ldots)$ form a bounding set in ℓ^∞.

EXERCISES

12.A. Let U be an open subset of E, and let B be a bounding subset of U. Show that each $f \in \mathcal{H}(U;F)$ is bounded on B.

12.B. Let U be an open subset of E, let $A \subset U$ and let $y \in \hat{A}_{\mathcal{H}(U)}$. Show that $\|f(y)\| \leq \sup_{x \in A} \|f(x)\|$ for every $f \in \mathcal{H}(U;F)$.

12.C. Let U be an open subset of E, and let $\mathcal{F}$ be a bounded subset of $(\mathcal{H}(U;F), \tau_c)$.

(a) Show that U is the union of an increasing sequence of open sets A_j such that the functions $f \in \mathcal{F}$ are uniformly bounded on each A_j.

(b) If B is a bounding subset of U, show that the functions $f \in \mathcal{F}$ are uniformly bounded on B.

(c) If B is a bounding subset of U with $d_U(B) > 0$, show that the functions $f \in \mathcal{F}$ are uniformly bounded on the set $B + B(0;\varepsilon)$, for a suitable $\varepsilon > 0$.

12.D. Let U be a balanced open subset of E. Show that the balanced hull of each bounding subset of U is also a bounding subset of U.

12.E. Let U be an open subset of E, and let V be an open subset of F. Given a bounding subset A of U, and a bounding subset B of V, show that $A \times B$ is a bounding subset of $U \times V$.

12.F. Let U and V be open subsets of E. Given a bounding subset A of U and a bounding subset B of V, show that $A + B$ is a bounding subset of $U + V$.

12.G. Let U be a convex open set in E. Show that $d_U(B) > 0$ for each bounding subset B of U.

NOTES AND COMMENTS

The proof of Theorem 10.9 based on the Baire Category Theorem is taken from the book of L. Nachbin [3]. The characterization of domains of holomorphy in Theorem 11.5 is due to H. Cartan and P. Thullen [1]. Theorem 11.4 is due to S. Dineen [1] and A. Hirschowitz [3], and represents an attempt to extend the Cartan-Thullen Theorem to infinite dimensional Banach spaces. Bounding sets were introduced by H. Alexander [1], who obtained Theorem 12.5 for separable Hilbert spaces. Theorem 12.5 is a special case of more general results obtained by S. Dineen [4] and A. Hirschowitz [2]. The proof of Theorem 12.5 given here is due to M. Schottenloher [2].

CHAPTER IV

DIFFERENTIABLE MAPPINGS

13. DIFFERENTIABLE MAPPINGS

This section is devoted to the study of differentiable mappings between Banach spaces. Unless stated otherwise, the letters E and F will represent Banach spaces over the same field $\mathbb{K}$.

13.1. DEFINITION. Let U be an open subset of E. A mapping $f : U \to F$ is said to be *differentiable* if for each point $a \in U$ there exists a mapping $A \in \mathcal{L}(E;F)$ such that

$$\lim_{x \to a} \frac{\| f(x) - f(a) - A(x - a)\|}{\| x - a\|} = 0.$$

13.2. REMARKS. Let U be an open subset of E. Then:

(a) Each differentiable mapping $f : U \to F$ is continuous.

(b) The mapping $A \in \mathcal{L}(E;F)$ that appears in Definition 13.1 is uniquely determined by f and a. It is called the *differential* of f at a and will be denoted by $Df(a)$. Thus a differentiable mapping $f : U \to F$ induces a mapping $Df : U \to \mathcal{L}(E;F)$.

(c) If E and F are complex Banach spaces then we have to distinguish between the complex differentiability of $f : U \subset E \to F$ and the real differentiability of $f : U \subset E_{\mathbb{R}} \to F_{\mathbb{R}}$. It is clear that complex differentiability implies real differentiability, but the converse is not true. Indeed, the function $z \in \mathbb{C} \to \bar{z} \in \mathbb{C}$ is $\mathbb{R}$-differentiable without being

$\mathbb{C}$-differentiable. We shall soon study the connection between $\mathbb{R}$-differentiable mappings and $\mathbb{C}$-differentiable mappings.

13.3. EXAMPLE. If $f : E \to F$ is a constant mapping then f is differentiable and $Df(x) = 0$ for every $a \in E$. If $A \in \mathcal{L}(E;F)$ then A is differentiable and $DA(a) = A$ for every $a \in E$.

13.4. EXAMPLE. Every $P \in \mathcal{P}(^mE;F)$ is differentiable. If $P = \hat{A}$ where $A \in \mathcal{L}^s(^mE;F)$ then $DP(a) = mAa^{m-1}$ for every $a \in E$.

PROOF. By the Newton Binomial Formula

$$P(x) = P(a) + mAa^{m-1}(x - a) + \sum_{j=2}^{m} \binom{m}{j} Aa^{m-j}(x - a)^j$$

for all $a, x \in E$. If we set

$$\varphi(x) = P(x) - P(a) - mAa^{m-1}(x - a) = \sum_{j=2}^{m} \binom{m}{j} Aa^{m-j}(x - a)^j$$

then it is clear that $\varphi(x)/\|x - a\| \to 0$ when $x \to a$.

13.5. EXAMPLE. Let E and F be complex Banach spaces and let U be an open subset of E. Then every $f \in \mathcal{H}(U;F)$ is $\mathbb{C}$-differentiable and $Df(a) = P^1 f(a)$ for every $a \in U$.

PROOF. Let $a \in U$ and let $0 < r < r_b f(a)$. Then we can write

$$f(x) = f(a) + P^1 f(a)(x - a) + \sum_{m=2}^{\infty} P^m f(a)(x - a)$$

for every $x \in B(a;r)$. If we set

$$\varphi(x) = f(x) - f(a) - P^1 f(a)(x - a) = \sum_{m=2}^{\infty} P^m f(a)(x - a)$$

then using the Cauchy inequalities one can readily show that $\varphi(x)/\|x - a\| \to 0$ when $x \to a$.

Additional examples of differentiable mappings are given in the exercises. Next we generalize the classical *Chain Rule*.

13.6. THEOREM. *Let E, F, G be Banach spaces over $\mathbb{K}$. Let $U \subset E$ and $V \subset F$ be two open sets and let $f : U \to F$ and $g : V \to G$ be two differentiable mappings with $f(U) \subset V$. Then the composite mapping $g \circ f : U \to G$ is differentiable as well and $D(g \circ f)(a) = Dg(f(a)) \circ Df(a)$ for every $a \in U$.*

PROOF. Let $a \in U$ and set $b = f(a) \in V$, $A = Df(a) \in \mathcal{L}(E;F)$ and $B = Dg(b) \in \mathcal{L}(F;G)$. Then for all $x \in U$ and $y \in V$ we can write

$$f(x) = f(a) + A(x - a) + \varphi(x)$$

and

$$g(y) = g(b) + B(y - b) + \psi(y)$$

where $\lim\limits_{x \to a} \dfrac{\|\varphi(x)\|}{\|x - a\|} = 0$ and $\lim\limits_{y \to b} \dfrac{\|\psi(y)\|}{\|y - b\|} = 0$. Hence

$$g(f(x)) = g(f(a)) + B(f(x) - f(a)) + \psi(f(x))$$

$$= g(f(a)) + B[A(x - a) + \varphi(x)] + \psi(f(x))$$

$$= g(f(a)) + B \circ A(x - a) + \rho(x)$$

where $\rho(x) = B(\varphi(x)) + \psi(f(x))$. Then

$$\frac{\|\rho(x)\|}{\|x - a\|} \leq \|B\| \frac{\|\varphi(x)\|}{\|x - a\|} + \frac{\|\psi(f(x))\|}{\|f(x) - f(a)\|} \frac{\|f(x) - f(a)\|}{\|x - a\|}$$

and since

$$\frac{\|f(x) - f(x)\|}{\|x - a\|} = \frac{\|A(x - a) + \varphi(x)\|}{\|x - a\|} \leq \|A\| + \frac{\|\varphi(x)\|}{\|x - a\|}$$

it follows that $\lim\limits_{x \to a} \dfrac{\|\rho(x)\|}{\|x - a\|} = 0$, completing the proof.

When $E = \mathbb{K}$ then the notion of differentiability takes the following, more familiar form.

13.7. **PROPOSITION.** *Let* U *be an open subset of* $\mathbb{K}$. *Then a mapping* $f : U \to F$ *is differentiable if and only if the derivative*

$$f'(\lambda) = \lim_{\mu \to \lambda} \frac{f(\mu) - f(\lambda)}{\mu - \lambda} \in F$$

exists for each $\lambda \in U$. *In this case* $Df(\lambda)(\mu) = \mu f'(\lambda)$ *for all* $\lambda \in U$ *and* $\mu \in \mathbb{K}$.

The proof of this proposition is straightforward and is left as an exercise to the reader. It will be useful in the proof of the next result, which generalizes the classical *Mean Value Theorem.*

13.8. **THEOREM.** *Let* U *be an open subset of* E, *and let* $f : U \to F$ *be a differentiable mapping. If a line segment* $[a, a + t]$ *is entirely contained in* U *then*

$$\|f(a + t) - f(a)\| \leq \|t\| \sup_{0 \leq \lambda \leq 1} \|Df(a + \lambda t)\|.$$

PROOF. In view of Remark 13.2(c) we may assume that $\mathbb{K} = \mathbb{R}$. Then for each $\psi \in F'$ we consider the function $g : [0,1] \to \mathbb{R}$ defined by $g(\lambda) = \psi \circ f(a + \lambda t)$. Then g is continuous on $[0,1]$, differentiable on $(0,1)$ and $g'(\lambda) = \psi[Df(a + \lambda t)(t)]$ for every $\lambda \in (0,1)$. By the classical Mean Value Theorem $|g(1) - g(0)| \leq \sup_{0<\lambda<1} |g'(\lambda)|$ and the desired conclusion follows.

13.9. **COROLLARY.** *Let* U *be an open subset of* E, *and let* $f : U \to F$ *be a differentiable mapping. If a line segment* $[a, a + t]$ *is entirely contained in* U *then*

$$\|f(a + t) - f(a) - Df(a)(t)\| \leq \|t\| \sup_{0 \leq \lambda \leq 1} \|Df(a + \lambda t) - Df(a)\|.$$

PROOF. It suffices to apply Theorem 13.8 to the mapping $g : U \to F$

defined by $g(x) = f(x) - Df(a)(x - a)$.

Next we give some applications of Theorem 13.8 and Corollary 13.9.

13.10. PROPOSITION. *Let U be an open subset of E and let $f: U \to F$ be a differentiable mapping. If U is connected and $Df(x) = 0$ for every $x \in U$ then f is a constant mapping:*

PROOF. Fix $a \in U$ and let A denote the set of all $x \in U$ such that $f(x) = f(a)$. Since f is continuous A is closed in U. To show that A is open, take $x \in A$ and choose $r > 0$ such that $B(x;r) \subset U$. By applying Theorem 13.8 to the line segment $[x, y]$ we see that $f(y) = f(x) = f(a)$ for every $y \in B(y;r)$. Thus A is open and closed in U and therefore $A = U$.

13.11. PROPOSITION. *Let $(e_1, \ldots, e_n)$ denote the canonical basis of $\mathbb{K}^n$ and let $\xi_1, \ldots, \xi_n$ denote the corresponding coordinate functionals. Let U be an open subset of $\mathbb{K}^n$ and let $f : U \to F$ be a mapping*

(a) *If f is differentiable then the partial derivative*

$$\frac{\partial f}{\partial \xi_j}(a) = \lim_{\lambda \to 0} \frac{f(a + \lambda e_j) - f(a)}{\lambda} \in F$$

exists and equals $Df(a)(e_j)$ for each $a \in U$ and $j = 1, \ldots, n$. Hence

$$Df(a)(t) = \sum_{j=1}^{n} t_j \frac{\partial f}{\partial \xi_j}(a)$$

for all $a \in U$ and $t \in \mathbb{K}^n$.

(b) *If all the partial derivatives $\partial f/\partial \xi_j$ exist and are continuous on U then f is differentiable there.*

PROOF. The proof of (a) is straightforward and is left as an exercise to the reader. To show (b) take $a \in U$. Then using

Corollary 13.9 we can write, for t near the origin:

$$f(a+t) - f(a)$$

$$= \sum_{j=1}^{n} [f(a + t_1 e_1 + \dots + t_j e_j) - f(a + t_1 e_1 + \dots + t_{j-1} e_{j-1})]$$

$$= \sum_{j=1}^{n} [t_j \frac{\partial f}{\partial \xi_j}(a) + \varphi_j(t)]$$

where

$$\|\varphi_j(t)\| \leq |t_j| \sup_{0\leq\lambda\leq 1} \|\frac{\partial f}{\partial \xi_j}(a + t_1 e_1 + \dots + t_{j-1} e_{j-1} + \lambda t_j e_j) -$$

$$\frac{\partial f}{\partial \xi_j}(a + t_1 e_1 + \dots + t_{j-1} e_{j-1})\|$$

for $j = 1,\dots,n$. Since the functions $\partial f/\partial \xi_j$ are continuous we see that $\|\varphi_j(t)\| / \|t\| \to 0$ when $t \to 0$ and the desired conclusion follows.

13.12. DEFINITION. Let U be an open subset of E. Then a mapping $f : U \to F$ is said to be *continuously differentiable* if f is differentiable and the mapping $Df : U \to \mathcal{L}(E;F)$ is continuous.

13.13. PROPOSITION. *Let U be an open subset of E, and let (f_i) be a net of continuously differentiable mappings from U into F. Assume that:*

(a) *The net (f_i) converges pointwise on U to a mapping f.*

(b) *The net (Df_i) converges uniformly on each compact subset of U to a mapping g.*

Then f is continuously differentiable on U and $Df = g$.

PROOF. Let $a \in U$ and choose $r > 0$ such that $B(a;r) \subset U$. Then by Corollary 13.9 for every $t \in B(0;r)$ and every i we have that

$$\|f_i(a+t) - f_i(a) - Df_i(a)(t)\| \leq \|t\| \sup_{0 \leq \lambda \leq 1} \|Df_i(a+\lambda t) - Df_i(a)\|.$$

Then using (a) and (b) we get that

$$\|f(a+t) - f(a) - g(a)(t)\| \leq \|t\| \sup_{0 \leq \lambda \leq 1} \|g(a+\lambda t) - g(a)\|.$$

Now, since (Df_i) converges to g uniformly on each compact subset of U, and since each Df_i is continuous, we conclude that $g|K$ is continuous for each compact subset K of U. Since U is a k-space we conclude that g is continuous. Hence, given $\varepsilon > 0$ we can find δ with $0 < \delta < r$ such that $\|g(x) - g(a)\| \leq \varepsilon$ whenever $\|x - a\| \leq \delta$. Using this we get that

$$\|f(a+t) - f(a) - g(a)(t)\| \leq \|t\| \varepsilon$$

whenever $\|t\| \leq \delta$. The desired conclusion follows.

13.14. PROPOSITION. *Let U be an open subset of E, and let μ be a $\mathbb{K}$-valued Borel measure on a compact Hausdorff space T. Let $\varphi \in C(U \times T; F)$ and let $f : U \to F$ be defined by*

$$f(x) = \int_T \varphi(x,t)\,d\mu(t)$$

for every $x \in U$. Then:

(a) *f is continuous.*

(b) *If the mapping $x \in U \to \varphi(x,t) \in F$ is differentiable for each $t \in T$, and the mapping $(x,t) \in U \times T \to D_x\varphi(x,t) \in \mathcal{L}(E;F)$ is continuous, then f is continuously differentiable and*

$$Df(x) = \int_T D_x\varphi(x,t)\,d\mu(t)$$

for every $x \in U$.

PROOF. (a) To show that f is continuous let $a \in U$ and $\varepsilon > 0$ be given. Since φ is continuous, for each $b \in T$ there is a neighborhood W_b of (a,b) in $U \times T$ such that $\|\varphi(x,t) - \varphi(a,b)\| \leq \varepsilon$ for every $(x,t) \in W_b$. Without loss of generality we may assume that $W_b = U_b \times V_b$ where U_b is a neighborhood of a in U and V_b is a neighborhood of b in T. Then, using the triangle inequality, we get that $\|\varphi(x,t) - \varphi(a,t)\| \leq 2\varepsilon$ for all $x \in U_b$ and $t \in V_b$. Since T is compact there are $b_1, \ldots, b_m \in T$ such that $T = V_{b_1} \cup \ldots \cup V_{b_m}$. Set $U_a = U_{b_1} \cap \ldots \cap U_{b_m}$. Then U_a is a neighborhood of a in U and

$$(13.1) \qquad \|\varphi(x,t) - \varphi(a,t)\| \leq 2\varepsilon$$

for all $x \in U_a$ and $t \in T$. Whence it follows that

$$\|f(x) - f(a)\| \leq \int_T \|\varphi(x,t) - \varphi(a,t)\| d|\mu|(t) \leq 2\varepsilon|\mu|(T)$$

for every $x \in U_a$. Hence f is continuous at a.

(b) Set $A(x) = \int_T D_x\varphi(x,t)d\mu(t)$ for every $x \in U$. By applying (a) to the mapping $\psi(x,t) = D_x\dot{\varphi}(x,t)$ we see that $A \in C(U;\mathcal{L}(E;F))$. We shall prove that f is differentiable and $Df(a) = A(a)$ for every $a \in U$. Let $a \in U$ and $\varepsilon > 0$ be given. By applying (13.1) to the mapping $\psi(x,t) = D_x\psi(x,t)$ we can find $\delta > 0$ such that

$$\|D_x\varphi(a + h,t) - D_x\varphi(a,t)\| \leq \varepsilon$$

whenever $\|h\| \leq \delta$ and $t \in T$. Then by Corollary 13.9 we have that

$$\|\varphi(a + h,t) - \varphi(a,t) - D_x\varphi(a,t)(h)\|$$

$$\leq \|h\| \sup_{0 \leq \lambda \leq 1} \|D_x\varphi(a + \lambda h,t) - D_x\varphi(a,t)\| \leq \|h\|\varepsilon$$

whenever $\|h\| \leq \delta$ and $t \in T$. Hence

$$\|f(a+h) - f(a) - A(a)(h)\|$$

$$\leq \int_T \|\varphi(a+h,t) - \varphi(a,t) - D_x\varphi(a,t)(h)\|\, d|\mu|(t)$$

$$\leq \|h\|\, \varepsilon |\mu|(T)$$

whenever $\|h\| \leq \delta$. The desired conclusion follows.

As we have already remarked, every $\mathbb{C}$-differentiable mapping is $\mathbb{R}$-differentiable. The next proposition tells us when an $\mathbb{R}$-differentiable mapping is $\mathbb{C}$-differentiable.

13.15 PROPOSITION. *Let E and F be complex Banach spaces, let U be an open subset of E, and let $f : U \to F$ be an $\mathbb{R}$-differentiable mapping. Let $Df(a)$ denote the real differential of f at a, and let $D'f(a)$ and $D''f(a)$ be defined by*

$$D'f(a)(t) = \frac{1}{2}[Df(a)(t) - i\, Df(a)(i\,t)]$$

and

$$D''f(a)(t) = \frac{1}{2}[Df(a)(t) + i\, Df(a)(i\,t)]$$

for every $t \in E$. Then f is $\mathbb{C}$-differentiable if and only if $D''f(a)(t) = 0$ for every $a \in U$ and $t \in E$. In this case $Df(a) = D'f(a)$ is also the complex differential of f at a.

PROOF. By Proposition 1.12 $Df(a)$ is $\mathbb{C}$-linear if and only if $D''f(a) = 0$. Thus it suffices to apply the definition of $\mathbb{C}$-differentiability and the uniqueness of the differential at a given point.

The distinction between $\mathbb{R}$-differentiability and $\mathbb{C}$-differentiability is emphasized by the following theorem.

13.16. THEOREM. *Let E and F be complex Banach spaces, and let U be an open subset of E. Then a mapping $f : U \to F$ is $\mathbb{C}$-differentiable if and only if it is holomorphic. In this case $Df(a) = P^1f(a)$ for every $a \in U$.*

PROOF. Suppose f is $\mathbb{C}$-differentiable. Then it follows from the Chain Rule 13.6 that the function $g(\lambda) = \psi \circ f(a + \lambda b)$ is $\mathbb{C}$-differentiable on the open set $\Lambda = \{\lambda \in U : a + \lambda b \in U\}$ for every $a \in U$, $b \in E$ and $\psi \in F'$. Hence g is holomorphic, by the corresponding result for functions of one complex variable. Thus f is G-holomorphic, by Theorem 8.12, and therefore holomorphic, by Theorem 8.7, since f is clearly continuous. Since the reverse implication was established in Example 13.5, the proof of the theorem is complete.

13.17. COROLLARY. *Let E and F be complex Banach spaces, and let U be an open subset of E. Then a mapping $f : U \to F$ is holomorphic if and only if f is $\mathbb{R}$-differentiable and $D''f$ is identically zero.*

EXERCISES

13.A. Let U be an open subset of E, and let $f : U \to F$ and $g : U \to F$ be two differentiable mappings.

(a) Show that $\alpha f + \beta g$ is differentiable for all $\alpha, \beta \in \mathbb{K}$ and $D(\alpha f + \beta g)(a) = \alpha Df(a) + \beta Dg(a)$ for every $a \in U$.

(b) Show that if $F = \mathbb{K}$ then the product fg is differentiable and $D(fg)(a) = g(a)Df(a) + f(a)Dg(a)$ for every $a \in U$.

13.B. Let E and F_j $(j = 1,\ldots,m)$ be Banach spaces, and let $F = F_1 \times \ldots \times F_m$. Let U be an open subset of E, let $f_j : U \to F_j$ $(j = 1,\ldots,m)$ and let $f = (f_1,\ldots,f_m) : U \to F$. Show that f is differentiable if and only if each f_j is differentiable. Show that in this case $Df(a) = (Df_1(a),\ldots,Df_m(a))$ for every $a \in U$.

13.C. Let E_j $(j = 1,\ldots,m)$ and F be Banach spaces, and let $A : E_1 \times \ldots \times E_m \to F$ be m-linear and continuous. Show that A is differentiable and

$$DA(a)(t) = \sum_{j=1}^{m} A(a_1,\ldots,a_{j-1},t_j,a_{j+1},\ldots,a_m)$$

for all $a = (a_1, \ldots, a_m)$ and $t = (t_1, \ldots, t_m)$ in $E_1 \times \ldots \times E_m$.

13.D. Let E, F_j $(j = 1, \ldots, m)$ and G be Banach spaces. Let U be an open subset of E, let $f_j : U \to F_j$ $(j = 1, \ldots, m)$ be differentiable, and let $A : F_1 \times \ldots \times F_m \to G$ be m-linear and continuous. Show that the mapping $g = A \circ (f_1, \ldots, f_m) : U \to G$ is differentiable and

$$Dg(a)(t) = \sum_{j=1}^{m} A(f_1(a), \ldots, f_{j-1}(a), Df_j(a)(t), f_{j+1}(a), \ldots, f_m(a))$$

for every $a \in U$ and $t \in E$.

13.E. Let E be a real or complex Hilbert space, with inner product $(x \mid y)$. Show that the function $f(x) = \|x\|^2$ is $\mathbb{R}$-differentiable on E and $Df(a)(t) = 2\mathrm{Re}(t|a)$ for all $a, t \in E$.

13.F. Let U be a connected open subset of E, and let $f : U \to F$ be a differentiable mapping whose differential $Df : U \to \mathcal{L}(E;F)$ is a constant mapping. Show the existence of $A \in \mathcal{L}(E;F)$ and $b \in F$ such that $f(x) = Ax + b$ for every $x \in U$.

13.G. Let $E = \mathbb{K}^m$ and $F = \mathbb{K}^n$, and let G be any Banach space over $\mathbb{K}$. Let $U \subset E$ and $V \subset F$ be two open sets, and let $f : U \to F$ and $g : V \to G$ be two differentiable mappings with $f(U) \subset V$. Let $\xi_1, \ldots, \xi_m$ denote the coordinate functionals of E, and let $\eta_1, \ldots, \eta_n$ denote the coordinate functionals of F. Show that if we write $f = (f_1, \ldots, f_n)$ then

$$\frac{\partial(g \circ f)}{\partial \xi_j}(a) = \sum_{k=1}^{n} \frac{\partial g}{\partial \eta_k}(f(a)) \frac{\partial f}{\partial \xi_j}(a)$$

for every $a \in U$ and $j = 1, \ldots, m$.

13.H. Let $z_1, \ldots, z_n$ denote the complex coordinate functionals of $\mathbb{C}^n$ and let $x_1, y_1, \ldots, x_n, y_n$ denote the corresponding real coordinate functionals. Let U be an open subset of $\mathbb{C}^n$, let F

be a complex Banach space, and let $f : U \to F$ be an $\mathbb{R}$-differentiable mapping. Let $\partial f/\partial x_j$ and $\partial f/\partial y_j$ denote the real partial derivatives in the sense of Proposition 13.11 and let $\partial f/\partial z_j$ and $\partial f/\partial \bar{z}_j$ be the functions defined by

$$\frac{\partial f}{\partial z_j}(a) = \frac{1}{2}\left[\frac{\partial f}{\partial x_j}(a) - i\frac{\partial f}{\partial y_j}(a)\right]$$

and

$$\frac{\partial f}{\partial \bar{z}_j}(a) = \frac{1}{2}\left[\frac{\partial f}{\partial x_j}(a) + i\frac{\partial f}{\partial y_j}(a)\right]$$

for every $a \in U$.

(a) Show that

$$D'f(a)(t) = \sum_{j=1}^{n} t_j \frac{\partial f}{\partial z_j}(a)$$

$$D''f(a)(t) = \sum_{j=1}^{n} \bar{t}_j \frac{\partial f}{\partial \bar{z}_j}(a)$$

for every $a \in U$ and $t \in \mathbb{C}^n$.

(b) Show that f is holomorphic if and only if $\partial f/\partial \bar{z}_j = 0$ on U for $j = 1,\ldots,n$. These are the *Cauchy-Riemann equations*.

Note that the operator $\partial/\partial z_j$ just defined coincides, when applied to holomorphic functions, with the complex partial derivative $\partial/\partial z_j$ in the sense of Proposition 13.11.

13.I. Let U be an open subset of E. Show that a function $f : U \to \mathbb{C}$ is $\mathbb{R}$-differentiable if and only if the conjugate function $\bar{f} : U \to \mathbb{C}$ is $\mathbb{R}$-differentiable.

13.J. Let U be an open subset of $\mathbb{C}^n$ and let $f : U \to \mathbb{C}$ be an $\mathbb{R}$-differentiable function. Show that

$$\left(\overline{\frac{\partial f}{\partial z_j}}\right) = \frac{\partial \bar{f}}{\partial \bar{z}_j} \quad \text{and} \quad \left(\overline{\frac{\partial f}{\partial \bar{z}_j}}\right) = \frac{\partial \bar{f}}{\partial z_j}$$

for every $j = 1, \ldots, n$.

14. DIFFERENTIABLE MAPPINGS OF HIGHER ORDER

This section is devoted to the study of differentiable mappings of higher order. Unless stated otherwise, the letters E and F will represent Banach spaces over the same field $\mathbb{K}$.

14.1. DEFINITION. Let U be an open subset of E. A mapping $f : U \to F$ is said to be *twice differentiable* if f is differentiable and the differential $Df : U \to \mathcal{L}(E;F)$ is differentiable as well.

14.2. REMARKS. Let U be an open subset of E and let $f : U \to F$ be a twice differentiable mapping. Then the differential of the mapping Df at a point $a \in U$ is called the *second differential* of f at a and will be denoted by $D^2f(a)$. Thus $D^2f(a)$ may be regarded as a member of $\mathcal{L}(E;\mathcal{L}(E;F))$ or $\mathcal{L}(^2E;F)$, in view of the canonical isomorphism given in Proposition 1.4. If the induced mapping $D^2f : U \to \mathcal{L}(^2E;F)$ is continuous, then f is said to be *twice continuously differentiable*.

The next result generalizes the classical *Schwarz' Theorem*.

14.3. THEOREM. *Let U be an open subset of E and let $f : U \to F$ be a twice differentiable mapping. Then the bilinear mapping $D^2f(a) \in \mathcal{L}(^2E;F)$ is symmetric for each $a \in U$. In other words, $D^2f(a)(s,t) = D^2f(a)(t,s)$ for all $s, t \in E$.*

PROOF. Let $a \in U$ and choose $r > 0$ such that $B(a;2r) \subset U$. For $s, t \in B(0;r)$ we define the mapping

$$g(s,t) = f(a + s + t) - f(a + s) - f(a + t) + f(a).$$

The idea of the proof is to estimate the difference $g(s,t) - D^2f(a)(s,t)$ for s and t near zero. First note that if we set

$$\varphi(x) = Df(x) - Df(a) - D^2f(a)(x - a)$$

for $x \in U$ then $\|\varphi(x)\| / \|x - a\| \to 0$ when $x \to a$. Hence we can write

$$\|g(s,t) - D^2f(a)(s,t)\|$$

$$= \|g(s,t) + \varphi(a + s)(t) - Df(a + s)(t) + Df(a)(t)\|$$

$$\leq \|\varphi(a + s)\| \, \|t\| + \|g(s,t) - Df(a + s)(t) + Df(a)(t)\|.$$

Now, fix $s \in B(0;r)$ and define $h(x) = f(x + s) - f(x)$ for each $x \in B(a;r)$. Noting that $Dh(x) = Df(x + s) - Df(x)$ and $g(s,t) = h(a + t) - h(a)$, and using Corollary 13.9 we get that

$$\|g(s,t) - Df(a + s)(t) + Df(a)(t)\|$$

$$= \|h(a + t) - h(a) - Dh(a)(t)\|$$

$$\leq \|t\| \sup_{0 \leq \lambda \leq 1} \|Dh(a + \lambda t) - Dh(a)\|.$$

Now,

$$Dh(a + \lambda t) - Dh(a)$$

$$= [Df(a + \lambda t + s) - Df(a + \lambda t)] - [Df(a + s) - Df(a)]$$

$$= [Df(a + \lambda t + s) - Df(a)] - [Df(a + \lambda t) - Df(a)] -$$

$$[Df(a + s) - Df(a)]$$

$$= [D^2f(a)(\lambda t + s) + \varphi(a + \lambda t + s)] - [D^2f(a)(\lambda t) + \varphi(a + \lambda t)] -$$

$$[D^2f(a)(s) + \varphi(a + s)]$$

$$= \varphi(a + \lambda t + s) - \varphi(a + \lambda t) - \varphi(a + s).$$

Thus we get the estimate

$$\|g(s,t) - D^2f(a)(s,t)\|$$

$$\leq \|\varphi(a+s)\| \, \|t\| + \|t\| \sup_{0\leq\lambda\leq 1} \|\varphi(a+\lambda t+s) - \varphi(a+\lambda t) - \varphi(a+s)\| .$$

Now, given $\varepsilon > 0$ we can find δ with $0 < \delta < r$ such that $\|\varphi(x)\| \,/\, \|x - a\| \leq \varepsilon$ for every $x \in B(a;2\delta)$. Then for all $s, t \in B(0;\delta)$ we get that

$$\|g(s,t) - D^2f(a)(s,t)\|$$

$$\leq \varepsilon \|s\| \, \|t\| + \|t\| \, \varepsilon \sup_{0\leq\lambda\leq 1} (\|\lambda t + s\| + \|\lambda t\| + \|s\|)$$

$$\leq \varepsilon \|t\| \, (3\|s\| + 2\|t\|) \leq 3\varepsilon \|t\| \, (\|s\| + \|t\|).$$

By interchanging the roles of s and t we get that

$$\|g(t,s) - D^2f(a)(t,s)\| \leq 3\varepsilon \, \|s\| \, (\|t\| + \|s\|)$$

for all $t, s \in B(0;\delta)$. And since $g(s,t) = g(t,s)$ for all $s, t \in B(0;r)$ we conclude that

$$\|D^2f(a)(s,t) - D^2f(a)(t,s)\| \leq 3\varepsilon \, (\|s\| + \|t\|)^2$$

for all $s, t \in B(0;\delta)$, but then, obviously for all $s, t \in E$. Since $\varepsilon > 0$ was arbitrary, we conclude that $D^2f(a)(s,t) = D^2f(a)(t,s)$ for all $s, t \in E$, as we wanted.

14.4. DEFINITION. Let U be an open subset of E. By induction on k we define a mapping $f : U \to F$ to be k times differentiable if f is $k-1$ times differentiable and its $(k-1)$th differential $D^{k-1}f : U \to \mathcal{L}(^{k-1}E;F)$ is differentiable. A mapping $f : U \to F$ is said to be *infinitely differentiable* if it is k times differentiable for each $k \in \mathbb{N}$.

14.5. REMARKS. Let U be an open subset of E, and let $f: U \to F$ be a k times differentiable mapping. Then the differential of the mapping $D^{k-1}f$ at a point $a \in U$ is called the *kth differential* of f at a and will be denoted by $D^kf(a)$. Thus $D^kf(a)$ may be regarded as a member of $\mathcal{L}(E;\mathcal{L}(^{k-1}E;F))$ or $\mathcal{L}(^kE;F)$, under the identification given by Proposition 1.4. If the induced mapping $D^k f : U \to \mathcal{L}(^k E;F)$ is continuous then f is said to be k *times continuously differentiable*. For convenience we also define $D^o f = f$.

Now we can easily generalize Theorem 14.3.

14.6. THEOREM. *Let U be an open subset of E and let $f: U \to F$ be a k times differentiable mapping. Then the k-linear mapping $D^k f(a) \in \mathcal{L}(^k E;F)$ is symmetric for each $a \in U$.*

PROOF. By induction on k. For $k = 2$ this is just Theorem 14.3. Let $k \geq 3$ and assume the theorem true for $k - 1$. If $a \in U$ then $D^k f(a)$ is the differential at a of the mapping $g = D^{k-1} f$, which, by the induction hypothesis, takes its values in $\mathcal{L}^s(^{k-1}E;F)$. Thus

$$D^k f(a) = Dg(a) \in \mathcal{L}(E;\mathcal{L}^s(^{k-1}E;F))$$

and hence $D^k f(a)(t_1,t_2,\ldots,t_k)$ is symmetric in the variables $t_2,\ldots,t_k$. On the other hand, $D^k f(a)$ is the second differential at a of the mapping $h = D^{k-2} f$, and if follows from Theorem 14.3 that

$$D^k f(a) = D^2 h(a) \in \mathcal{L}^s(^2E;\mathcal{L}(^{k-2}E;F)).$$

Hence $D^k f(a)(t_1,t_2,\ldots,t_k)$ is symmetric in the variables t_1 and t_2, and the desired conclusion follows.

The next theorem strengthens the conclusion in Theorem 13.16.

14.7. THEOREM. *Let E and F be complex Banach spaces, and let U be an open subset of E. Then for each mapping $f: U \to F$ the following conditions are equivalent:*

(a) *f is holomorphic.*

(b) *f is $\mathbb{C}$-differentiable.*

(c) *f is infinitely $\mathbb{C}$-differentiable.*

If these conditions are satisfied then $D^k f(a) = k!\, A^k f(a)$ for every $a \in U$ and $k \in \mathbb{N}$.

PROOF. In view of Theorem 13.16 it is sufficient to show that if $f : U \to F$ is holomorphic, then f is k times $\mathbb{C}$-differentiable and $D^k f(a) = k!\, A^k f(a)$ for every $a \in U$ and $k \in \mathbb{N}$. We proceed by induction on k, the statement being true for $k = 1$ by Theorem 13.16. Let $k \geq 2$ and assume f is $k-1$ times $\mathbb{C}$-differentiable and $D^{k-1} f = (k-1)!\, A^{k-1} f$. By Theorem 7.17, $P^{k-1} f \in \mathcal{H}(U; P({}^{k-1}E;F))$ and $P^1(P^{k-1} f)(a) = P^{k-1}(P^k f(a))$ for every $a \in U$. Hence $D^{k-1} f = (k-1)!\, A^{k-1} f \in \mathcal{H}(U; \mathcal{L}^s({}^{k-1}E;F))$, and by Theorem 13.16 we may conclude that $D^{k-1} f$ is $\mathbb{C}$-differentiable and $D^k f(a) = P^1(D^{k-1} f)(a) = (k-1)!\, P^1(A^{k-1} f)(a)$ for every $a \in U$. Thus f is k times $\mathbb{C}$-differentiable and

$$\begin{aligned} D^k f(a) t^k &= (k-1)!\, P^1(A^{k-1} f)(a)(t) t^{k-1} \\ &= (k-1)!\, P^1(P^{k-1} f)(a)(t)(t) \\ &= (k-1)!\, P^{k-1}(P^k f(a))(t)(t) \\ &= (k-1)! \binom{k}{k-1} A^k f(a) t\, t^{k-1} \\ &= k!\, A^k f(a) t^k. \end{aligned}$$

Since $D^k f(a)$ and $A^k f(a)$ are both symmetric, we conclude that $D^k f(a) = k!\, A^k f(a)$, as asserted.

14.8. DEFINITION. Let U be an open subset of E. For each $k \in \mathbb{N}$ we shall denote by $C^k(U;F)$ the vector space of all mappings $f : U \to F$ which are k times continuously $\mathbb{R}$-differentiable. We shall denote by $C^\infty(U;F)$ the vector space of all mappings $f : U \to F$ which belong to $C^k(U;F)$ for every $k \in \mathbb{N}$. For convenience we also define $C^0(U;F) = C(U;F)$. The members of $C^k(U;F)$, where $k \in \mathbb{N}_0 \cup \{\infty\}$, are often called mappings of class C^k. When $F = \mathbb{K}$ then we shall write for short $C^k(U) = C^k(U;\mathbb{K})$.

14.9. THEOREM. *Let E, F, G be Banach spaces over $\mathbb{K}$. Let $U \subset E$ and $V \rightarrow F$ be two open sets, and let $f : U \rightarrow F$ and $g : V \rightarrow G$ be two mappings of class C^k, with $f(U) \subset V$. Then the composite mapping $g \circ f : U \rightarrow G$ is also of class C^k.*

PROOF. We proceed by induction on k, the theorem being obvious for $k = 0$. Let $k \geq 1$ and assume the theorem true for $k - 1$. By the Chain Rule 13.6 the mapping $g \circ f : U \rightarrow G$ is $\mathbb{R}$-differentiable and $D(g \circ f)(x) = Dg(f(x)) \circ Df(x)$ for every $x \in U$. Thus the mapping $D(g \circ f) : U \rightarrow \mathcal{L}(E;G)$ is obtained by composition of the mappings S and T defined by

$$S : x \in U \rightarrow (Df(x), Dg(f(x))) \in \mathcal{L}(E;F) \times \mathcal{L}(F;G)$$

$$T : (A,B) \in \mathcal{L}(E;F) \times \mathcal{L}(F;G) \rightarrow B \circ A \in \mathcal{L}(E;G).$$

The mapping T is bilinear and continuous, and hence is of class C^∞, by Exercise 14.B. On the other hand, the mapping S takes its values in a product and its components are the mappings $Df : U \rightarrow \mathcal{L}(E;F)$ and $Dg \circ f : U \rightarrow \mathcal{L}(F;G)$. By the induction hypothesis the mappings Df and $Dg \circ f$ are both or class C^{k-1}, and then it follows from Exercise 14.A that the mapping S is of class C^{k-1} too. Then, again by the induction hypothesis, $D(g \circ f)$ is of class C^{k-1}. Hence $g \circ f$ is of class C^k and the proof is complete.

EXERCISES

14.A. Let E and F_j $(j = 1,\ldots,m)$ be Banach spaces, and let $F = F_1 \times \ldots \times F_m$. Let U be an open subset of E, let $f_j : U \rightarrow F_j$ $(j = 1,\ldots,m)$ and let $f = (f_1,\ldots,f_m) : U \rightarrow F$. Show that f is k times (continuously) differentiable if and only if each f_j is k times (continuously) differentiable.

14.B. Let E_j $(j = 1,\ldots,m)$ and F be Banach spaces, and let $A: E_1 \times \ldots \times E_m \to F$ be m-linear and continuous. Show that A is infinitely differentiable and $D^k A = 0$ for every $k \geq m+1$.

14.C. Show that if E is a real or complex Hilbert space then the function $x \to \|x\|^2$ is of class C^∞ on E.

14.D. Let U be an open subset of E, let $f : U \to F$ be a k times differentiable mapping, and let $[a, a+t]$ be a line segment which is entirely contained in U.

(a) Using the classical *Taylor's formula* show that

$$\|f(a+t) - \sum_{j=0}^{k-1} \frac{1}{j!} D^j f(a) t^j\| \leq \frac{\|t\|^k}{k!} \sup_{0 \leq \lambda \leq 1} \|D^k f(a+\lambda t)\|.$$

(b) Applying (a) to the mapping $g(x) = f(x) - \frac{1}{k!} D^k f(a)(x-a)^k$ show that

$$\|f(a+t) - \sum_{j=0}^{k} \frac{1}{j!} D^j f(a) t^j\| \leq \frac{\|t\|^k}{k!} \sup_{0 \leq \lambda \leq 1} \|D^k f(a+\lambda t) - D^k f(a)\|.$$

14.E. Let U be an open subset of E, and consider the vector space $C^\infty(U;F)$, endowed with the locally convex topology generated by all the seminorms of the form $f \to \sup_{x \in K} \|D^j f(x)\|$, where $j \in \mathbb{N}_0$ and K is a compact subset of U. Using Proposition 13.13 show that the locally convex space $C^\infty(U;F)$ is always complete. Show that if E is finite dimensional then $C^\infty(U;F)$ is a Fréchet space.

14.F. Let $(e_1,\ldots,e_n)$ denote the canonical basis of $\mathbb{K}^n$ and let $\xi_1,\ldots,\xi_n$ denote the corresponding coordinate functionals. Let U be an open subset of $\mathbb{K}^n$ and let $f : U \to F$ be a k times differentiable mapping.

(a) Show that the partial derivative $\partial^k f(a)/\partial\xi_{j_1} \ldots \partial\xi_{j_k}$ exists and equals $D^k f(a)(e_{j_1},\ldots,e_{j_k})$ for every $a \in U$ and $j_1,\ldots,j_k \in \{1,\ldots,n\}$.

(b) Show that $\partial^k f/\partial\xi_{j_1} \ldots \partial\xi_{j_k}$ is a symmetric function of the indices $j_1,\ldots,j_k$.

(c) Using the Leibniz Formula 1.8 show that

$$D^k f(a)t^k = \sum_{|\alpha|=k} \frac{k!}{\alpha!} t_1^{\alpha_1} \ldots t_n^{\alpha_n} \frac{\partial^k f(a)}{\partial\xi_1^{\alpha_1} \ldots \partial\xi_n^{\alpha_n}}$$

for every $a \in U$ and $t = (t_1,\ldots,t_n) \in \mathbb{K}^n$. For short we shall use the notation $\partial^\alpha f/\partial\xi^\alpha = \partial^k f/\partial\xi_1^{\alpha_1} \ldots \partial\xi_n^{\alpha_n}$ for every multi-index α with $|\alpha| = k$.

14.G. Let U be an open subset of $\mathbb{R}^n$, and let $x_1, \ldots, x_n$ denote the coordinate functionals of $\mathbb{R}^n$. Show that a mapping $f : U \to F$ is of class C^k if and only if all the partial derivatives $\partial^\alpha f/\partial x^\alpha$ of order $|\alpha| \leq k$ exist and are continuous on U.

15. PARTITIONS OF UNITY

In this section we introduce partitions of unity, a fundamental tool which in many different situations is used to construct objects with certain global properties by patching together objects which have locally the same properties. Before stating the main result we give two preparatory lemmas.

15.1. LEMMA. *If $a < b$ then there is a function $\varphi \in C^\infty(\mathbb{R})$ such that $\varphi(x) = 0$ if $x \leq a$, $\varphi(x) = 1$ if $x \geq b$ and $\varphi(x)$ is strictly increasing if $a \leq x \leq b$.*

PROOF. Start with the function f defined by $f(x) = e^{-1/x}$ if $x > 0$ and $f(x) = 0$ if $x \leq 0$. Then $f \in C^\infty(\mathbb{R})$ by Exercise 15.A, f is strictly positive for $x > 0$ and f is identically zero for $x \leq 0$. Then the function $g(x) = f[(x - a)(b - x)]$ is also of class C^∞ on $\mathbb{R}$, is strictly positive on the interval $a < x < b$, and is identically zero outside that interval. Then the function $\varphi(x) = \int_a^x g(t)dt / \int_a^b g(t)dt$ is identically zero for $x \leq a$, and identically one for $x \geq b$. Since $\varphi'(x) = g(x) / \int_a^b g(t)dt$ we conclude that φ is of class C^∞ on $\mathbb{R}$ and that φ is strictly increasing for $a \leq x \leq b$.

15.2. LEMMA. *Let E be a Hilbert space, and let $0 < r < R$. Then there is a function $\varphi \in C^\infty(E)$ such that $\varphi(x) = 1$ if $\|x\| \leq r$, $\varphi(x) = 0$ if $\|x\| \geq R$ and $0 < \varphi(x) < 1$ if $r < \|x\| < R$.*

PROOF. By Exercise 14.C the function $f(x) = - \|x\|^2$ is of class C^∞ on E. By Lemma 15.1 there is a function $g \in C^\infty(\mathbb{R})$ such that $g(t) = 0$ if $t \leq - R^2$, $g(t) = 1$ if $t \geq - r^2$ and $0 < g(t) < 1$ if $-R^2 < t < -r^2$. Then the function $\varphi(x) = g \circ f(x) = g(- \|x\|^2)$ has the required properties.

15.3. DEFINITION. Let X be a topological space and let F be a Banach space.

(a) The *support* of a mapping $f : X \to F$ is the closure of the set $\{x \in X : f(x) \neq 0\}$. The support of f will be denoted by $supp\, f$.

(b) A collection $(f_i)_{i\in I}$ of mappings from X into F is said to be *locally finite* if each point of X has a neighborhood which meets only finitely many of the sets $supp\, f_i$.

(c) A *partition of unity* on X is a locally finite collection $(\varphi_i)_{i\in I}$ of continuous functions from X into $[0, 1]$ such that $\sum_{i \in I} \varphi_i(x) = 1$ for every $x \in X$.

(d) A partition of unity $(\varphi_i)_{i\in I}$ on X is said to be *subordinated* to an open cover $(U_i)_{i\in I}$ of X if $supp\, \varphi_i \subset U_i$ for every $i \in I$.

If X is an open subset of a Banach space then it makes sense to talk about C^∞ *partitions of unity* on X: this means of course that each member of the partition of unity is a C^∞ function. Actually this is the situation we shall be primarily interested in. Indeed, we have the following theorem.

15.4. THEOREM. *Let U be an open subset of a separable Hilbert space E. Then for each open cover $(U_i)_{i\in I}$ of U there is a C^∞ partition of unity $(\varphi_i)_{i\in I}$ on U which is subordinated to $(U_i)_{i\in I}$.*

PROOF. Since U is a Lindelöf space there is a sequence of open balls $B(a_n;r_n)$ whose union is U and such that each $\overline{B}(a_n;2r_n)$ is contained in some U_i. By the axiom of choice there is a function $\tau : \mathbb{N} \to I$ such that $\overline{B}(a_n;2r_n) \subset U_{\tau(n)}$ for every $n \in \mathbb{N}$. By Lemma 15.2 there is a sequence (f_n) in $C^\infty(E)$ such that $f_n(x) = 1$ if $\|x - a_n\| \leq r_n$, $f_n(x) = 0$ if $\|x-a_n\| \geq 2r_n$ and $0 < f_n(x) < 1$ if $r_n < \|x - a_n\| < 2r_n$. Define another sequence (g_n) in $C^\infty(E)$ by $g_1 = f_1$ and

$$g_n = (1 - f_1) \ldots (1 - f_{n-1})f_n$$

if $n \geq 2$. Then it is clear that $0 \leq g_n \leq 1$ on E and that $\operatorname{supp} g_n \subset \operatorname{supp} f_n \subset \overline{B}_n(a_n;2r_n) \subset U_{\tau(n)}$ for every n. Furthermore, one can readily prove by induction that

$$(15.1) \qquad g_1 + \ldots + g_n = 1 - (1 - f_1) \ldots (1 - f_n)$$

for every n. Since $f_n = 1$ on $\overline{B}(a_n;r_n)$ it follows from (15.1) that

$$(15.2) \qquad g_1 + \ldots + g_n = 1 \quad \text{on} \quad \overline{B}(a_n;r_n)$$

and

$$(15.3) \qquad g_j = 0 \quad \text{on} \quad \overline{B}(a_n;r_n) \quad \text{for every} \quad j > n.$$

Thus (15.3) guarantees that the sequence (g_n) is locally finite in U, whereas (15.2) guarantees that $\sum_{n\in\mathbb{N}} g_n(x) = 1$ for every $x \in U$. Finally we define $\varphi_i(x) = \sum_{\tau(n)=i} g_n(x)$ for each $i \in \tau(\mathbb{N})$ and $x \in U$, and $\varphi_i(x) = 0$ for each $i \in I \setminus \tau(\mathbb{N})$

and $x \in U$. Since the sequence (g_n) is locally finite in U, each φ_i is well defined and belongs to $C^\infty(U)$. Furthermore, the set $\bigcup_{\tau(n)=i} supp\, g_n$ is closed in U and therefore $supp\, \varphi_i \subset \bigcup_{\tau(n)=i} supp\, g_n \subset U_i$. Since $\sum_{i \in I} \varphi_i(x) = \sum_{n \in \mathbb{N}} g_n(x)$ for every $x \in U$, $(\varphi_i)_{i \in I}$ is the required partition of unity.

15.5. COROLLARY. *Let U be an open subset of a separable Hilbert space E. Let A and B be two disjoint closed subsets of U. Then there is a function $\varphi \in C^\infty(U)$ such that $0 \leq \varphi \leq 1$ on U, $\varphi = 1$ on a neighborhood of A in U, and $\varphi = 0$ on a neighborhood of B in U.*

PROOF. By Theorem 15.4 there are two nonnegative functions φ, ψ in $C^\infty(U)$ such that $supp\, \varphi \subset U \setminus B$, $supp\, \psi \subset U \setminus A$ and $\varphi + \psi = 1$ on U. Then $\varphi = 1$ on the open set $U \setminus supp\, \psi$, which contains A, whereas $\varphi = 0$ on the open set $U \setminus supp\, \varphi$, which contains B.

EXERCISES

15.A. Consider the function f defined by $f(x) = e^{-1/x}$ for $x > 0$ and $f(x) = 0$ for $x \leq 0$.

(a) Show that $f^{(k)}(x) = e^{-1/x} P_{2k}(1/x)$ for every $x > 0$ and $k \in \mathbb{N}$, where $P_{2k}(t)$ is a polynomial in t of degree $2k$.

(b) Using L'Hospital rule show that $f^{(k)}(0)$ exists and equals zero for every $k \in \mathbb{N}$. Conclude that $f \in C^\infty(\mathbb{R})$.

15.B. Find an increasing sequence of convex, increasing C^∞ functions $\varphi_n : \mathbb{R} \to \mathbb{R}$ such that $\varphi_n(x) = 0$ for every $x \leq 0$ and $n \in \mathbb{N}$, and $\lim_{n \to \infty} \varphi_n(x) = \infty$ for every $x > 0$.

15.C. Let U be an open subset of a separable Hilbert space. Let (U_n) be an increasing sequence of open sets whose union is U, and let (c_n) be an increasing sequence of real numbers. Using

Theorem 15.4 find a function $\varphi \in C^{\infty}(U; \mathbb{R})$ such that $\varphi \geq c_1$ on U_1 and $\varphi \geq c_n$ on $U_n \setminus U_{n-1}$ for every $n \geq 2$.

15.D. Let U be an open subset of a separable Hilbert space. Given a function $f \in C(U;\mathbb{R})$ find a function $g \in C^{\infty}(U; \mathbb{R})$ such that $g \geq f$ on U.

15.E. Let $f : \mathbb{R} \to \mathbb{R}$ be a function which is bounded above on each interval of the form $(-\infty, b)$. Find a function $g \in C^{\infty}(\mathbb{R};\mathbb{R})$ such that $g(x) =$ constant for $x \leq 0$ and $g(x) \geq f(x)$ for every $x \in \mathbb{R}$.

15.F. Let K be a compact subset of a Hilbert space E, and let U be an open neighborhood of K. By adapting the proof of Theorem 15.4 find a nonnegative function $\varphi \in C^{\infty}(E)$ such that $supp\ \varphi \subset U$, $\varphi \leq 1$ on E and $\varphi = 1$ on a neighborhood of K.

15.G. Let K be a compact subset of a Hilbert space E, and let $U_1,\ldots,U_n$ be open subsets of E which cover K. Using Exercises 9.B and 15.F find nonnegative functions $\varphi_1, \ldots, \varphi_n \in C^{\infty}(E)$ such that $supp\ \varphi_j \subset U_j$, $\varphi_1 + \ldots + \varphi_n \leq 1$ on E, and $\varphi_1 + \ldots + \varphi_n = 1$ on a neighborhodd of K.

15.H. Let K be a compact subset of a completely regular Hausdorff space X, and let $U_1,\ldots,U_n$ be open subsets of X which cover K. Find nonnegative functions $\varphi_1,\ldots,\varphi_n \in C(X)$ such that $supp\ \varphi_j \subset U_j$, $\varphi_1 + \ldots + \varphi_n \leq 1$ on X, and $\varphi_1 + \ldots + \varphi_n = 1$ on a neighborhood of K.

16. TEST FUNCTIONS

In this section we introduce the space of test functions and establish some properties that will be of frequent use in this book.

16.1. DEFINITION. We shall denote by $\mathcal{D}(\mathbb{R}^n)$ the vector space of all $f \in C^{\infty}(\mathbb{R}^n)$ which have compact support. Each $f \in \mathcal{D}(\mathbb{R}^n)$

is called a *test function*. If U is an open subset of $\mathbb{R}^n$ then we shall denote by $\mathcal{D}(U)$ the vector space of all $f \in \mathcal{D}(\mathbb{R}^n)$ such that $supp\, f \subset U$. Likewise, if K is a compact subset of $\mathbb{R}^n$ then we shall denote by $\mathcal{D}(K)$ the vector space of all $f \in \mathcal{D}(\mathbb{R}^n)$ such that $supp\, f \subset K$.

It is not at all obvious that test functions exist, apart from the zero function. But the results in the preceding section guarantee the existence of large collections of test functions. Indeed, let U be an open subset of $\mathbb{R}^n$, let (U_i) be an open cover of U, and let (φ_i) be a C^∞ partition of unity on U, subordinated to the cover (U_i). If each U_i is relatively compact in U then the entire collection (φ_i) is contained in $\mathcal{D}(U)$. Another example of a test function. which is very useful for it serves to generate new test functions, is the following.

16.2. EXAMPLE. Let $\rho : \mathbb{R}^n \to \mathbb{R}$ be defined by $\rho(x) = ke^{-1/(1-\|x\|^2)}$ if $\|x\| < 1$ and $\rho(x) = 0$ if $\|x\| \geq 1$, where the constant $k > 0$ is chosen so that $\int_{\mathbb{R}^n} \rho d\lambda = 1$, and the letter λ stands for n dimensional Lebesgue measure. Then $\rho \in C^\infty(\mathbb{R}^n)$, by Exercise 15.A, and $supp\, \rho = \overline{B}(0;1)$. More generally, for each $\delta > 0$ let $\rho_\delta \in \mathcal{D}(\mathbb{R}^n)$ be defined by $\rho_\delta(x) = \delta^{-n}\rho(x/\delta)$ for every $x \in \mathbb{R}^n$. Then $\int_{\mathbb{R}^n} \rho_\delta d\lambda = 1$ and $supp\, \rho_\delta = \overline{B}(0;\delta)$.

If U is an open subset of $\mathbb{R}^n$ then we shall denote by $L^1(U)$ the Banach space of all equivalent classes of Lebesgue integrable functions on U. We shall denote by $L^1(U, loc)$ the vector space of all equivalent classes of Lebesgue measurable functions on U which are integrable over each compact subset of U.

16.3. DEFINITION. Let U be an open subset of $\mathbb{R}^n$, let $\delta > 0$ and let $U_\delta = \{x \in U : d_U(x) > \delta\}$. Given $f \in L^1(U, loc)$ and $\varphi \in \mathcal{D}(\overline{B}(0;\delta))$ we define their *convolution* $f * \varphi : U_\delta \to \mathbb{K}$ by

$$(f*\varphi)(x) = \int_{\overline{B}(0;\delta)} f(x-y)\varphi(y)d\lambda(y) = \int_{\overline{B}(x;\delta)} f(y)\varphi(x-y)d\lambda(y)$$

for every $x \in U_\delta$.

16.4. **PROPOSITION.** *Let U be an open subset of $\mathbb{R}^n$, let $f \in L^1(U, loc)$ and let $\varphi \in \mathcal{D}(\overline{B}(0;\delta))$. Then:*

(a) $f * \varphi \in C^\infty(U_\delta)$.

(b) $\dfrac{\partial^\alpha}{\partial x^\alpha}(f * \varphi) = f * \dfrac{\partial^\alpha \varphi}{\partial x^\alpha}$ *for each multi-index* α.

(c) $supp(f * \varphi) \subset supp\ f + \overline{B}(0;\delta)$.

In particular $f * \varphi \in \mathcal{D}(U_\delta)$ if the support of f is a compact subset of $U_{2\delta}$.

PROOF. For each $x \in U_\delta$ we can write

$$(f * \varphi)(x) = \int_U f(y)\varphi(x - y)d\lambda(y).$$

By differentiation under the integral sign we get that $f * \varphi \in C^1(U_\delta)$ and

$$\frac{\partial}{\partial x_j}(f * \varphi)(x) = \int_U f(y)\frac{\partial \varphi}{\partial x_j}(x - y)d\lambda(y) = (f * \frac{\partial \varphi}{\partial x_j})(x)$$

for every j. Then (a) and (b) follow by induction. Since (c) is clear, the proof of the proposition is complete.

16.5. **PROPOSITION.** *Let U be an open subset of $\mathbb{R}^n$ and let $f \in C(U)$ be a function with compact support. Then $f * \rho_\delta$ converges to f uniformly on U when $\delta \to 0$.*

PROOF. Since f is continuous and has compact support, it is uniformly continuous on U. Hence, given $\varepsilon > 0$ we can find $\delta_0 > 0$ such that $supp\ f \subset U_{2\delta_0}$ and $|f(x - y) - f(x)| \leq \varepsilon$ for every $x \in U_{\delta_0}$ and $y \in \overline{B}(0;\delta_0)$. If $0 < \delta < \delta_0$ then for every $x \in U_\delta$ we have that

$$|f * \rho_\delta(x) - f(x)| \leq \int_{\overline{B}(0;\delta)} |f(x - y) - f(x)|\rho_\delta(y)d\lambda(y) \leq \varepsilon.$$

16.6. PROPOSITION. *Let U be an open subset of $\mathbb{R}^n$ and let $f \in L^1(U)$ be a function with compact support. Then:*

(a) $\displaystyle\int_U |f * \rho_\delta| d\lambda \leq \int_U |f| d\lambda$ *whenever* $\operatorname{supp} f \subset U_{2\delta}$.

(b) $\displaystyle\int_U |f * \rho_\delta - f| d\lambda \to 0$ *when* $\delta \to 0$.

PROOF. (a) Since $f * \rho_\delta$ vanishes outside U_δ an application of the Fubini Theorem shows that

$$\int_U |f * \rho_\delta(x)| d\lambda(x) \leq \int_{U_\delta} d\lambda(x) \int_{\overline{B}(0;\delta)} |f(x-y)| \rho_\delta(y) d\lambda(y)$$

$$= \int_{\overline{B}(0;\delta)} d\lambda(y) \int_{U_\delta} |f(x-y)| \rho_\delta(y) d\lambda(x) = \int_U |f(x)| d\lambda(x).$$

(b) Since the continuous functions with compact support are dense in $L^1(U)$, given $\varepsilon > 0$ we can find a function $g \in C(U)$, with compact support, such that $\int_U |g - f| d\lambda \leq \varepsilon$. Let $0 < 2r < d_U(\operatorname{supp} g)$ and let $K = \sup g + \overline{B}(0;r)$. Then by Proposition 16.5 we can find δ_o with $0 < \delta_o < r$ such that $|g * \rho_\delta - g| \leq \varepsilon/\lambda(K)$ whenever $0 < \delta < \delta_o$. Then, using part (a) we get for $0 < \delta < \delta_o$ that

$$\int_U |f * \rho_\delta - f| d\lambda$$

$$\leq \int_U |f * \rho_\delta - g * \rho_\delta| d\lambda + \int_U |g * \rho_\delta - g| d\lambda + \int_U |g - f| d\lambda$$

$$\leq 2 \int_U |f - g| d\lambda + \int_U |g * \rho_\delta - g| d\lambda \leq 3\varepsilon.$$

16.7. COROLLARY. *If U is an open subset of $\mathbb{R}^n$ then $\mathcal{D}(U)$ is dense in $L^1(U)$.*

PROOF. Let *(K_j) be an increasing sequence of compact subsets of U which cover U. If $f \in L^1(U)$ then* $\int_U |f\chi_{K_j} - f| d\lambda \to 0$

when $j \to \infty$, by the Dominated Convergence Theorem. Since each of the functions $f\chi_{K_j}$ has compact support, it suffices to apply Proposition 16.6.

16.8. DEFINITION. Let K be a compact subset of $\mathbb{R}^n$. Then the vector space $\mathcal{D}(K)$ will be always endowed with the locally convex topology generated by the seminorms $f \to \sup_{x \in K} \| D^k f(x) \|$, where k varies over all nonnegative integers.

Using Exercise 14.F we see that the topology of $\mathcal{D}(K)$ is also generated by the seminorms of the form $f \to \sup_K |\partial^\alpha f/\partial x^\alpha|$, where α varies over all multi-indices in $\mathbb{N}_o^n$. It follows from Proposition 13.13 that $\mathcal{D}(K)$ is a Fréchet space. Actually $\mathcal{D}(K)$ is a closed vector subspace if $C^\infty(\mathbb{R}^n)$, which is also a Fréchet space, by Exercise 14.E.

16.9. DEFINITION. Let U be an open subset of $\mathbb{R}^n$. Then the vector space $\mathcal{D}(U)$ will be endowed with the finest locally convex topology such that the inclusion mapping $\mathcal{D}(K) \hookrightarrow \mathcal{D}(U)$ is continuous for each compact subset K of U. In other words, a seminorm p on $\mathcal{D}(U)$ is continuous if and only if its restriction to the Fréchet space $\mathcal{D}(K)$ is continuous for each compact subset K of U.

It is clear from the description of the topology of $\mathcal{D}(U)$ that a linear mapping $T : \mathcal{D}(U) \to Y$ from $\mathcal{D}(U)$ into a locally convex space Y is continuous if and only if the restriction of T to each $\mathcal{D}(K)$ is continuous.

EXERCISES

16.A. Let U be an open subset of $\mathbb{R}^n$, let $1 \le p < \infty$, and let $L^p(U)$ and $L^p(U, loc)$ be defined in the obvious way. Given a function $f \in L^p(U)$ with compact support show that:

(a) $\int_U |f * \rho_\delta|^p d\lambda \le \int_U |f|^p d\lambda$ whenever $supp\ f \subset U_{2\delta}$.

(b) $\int_U |f * \rho_\delta - f|^p d\lambda \to 0$ when $\delta \to 0$.

16.B. If U is an open subset of $\mathbb{R}^n$ show that $L^p(U, loc) \subset L^1(U, loc)$ for every $p \geq 1$.

16.C. Let U be an open subset of $\mathbb{R}^n$. For each $\varphi \in C^\infty(U;\mathbb{R})$ let $L^p(U,\varphi)$ denote the Banach space of all equivalent classes of Lebesgue measurable functions $f : U \to \mathbb{K}$ such that $\int_U |f|^p e^{-\varphi} d\lambda < \infty$. Using Exercise 15.C show that $L^p(U, loc)$ is the union of the spaces $L^p(U,\varphi)$, with $\varphi \in C^\infty(U;\mathbb{R})$.

16.D. Show that if U is an open subset of $\mathbb{R}^n$ then the mapping $\varphi \in \mathcal{D}(U) \to \partial^\alpha\varphi/\partial x^\alpha \in \mathcal{D}(U)$ is continuous for each multi-index α.

16.E. Let $\varphi \in \mathcal{D}(\mathbb{R}^n)$, let $\varepsilon > 0$ and let $K = supp\, \varphi + \overline{B}(0;\varepsilon)$. For $0 < |\lambda| < \varepsilon$ define $\varphi_\lambda \in \mathcal{D}(K)$ by

$$\varphi_\lambda(x) = \frac{\varphi(x + \lambda e_j) - \varphi(x)}{\lambda}.$$

(a) Using Corollary 13.9 show that (φ_λ) converges to $\partial\varphi/\partial x_j$ uniformly on K when $\lambda \to 0$.

(b) Show that (φ_λ) converges to $\partial\varphi/\partial x_j$ in $\mathcal{D}(K)$ when $\lambda \to 0$.

16.F. Let U be an open subset of $\mathbb{R}^n$.

(a) Show that $\frac{\partial^\alpha}{\partial x^\alpha}(f * \psi) = \frac{\partial^\alpha f}{\partial x^\alpha} * \psi$ on U_δ for every $f \in C^\infty(U)$ and $\psi \in \mathcal{D}(\overline{B}(0;\delta))$.

(b) Using (a) and Proposition 16.5 show that $(\varphi * \rho_\delta)$ converges to φ in $\mathcal{D}(U)$ when $\delta \to 0$ for every $\varphi \in \mathcal{D}(U)$.

17. DISTRIBUTIONS

In this section we introduce distributions and establish

some properties that will be frequently used in this book.

The idea is to define a new class of objects, called distributions, which should include all continuous functions on $\mathbb{R}^n$, and should have, in some sense, partial derivatives of all orders. The partial derivatives of distributions should again be distributions, and in the case of C^∞ functions, the partial derivatives in the sense of distributions should coincide with the partial derivative in the classical sense.

17.1. DEFINITION. Let U be an open subset of $\mathbb{R}^n$. Then a *distribution* on U is a continuous linear functional on $\mathcal{D}(U)$. In other words, a distribution on U is a linear functional on $\mathcal{D}(U)$ whose restriction to $\mathcal{D}(K)$ is continuous for each compact subset K of U. Thus we shall denote by $\mathcal{D}'(U)$ the vector space of all distributions on U.

17.2. EXAMPLE. Let U be an open subset of $\mathbb{R}^n$. Then each function $f \in C(U)$ defines a distribution $T_f \in \mathcal{D}'(U)$ by $T_f(\varphi) = \int_U \varphi f d\lambda$ for every $\varphi \in \mathcal{D}(U)$.

Using Proposition 16.5 one can readily show that the mapping $f \in C(U) \to T_f \in \mathcal{D}'(U)$ is injective. Hence we shall identify each $f \in C(U)$ with its image $T_f \in \mathcal{D}'(U)$ and speak of the distribution f.

17.3. EXAMPLE. Let U be an open subset of $\mathbb{R}^n$. Then more generally, each $f \in L^1(U, loc)$ defines a distribution $T_f \in \mathcal{D}'(U)$ by $T_f(\varphi) = \int_U \varphi f d\lambda$ for every $\varphi \in \mathcal{D}(U)$. Using Proposition 16.6 one can readily show that the mapping $f \in L^1(U, loc) \to T_f \in \mathcal{D}'(U)$ is injective. Hence we shall identify each $f \in L^1(U, loc)$ with its image $T_f \in \mathcal{D}'(U)$ and speak of the distribution f.

Thus we see that in particular we can identify $C^\infty(U)$ with a vector subspace of $\mathcal{D}'(U)$. We would like to extend the differential operators $\partial^\alpha/\partial x^\alpha$ from $C^\infty(U)$ to all of $\mathcal{D}'(U)$. To motivate the definition take $f \in C^\infty(U)$ and $\varphi \in \mathcal{D}(U)$. Since φ

has compact support it follows from the Fubini Theorem and the integration by parts formula that

$$\int_U \varphi \frac{\partial f}{\partial x_j} d\lambda = - \int_U \frac{\partial \varphi}{\partial x_j} f d\lambda$$

for every j. Then by induction we get that

$$\int_U \varphi \frac{\partial^\alpha f}{\partial x^\alpha} d\lambda = (-1)^{|\alpha|} \int_U \frac{\partial^\alpha \varphi}{\partial x^\alpha} f d\lambda$$

for every multi-index α. This motivates the following definition.

17.4. DEFINITION. Let U be an open subset of $\mathbb{R}^n$ and let $T \in \mathcal{D}'(U)$. For each multi-index α we define $\partial^\alpha T/\partial x^\alpha \in \mathcal{D}'(U)$ by

$$\frac{\partial^\alpha T}{\partial x^\alpha}(\varphi) = (-1)^{|\alpha|} T\left(\frac{\partial^\alpha \varphi}{\partial x^\alpha}\right)$$

for every $\varphi \in \mathcal{D}(U)$.

Since the mapping $\varphi \in \mathcal{D}(U) \to \partial^\alpha \varphi/\partial x^\alpha \in \mathcal{D}(U)$ is linear and continuous, it is clear that $\partial^\alpha T/\partial x^\alpha$ is indeed a distribution on U. It is also clear that if $f \in C^\infty(U)$ then the derivative $\partial^\alpha f/\partial x^\alpha$ in the sense of distributions coincides with the derivative $\partial^\alpha f/\partial x^\alpha$ in the classical sense.

Next we introduce a locally convex topology on $\mathcal{D}'(U)$. The topology we choose is very simple and is sufficient for all our needs in this book.

17.5. DEFINITION. Let U be an open subset of $\mathbb{R}^n$. Then the vector space $\mathcal{D}'(U)$ will be always endowed with the *topology of pointwise convergence*, that is, with the locally convex topology generated by the seminorms of the form $T \to \sup_{\varphi \in \Phi} |T(\varphi)|$, where Φ varies over all finite subsets of $\mathcal{D}(U)$.

Then the following result is clear.

17.6. PROPOSITION. *Let* U *be an open subset of* $\mathbb{R}^n$. *Then the mapping* $T \in \mathcal{D}'(U) \to \partial^\alpha T/\partial x^\alpha \in \mathcal{D}'(U)$ *is continuous for each multi-index* α.

In the preceding section we proved that $\mathcal{D}(U)$ is dense in $L^1(U)$, and using Exercise 17.A one can actually prove that $\mathcal{D}(U)$ is dense in $L^p(U)$ for every $p \geq 1$. Now we would like to obtain a similar result for distributions, namely we would like to prove that $\mathcal{D}(U)$ is dense in $\mathcal{D}'(U)$. With this aim in wind we shall introduce two new operations with distributions. First we shall define the multiplication of a distribution and a C^∞ function, and afterwards we shall define the convolution of a distribution and a test function.

17.7. DEFINITION. Let U be an open subset of $\mathbb{R}^n$. Given $f \in C^\infty(U)$ and $T \in \mathcal{D}'(U)$ we define the product $fT \in \mathcal{D}'(U)$ by

$$(fT)(\varphi) = T(f\varphi)$$

for every $\varphi \in \mathcal{D}(U)$.

Since the mapping $\varphi \in \mathcal{D}(U) \to \varphi f \in \mathcal{D}(U)$ is linear and continuous, it is clear that fT is indeed a distribution on U. If $g \in L^1(U, loc)$ then it is also clear that the product fg in the sense of distributions coincides with the pointwise product fg.

17.8. PROPOSITION. *Let* U *be an open subset of* $\mathbb{R}^n$, *let* $f \in C^\infty(U)$ *and let* $T \in \mathcal{D}'(U)$. *Then:*

$$\frac{\partial(fT)}{\partial x_j} = \frac{\partial f}{\partial x_j} T + f \frac{\partial T}{\partial x_j}$$

for every $j = 1, \ldots, n$.

PROOF. For every $\varphi \in \mathcal{D}(U)$ we have that

$$\frac{\partial(fT)}{\partial x_j}(\varphi) = -(fT)\left(\frac{\partial \varphi}{\partial x_j}\right) = -T\left(\frac{\partial \varphi}{\partial x_j} f\right)$$

$$= - T\left[\frac{\partial(\varphi f)}{\partial x_j} - \varphi \frac{\partial f}{\partial x_j}\right] = \frac{\partial T}{\partial x_j}(\varphi f) + T\left(\varphi \frac{\partial f}{\partial x_j}\right)$$

$$= \left(f \frac{\partial T}{\partial x_j}\right)(\varphi) + \left(\frac{\partial f}{\partial x_j} T\right)(\varphi).$$

Let $V \subset U$ be two open subsets of $\mathbb{R}^n$. Then $\mathcal{D}(V) \subset \mathcal{D}(U)$ and every distribution on U defines by restriction a distribution on V. If $T \in \mathcal{D}'(U)$ then we shall say that $T = 0$ on V if $T(\varphi) = 0$ for every $\varphi \in \mathcal{D}(V)$. If follows from the next proposition that for every $T \in \mathcal{D}'(U)$ there exists a largest open set $V \subset U$ (possibly empty) where $T = 0$.

17.9. PROPOSITION. *Let U be an open subset of $\mathbb{R}^n$ and let $T \in \mathcal{D}'(U)$. Let $(V_i)_{i\in I}$ be a collection of open subsets of U such that $T = 0$ on V_i for every $i \in I$. Then $T = 0$ on the open set $V = \bigcup_{i\in I} V_i$.*

PROOF. Let $(\psi_i)_{i\in I}$ be a C^∞ partition of unity on V, subordinated to the cover $(V_i)_{i\in I}$. If $\varphi \in \mathcal{D}(V)$ then $\varphi = \Sigma\, \varphi\psi_i$ and the sum has only finitely many nonzero terms, since φ has compact support and the collection (ψ_i) is locally finite. It is also clear that $\varphi\psi_i \in \mathcal{D}(V_i)$ for every i. Hence $T(\varphi) = \Sigma\, T(\varphi\psi_i) = 0$ and the proof is complete.

17.10. DEFINITION. Let U be an open subset of $\mathbb{R}^n$ and let $T \in \mathcal{D}'(U)$. If V is the largest open subset of U where $T = 0$ then the set $U \setminus V$ is called the *support* of T and will be denoted by $\operatorname{supp} T$.

17.11. EXAMPLE. Let U be an open subset of $\mathbb{R}^n$ and let $f \in C(U)$. Then $\operatorname{supp} T_f = \operatorname{supp} f$.

17.12. PROPOSITION. *Let U be an open subset of $\mathbb{R}^n$. Then the distributions on U with compact support are dense in $\mathcal{D}'(U)$.*

PROOF. Let (K_j) be a sequence of compact subsets of U such that

$U = \bigcup_{j=1}^{\infty} K_j$ and $K_j \subset \overset{\circ}{K}_{j=1}$ for every j. By Corollary 15.5 there is a sequence (ψ_i) in $\mathcal{D}(U)$ such that $\psi_j = 1$ on a neighborhood of K_j and $supp\, \psi_j \subset \overset{\circ}{K}_{j+1}$ for every j. If $\varphi \in \mathcal{D}(K_j)$ then $\varphi\psi_k = \varphi$ on a neighborhood of K_j for every $k \geq j$, and hence the sequence $(\varphi\psi_k)$ converges to φ in $\mathcal{D}(K_j)$. Hence the sequence $(\psi_k T)$ converges to T in $\mathcal{D}'(U)$ for every $T \in \mathcal{D}'(U)$. Since $supp(\psi_k T) \subset supp\, \psi_k \subset \overset{\circ}{K}_{k+1}$, the proof is complete.

Next we want to define the convolution of a distribution and a test function. We recall that if $f \in L^1(U, loc)$ and $\varphi \in \mathcal{D}(\overline{B}(0;\delta))$ then the convolution $f * \varphi : U_\delta \to \mathbb{K}$ is given by

$$(f * \varphi)(x) = \int_U f(y)\varphi(x - y)\, d\lambda(y)$$

for every $x \in U_\delta$. This motivates the following definition.

17.13. DEFINITION. Let U be an open subset of $\mathbb{R}^n$. Given $T \in \mathcal{D}'(U)$ and $\varphi \in \mathcal{D}(\overline{B}(0;\delta))$ we define their *convolution* $T * \varphi : U_\delta \to \mathbb{K}$ by

$$(T * \varphi)(x) = T_y\, [\varphi(x - y)]$$

for every $x \in U_\delta$. Here $T_y\, [\varphi(x - y)]$ means that the distribution T is applied to the test function $\varphi(x - y)$ regarded as a function of y for x fixed.

If $f \in L^1(U, loc)$ then it is clear that the convolution $f * \varphi$ in the sense of distributions coincides with the convolution $f * \varphi$ in the sense of functions.

17.14. PROPOSITION. *Let U be an open subset of $\mathbb{R}^n$, let $T \in \mathcal{D}'(U)$ and let $\varphi \in \mathcal{D}(\overline{B}(0;\delta))$. Then:*

(a) $T * \varphi \in C^\infty(U_\delta)$.

(b) $$\frac{\partial^\alpha}{\partial x^\alpha}(T * \varphi) = T * \frac{\partial^\alpha \varphi}{\partial x^\alpha} = \frac{\partial^\alpha T}{\partial x^\alpha} * \varphi.$$

(c) $supp(T * \varphi) \subset supp\ T + \overline{B}(0;\delta)$.

In particular $T * \varphi \in \mathcal{D}(U_\delta)$ *if the support of* T *is a compact subset of* $U_{2\delta}$.

PROOF. First we show that $T * \varphi$ is continuous at each $a \in U_\delta$. Indeed, choose $\varepsilon < 0$ such that $\overline{B}(a;r) \subset U_\delta$ and let $K = \overline{B}(a;\varepsilon + \delta)$. If we set $\psi_x(y) = \varphi(x - y)$ for each $x \in \overline{B}(a;\varepsilon)$ and $y \in \mathbb{R}^n$ then it is clear that $\psi_x \to \psi_a$ in $\mathcal{D}(K)$ and therefore $(T * \varphi)(x) = T(\psi_x) \to T(\psi_a) = (T * \varphi)(a)$, as we wanted.

Next we show that $\frac{\partial \varphi}{\partial x_j}(T * \varphi)(a)$ exists and equals $(T * \frac{\partial \varphi}{\partial x_j})(a)$ for every $j = 1,\dots,n$. Indeed, first we observe that

$$\frac{(T * \varphi)(a + \lambda e_j) - (T * \varphi)(a)}{\lambda} = (T * \varphi_\lambda)(a),$$

for $0 < |\lambda| < \varepsilon$, where

$$\varphi_\lambda(x) = \frac{\varphi(x + \lambda e_j) - \varphi(x)}{\lambda}.$$

Now, by Exercise 16.E $\varphi_\lambda \to \partial\varphi/\partial x_j$ in $\mathcal{D}(\overline{B}(0;\varepsilon + \delta))$ and whence it follows that $(T * \varphi_\lambda)(a) = T_y[\varphi_\lambda(a - y)] \to T_y[\frac{\partial \varphi}{\partial x_j}(a - y)]$ $= (T * \frac{\partial \varphi}{\partial x_j})(a)$. Hence $\frac{\partial}{\partial x_j}(T * \varphi)(a)$ exists and equals $(T * \frac{\partial \varphi}{\partial x_j})(a)$, as we wanted.

From the first part of the proof we conclude that $T * \varphi \in C^1(U_\delta)$ and $\frac{\partial}{\partial x_j}(T * \varphi) = T * \frac{\partial \varphi}{\partial x_j}$ on U_δ for $j = 1,\dots,n$. Then it follows by induction that $T * \varphi \in C^\infty(U_\delta)$ and $\frac{\partial^\alpha}{\partial x^\alpha}(T * \varphi) = T * \frac{\partial^\alpha \varphi}{\partial x^\alpha}$ for every multi-index α. To complete the proof of (b) we first observe that

$$(T * \frac{\partial \varphi}{\partial x_j})(a) = T_y\left[\frac{\partial \varphi}{\partial x_j}(a - y)\right] = T_y\left[-\frac{\partial}{\partial x_j}(\varphi(a - y))\right]$$

$$= \left(\frac{\partial T}{\partial x_j}\right)_y[\varphi(a - y)] = \left(\frac{\partial T}{\partial x_j} * \varphi\right)(a)$$

for $j = 1,\ldots,n$. Whence it follows by induction that $T * \frac{\partial^\alpha \varphi}{\partial x^\alpha} = \frac{\partial^\alpha T}{\partial x^\alpha} * \varphi$ for every multi-index α. Since (c) is clear from the definitions, the proof of the proposition is complete.

If $\psi \in \mathcal{D}(\mathbb{R}^n)$ then $\check{\psi} \in \mathcal{D}(\mathbb{R}^n)$ will denote the test function defined by $\check{\psi}(x) = \psi(-x)$ for every $x \in \mathbb{R}^n$. With this notation we have the following lemma.

17.15. LEMMA. *Let U be an open subset of $\mathbb{R}^n$, let $T \in \mathcal{D}'(U)$ and let $\psi \in \mathcal{D}(\overline{B}(0;\delta))$. Then $(T * \psi)(\varphi) = T(\varphi * \check{\psi})$ for every $\varphi \in \mathcal{D}(U_{2\delta})$.*

PROOF. Note that $T * \psi \in C^\infty(U_\delta)$ and $\varphi * \check{\psi} \in \mathcal{D}(U_\delta)$. Let $K = \operatorname{supp} \varphi$ and let $L = K + \overline{B}(0;\delta)$. Then

$$(T * \psi)(\varphi) = \int_K \varphi(x)(T * \psi)(x)\,d\lambda(x) = \int_K \varphi(x)\, T_y[\psi(x-y)]\,d\lambda(x)$$

$$= \int_K T_y[\varphi(x)\psi(x-y)]\,d\lambda(x) = \int_K T(f(x))\,d\lambda(x),$$

where $f : K \to \mathcal{D}(L)$ denotes the continuous mapping defined by $f(x)(y) = \varphi(x)\psi(x-y)$ for every $x \in K$ and $y \in \mathbb{R}^n$. We would like to interchange T and the integral sign in the last expression for $(T * \psi)(\varphi)$, but since $\mathcal{D}(L)$ is not a Banach space we cannot apply Proposition 6.5 to guarantee that f is Bochner integrable. However, since $\mathcal{D}(L)$ is a Fréchet space, the closed convex hull of each compact subset of $\mathcal{D}(L)$ is compact as well, and then the argument in Exercise 6.F applies. Thus there exists a vector $\theta \in \mathcal{D}(L)$ such that $S(\theta) = \int_K S(f(x))\,d\lambda(x)$ for every $S \in \mathcal{D}'(L)$. Then on one hand we have that

$$(T * \psi)(\varphi) = \int_K T(f(x))\,d\lambda(x) = T(\theta),$$

and on the other hand we have that

$$\theta(y) = \int_K f(x)(y)\,d\lambda(x) = \int_K \varphi(x)\psi(x-y)\,d\lambda(x) = (\varphi * \check{\psi})(y)$$

for every $y \in \mathbb{R}^n$. Thus $\theta = \varphi * \check{\psi}$ and $(T * \psi)(\varphi) = T(\varphi * \check{\psi})$, as asserted.

Now it is easy to prove the following result.

17.16. **PROPOSITION.** *Let U be an open subset of $\mathbb{R}^n$, and let $T \in \mathcal{D}'(U)$ be a distribution with compact support. Then $T * \rho_\delta \in \mathcal{D}(U)$ when $\delta > 0$ is sufficiently small, and $T * \rho_\delta \to T$ in $\mathcal{D}'(U)$ when $\delta \to 0$.*

PROOF. Given $\varphi \in \mathcal{D}(U)$ choose $r > 0$ such that $\operatorname{supp} T \subset U_{2r}$ and $\operatorname{supp} \varphi \subset U_{2r}$. Then $T * \rho_\delta \in \mathcal{D}(U_r)$ and $\varphi * \rho_\delta \in \mathcal{D}(U_r)$ whenever $0 < \delta < r$. Using Lemma 17.15 and Exercise 16.F, and observing that $\rho_\delta = \check{\rho}_\delta$, we get that $(T * \rho_\delta)(\varphi) = T(\varphi * \rho_\delta) \to T(\varphi)$ when $\delta \to 0$.

From Propositions 17.12 and 17.16 we get at once the following corollary.

17.17. **COROLLARY.** *If U is an open subset of $\mathbb{R}^n$ then $\mathcal{D}(U)$ is dense in $\mathcal{D}'(U)$.*

We end this section with the following result, which will be needed later on.

17.18. **PROPOSITION.** *Let U be an open subset of $\mathbb{R}^n$. Let $f: U \to \mathbb{K}$ be a function which is Lipschitz continuous, that is, there is a constant $k > 0$ such that $|f(x) - f(y)| \leq k \|x - y\|$ for all $x, y \in U$. Then $\partial f / \partial x_j \in L^p(U; loc)$ for every $j = 1, \ldots, n$ and $p \geq 1$.*

To prove this proposition we need the following lemma.

17.19. **LEMMA.** *Let (f_j) be a sequence of $\mathbb{K}$-valued measurable functions defined on a measurable space X. Then the set of points $x \in X$ for which the sequence $(f_j(x))$ converges is measurable.*

PROOF. For each $m, n \in \mathbb{N}$ consider the measurable set

$$A_{mn} = \{x \in X : |f_j(x) - f_k(x)| < 1/m \quad \text{for all} \quad j, k \geq n\}.$$

Since the set of points $x \in X$ for which $(f_j(x))$ converges is $\bigcap_{m=1}^{\infty} \bigcup_{n=1}^{\infty} A_{mn}$, the desired conclusion follows.

PROOF OF PROPOSITION 17.18. First suppose $n = 1$. Without loss of generality we may assume that U is a bounded open interval. Then the Lipschitz condition implies that f is absolutely continuous, and in particular of bounded variation. By a theorem of Lebesgue the derivative $f'(a)$ exists for almost every $a \in U$, and clearly $|f'(a)| \leq k$ wherever it exists. Now, since f is absolutely continuous, the integration by parts formula $\int \varphi f' dx = - \int \varphi' f dx$ is valid for every $\varphi \in \mathcal{D}(U)$, and hence the derivative of f in the sense of distributions coincides with the classical derivative. The desired conclusion follows.

Next suppose $n \geq 2$. We shall prove that $\partial f/\partial x_1 \in L^p(U, loc)$ for every $p \geq 1$. The proof for $\partial f/\partial x_j$ is analogous. Without loss of generality we may assume that $U = A \times B$, where A is a bounded open interval in $\mathbb{R}$ and B is a bounded open set in $\mathbb{R}^{n-1}$. Let C denote the set of all points $(a,b) \in A \times B$ such that the partial derivative $\frac{\partial f}{\partial x_1}(a,b)$ exists. Then C is a measurable subset of $A \times B$, by Lemma 17.19. For each $b \in B$ let C_b denote the set of all points $a \in A$ such that $\frac{\partial f}{\partial x_1}(a,b)$ exists. Then, again by Lemma 17.19, each C_b is a measurable subset of A, and clearly $C = \bigcup_{b \in B} (C_b \times \{b\})$. Let C' denote the complement of C in $A \times B$, and let C_b' denote the complement of C_b in A, for each $b \in B$. Then $C' = \bigcup_{b \in B} (C_b' \times \{b\})$. Now, since f is Lipschitz continuous, it is clear that for each $b \in B$ the function $f(x_1, b)$ is absolutely continuous, and in particular of bounded variation on A. Then it follows again from Lebesgue's Theorem that the set C_b' has one dimensional Lebesgue measure zero for each $b \in B$. Then a direct application of the Fubini Theorem shows that the set C has n dimensional Lebesgue measure zero. Thus the partial derivative $\frac{\partial f}{\partial x_1}(a,b)$ exists for almost every $(a,b) \in U$, and clearly $|\frac{\partial f}{\partial x_1}(a,b)| \leq k$ wherever it exists. As before, it follows from the absolute continuity of $f(x_1,b)$ that the derivative $\partial f/\partial x_1$ in the sense of distributions coincides with the derivative $\partial f/\partial x_1$ in the classical sense. This completes the proof.

EXERCISES

17.A. Let U be an open subset of $\mathbb{R}^n$, let $a \in U$ and let $\delta_a : \mathcal{D}(U) \to \mathbb{K}$ be defined by $\delta_a(\varphi) = \varphi(a)$ for every $\varphi \in \mathcal{D}(U)$. Show that δ_a is a distribution on U, called the *Dirac measure* at a.

17.B. Let $Y : \mathbb{R} \to \mathbb{R}$ be the *Heaviside function*, which is defined by $Y(x) = 0$ if $x < 0$, and $Y(x) = 1$ if $x \geq 0$. Show that Y defines a distribution on $\mathbb{R}$. Show that the derivative of Y is the Dirac measure δ_0.

17.C. Let U be an open subset of $\mathbb{R}^n$. Show that a distribution $T \in \mathcal{D}'(U)$ has compact support if and only if T extends as a continuous linear functional to $C^\infty(U)$.

17.D. Let $(U_i)_{i\in I}$ be an open cover of an open subset U of $\mathbb{R}^n$. Suppose that for each $i \in I$ there exists a distribution $T_i \in \mathcal{D}'(U_i)$ with the property that $T_i(\varphi) = T_j(\varphi)$ for every $\varphi \in \mathcal{D}(U_i \cap U_j)$, whenever $U_i \cap U_j \neq \emptyset$. Using Theorem 15.4 find a distribution $T \in \mathcal{D}'(U)$ such that $T(\varphi) = T_i(\varphi)$ for every $\varphi \in \mathcal{D}(U_i)$ and every $i \in I$. Show that T is unique.

17.E. Let U be an open subset of $\mathbb{R}^n$ and let $f, g \in C(U)$. Show that if $\partial f/\partial x_j = g$ for some j $(j = 1,\dots,n)$ in the sense of distributions, then $\partial f/\partial x_j = g$ in the classical sense.

NOTES AND COMMENTS

The material in this chapter can be found in many standard

books. We have included only that material which is required for the rest of the book. The material in Sections 13 and 14 can be found, for instance, in the texts of J. Dieudonné [1], H. Cartan [2] and L. Nachbin [4]. The remaining three sections constitute a brief introduction to the theory of distributions. The standard reference for the theory of distributions is of course the book of L. Schwartz [1]. The book of L. Hörmander [1] contains a concise introduction to the subject. The proof of Proposition 17.18 given here is taken from the book of S. M. Nikol'skii [1].

CHAPTER V

DIFFERENTIAL FORMS

18. ALTERNATING MULTILINEAR FORMS

In this section we introduce the exterior product of alternating multilinear forms. This is indispensable for the study of differential forms, to begin in the next section.

18.1. DEFINITION. Given $A \in \mathcal{L}^a(^mE)$ and $B \in \mathcal{L}^a(^nE)$ their *exterior product* or *wedge product* $A \wedge B \in \mathcal{L}^a(^{m+n}E)$ is defined by

$$A \wedge B = \frac{(m+n)!}{m!\,n!} (A \otimes B)^a.$$

In other words,

$$(A \wedge B)(x_1, \ldots, x_{m+n})$$

$$= \frac{1}{m!\,n!} \sum_{\sigma \in S_{m+n}} (-1)^\sigma A(x_{\sigma(1)}, \ldots, x_{\sigma(m)}) B(x_{\sigma(m+1)}, \ldots, x_{\sigma(m+n)}).$$

The exterior product $A \wedge B$ can also be defined if one of the mappings A or B takes values in a Banach space. Clearly the mapping $(A, B) \to A \wedge B$ is bilinear and continuous.

18.2. PROPOSITION. *If* $A \in \mathcal{L}^a(^mE)$ *and* $B \in \mathcal{L}^a(^nE)$ *then* $B \wedge A = (-1)^{mn} A \wedge B$.

PROOF: If α denotes the permutation taking $(1, \ldots, m+n)$ into $(n+1, \ldots, n+m, 1, \ldots, n)$ then it is clear that $(-1)^\alpha = (-1)^{mn}$. Hence

$$\frac{m!\,n!}{(m+n)!}(B \wedge A)(x_1,\ldots,x_{m+n})$$

$$= \sum_{\sigma\in S_{m+n}} (-1)^{\sigma} B(x_{\sigma(1)},\ldots,x_{\sigma(n)})A(x_{\sigma(n+1)},\ldots,x_{\sigma(n+m)})$$

$$= (-1)^{\alpha} \sum_{\sigma\in S_{m+n}} (-1)^{\sigma\alpha} A(x_{\sigma\alpha(1)},\ldots,x_{\sigma\alpha(m)})B(x_{\sigma\alpha(m+1)},\ldots,x_{\sigma\alpha(m+n)})$$

$$= (-1)^{mn} \frac{m!\,n!}{(m+n)!}(A \wedge B)(x_1,\ldots,x_{m+n}).$$

18.3. PROPOSITION. *If* $A \in \mathcal{L}^a({}^mE)$, $B \in \mathcal{L}^a({}^nE)$ *and* $C \in \mathcal{L}^a({}^pE)$ *then*

$$(A \wedge B) \wedge C = A \wedge (B \wedge C) = \frac{(m+n+p)!}{m!\,n!\,p!}(A \otimes B \otimes C)^a.$$

This proposition is an immediate consequence of the following lemma.

18.4. LEMMA. *Let* $A \in \mathcal{L}({}^mE)$, $B \in \mathcal{L}({}^nE)$ *and* $C \in \mathcal{L}({}^pE)$. *Then:*

(a) $(A \otimes B)^a = 0$ *whenever* $A^a = 0$ *or* $B^a = 0$.

(b) $(A^a \otimes B)^a = (A \otimes B^a)^a = (A \otimes B)^a$.

(c) $[(A \otimes B)^a \otimes C]^a = [A \otimes (B \otimes C^a]^a = (A \otimes B \otimes C)^a$.

PROOF. (a) We show that $(A \otimes B)^a = 0$ whenever $A^a = 0$. The other statement is proved similarly. Let T denote the subgroup of all $\tau \in S_{m+n}$ which leave $m+1,\ldots,m+n$ fixed. Then T is isomorphic to S_m and

$$\sum_{\tau\in T} (-1)^{\tau} A(x_{\tau(1)},\ldots,x_{\tau(m)})B(x_{\tau(m+1)},\ldots,x_{\tau(m+n)})$$

$$= \sum_{\tau\in S_m} (-1)^{\tau} A(x_{\tau(1)},\ldots,x_{\tau(m)})B(x_{m+1},\ldots,x_{m+n})$$

$$= m!\, A^a(x_1,\ldots,x_m)(B(x_{m+1},\ldots,x_{m+n}) = 0.$$

Given $\alpha \in S_{m+n}$ and $x_1,\ldots,x_{m+n} \in E$ set $y_j = x_{\alpha(j)}$ for

$j = 1,\ldots,m+n$. Then

$$\sum_{\tau\in T} (-1)^{\alpha\tau} A(x_{\alpha\tau(1)},\ldots,x_{\alpha\tau(m)})\; B(x_{\alpha\tau(m+1)},\ldots,x_{\alpha\tau(m+n)})$$

$$= (-1)^{\alpha} \sum_{\tau\in T} (-1)^{\tau} A(y_{\tau(1)},\ldots,y_{\tau(m)})\; B(y_{\tau(m+1)},\ldots,y_{\tau(m+n)}) = 0.$$

Since the group S_{m+n} is the union of the cosets αT, and since the cosets αT are either disjoint or identical, we get that

$$(m+n)!\;(A\otimes B)^a(x_1,\ldots,x_{m+n})$$

$$= \sum_{\sigma\in S_{m+n}} (-1)^{\sigma} A(x_{\sigma(1)},\ldots,x_{\sigma(m)}) B(x_{\sigma(m+n)},\ldots,x_{\sigma(m+n)}) = 0.$$

(b) Since $(A^a - A)^a = 0$, (a) implies that $[(A^a - A)\otimes B]^a = 0$ and therefore $(A^a \otimes B)^a = (A\otimes B)^a$. The equality $(A\otimes B^a)^a = (A\otimes B)^a$ is proved similarly.

(c) follows at once from (b).

If $A\in \mathcal{L}^a({}^mE)$ and $B\in \mathcal{L}^a({}^nE)$ then the tensor product $A\otimes B$ is alternating in the first m variables and alternating in the last n variables. This motivates the following definition.

18.5. DEFINITION. We shall denote by $\mathcal{L}^{amn}({}^{m+n}E;F)$ the subspace of all $A\in\mathcal{L}({}^{m+n}E;F)$ which are alternating in the first m variables and alternating in the last n variables.

Let S_{mn} denote the set of all permutations $\sigma\in S_{m+n}$ such that $\sigma(1)<\ldots<\sigma(m)$ and $\sigma(m+1)<\ldots<\sigma(m+n)$. Note that S_{mn} has $(m+n)!/m!\,n!$ elements. Then we have the following proposition, whose proof is straightforward and is left as an exercise to the reader.

18.6. PROPOSITION. *For each* $A\in\mathcal{L}({}^{m+n}E;F)$ *let* $A^{amn}\in\mathcal{L}({}^{m+n}E;F)$ *be defined by*

$$A^{amn}(x_1,\ldots,x_{m+n}) = \frac{m!\,n!}{(m+n)!} \sum_{\sigma\in S_{mn}} (-1)^{\sigma} A(x_{\sigma(1)},\ldots,x_{\sigma(m+n)}).$$

Then $A^{amn} = A^a$ *for every* $A \in \mathcal{L}^{amn}({}^{m+n}E)$. *In particular the mapping* $A \to A^{amn}$ *induces a continuous projection from* $\mathcal{L}^{amn}({}^{m+n}E;F)$ *onto* $\mathcal{L}^a({}^{m+n}E;F)$.

As an immediate consequence we get another, somewhat different formula for the exterior product.

18.7. PROPOSITION. *If* $A \in \mathcal{L}^a({}^mE)$ *and* $B \in \mathcal{L}^a({}^nE)$ *then*

$$A \wedge B = \frac{(m+n)!}{m!\,n!} (A \otimes B)^{amn}.$$

In other words,

$$(A \wedge B)(x_1,\ldots,x_{m+n})$$

$$= \sum_{\sigma\in S_{mn}} (-1)^{\sigma} A(x_{\sigma(1)},\ldots,x_{\sigma(m)}) B(x_{\sigma(m+1)},\ldots,x_{\sigma(m+n)}).$$

We end this section with the following result, which parallels Theorem 1.15.

18.8. THEOREM. *If* E *and* F *are complex Banach spaces then* $\mathcal{L}^a({}^mE_{\mathbb{R}}; F_{\mathbb{R}})$ *is the topological direct sum of the subspaces* $\mathcal{L}^a({}^{pq}(E;F)$ *with* $p + q = m$.

PROOF. By Theorem 1.15 each $A \in \mathcal{L}({}^mE_{\mathbb{R}}; F_{\mathbb{R}})$ can be uniquely written in the form $A = A_0 + A_1 + \ldots + A_m$ where $A_k \in \mathcal{L}({}^{m-k,k}E;F)$ for $k = 0,1,\ldots,m$. To prove the theorem it is sufficient to show that if A is alternating then each A_k is alternating as well. To show this pick $m + 1$ distinct numbers $\theta_0,\ldots,\theta_m$ in the interval $(0,\pi)$ and set $\lambda_j = e^{i\theta_j}$ and $\zeta_j = \bar{\lambda}_j/\lambda_j = e^{-2i\theta_j}$ for $j = 0,\ldots,m$. Then, for $j = 0,\ldots,m$ and arbitrary vectors $x_1,\ldots,x_m \in E$ we have that

$$(18.1)\qquad A(\lambda_j x_1,\dots,\lambda_j x_m) = \sum_{k=0}^{m} \lambda_j^{m-k}\bar{\lambda}_j^{k} A_k(x_1,\dots,x_m)$$

$$= \lambda_j^m \sum_{k=0}^{m} \zeta_j^k A_k(x_1,\dots,x_m).$$

Since the complex numbers $\zeta_o,\dots,\zeta_m$ are distinct, the determinant of the coefficients of the system of equations (18.1) equals $(\lambda_o \cdots \lambda_m)^m$ times a nonzero Vandermonde determinant. Then it follows from Cramer's Rule that each $A_k(x_1,\dots,x_m)$ can be written in the form

$$A_k(x_1,\dots,x_m) = \sum_{j=0}^{m} c_{kj} A(\lambda_j x_1,\dots,\lambda_j x_m),$$

where the complex numbers c_{kj} are independent from the vectors $x_1,\dots,x_m$. Whence it is clear that each A_k is alternating if A is alternating.

EXERCISES

18.A. Let E and F be Banach spaces over $\mathbb{K}$, with E finite dimensional. Let $(e_1,\dots,e_n)$ be a basis for E and let $\xi_1,\dots,\xi_n$ be the corresponding coordinate functionals.

(a) Show that $\mathcal{L}^a(^mE;F) = \{0\}$ whenever $m > n$.

(b) Show that if $1 \le m \le n$ then each $A \in \mathcal{L}^a(^mE;F)$ can be uniquely represented as a sum

$$A = \sum c_{j_1\dots j_m}\, \xi_{j_1} \wedge \dots \wedge \xi_{j_m}$$

where $c_{j_1\dots j_m} \in F$ and the summation is taken over all multi-indices $(j_1,\dots,j_m)$ such that $1 \le j_1 < \dots < j_m \le n$.

(c) Show that $\mathcal{L}^a(^mE)$ has dimension $\binom{n}{m}$ whenever $0 \le m \le n$.

(d) Given $A \in \mathcal{L}^a(^nE;F)$ and $x_j = \sum_{j=1}^{n} x_{jk}e_k \in E \quad (j = 1,\dots,n)$

show that

$$A(x_1,\ldots,x_n) = det(x_{jk})A(e_1,\ldots,e_n).$$

18.B. Let $z_1,\ldots,z_n$ be the complex coordinate functionals of $\mathbb{C}^n$ and let F be a complex Banach space.

(a) Show that $\mathcal{L}^a({}^{pq}\mathbb{C}^n;F) = \{0\}$ if $p > n$ or $q > n$.

(b) Show that if $0 \leq p \leq n$, $0 \leq q \leq n$ and $p + 1 \geq 1$ then each $A \in \mathcal{L}^a({}^{pq}\mathbb{C}^n;F)$ can be uniquely represented as a sum

$$A = \Sigma\, c_{j_1\ldots j_p k_1\ldots k_q} z_{j_1} \wedge \ldots \wedge z_{j_p} \wedge \bar{z}_{k_1} \wedge \ldots \wedge \bar{z}_{k_q}$$

where $c_{j_1\ldots j_p k_1\ldots k_q} \in F$ and the summation is taken over all multi-indices $(j_1,\ldots,j_p,k_1,\ldots,k_q)$ such that $1 \leq j_1 < \ldots < j_p \leq n$ and $1 \leq k_1 < \ldots < k_q \leq n$.

(c) Show that if $0 \leq p \leq n$ and $0 \leq q \leq n$ then $\mathcal{L}^a({}^{pq}\mathbb{C}^n)$ has dimension $\binom{n}{p}\binom{n}{q}$.

19. DIFFERENTIAL FORMS

In this section we introduce differential forms in Banach spaces. We define the exterior differential d and derive its elementary properties.

Throughout this section the letters E and F will represent real Banach spaces.

19.1. DEFINITION. If U is an open subset of E then we shall denote by $\Omega_m(U;F)$ the vector space of all mappings

$$f : U \to \mathcal{L}^a({}^mE;F).$$

Each $f \in \Omega_m(U;F)$ is called an *F-valued differential form of degree m on U*. If $k \in \mathbb{N}_o \cup \{\infty\}$ then we shall denote by

$C^k_m(U;F)$ the vector subspace of all $f \in \Omega_m(U;F)$ which are of class C^k. In other words,

$$C^k_m(U;F) = C^k(U;\mathcal{L}^a(^mE;F)).$$

If F is $\mathbb{R}$ or $\mathbb{C}$ (regarded as a real Banach space) then we shall write $\Omega_m(U;F) = \Omega_m(U)$ and $C^k_m(U;F) = C^k_m(U)$.

19.2. DEFINITION. Let U be an open subset of E. Given differential forms $f \in \Omega_m(U)$ and $g \in \Omega_n(U)$, their *exterior product* or *wedge product* $f \wedge g \in \Omega_{m+n}(U)$ is defined by $(f \wedge g)(x) = f(x) \wedge g(x)$ for every $x \in U$.

The exterior product $f \wedge g$ can also be defined if one of the differential forms f or g is Banach-valued. Clearly the mapping $(f,g) \to f \wedge g$ is bilinear and continuous.

19.3. PROPOSITION. *Let U be an open subset of E. Let $f \in \Omega_m(U)$, $g \in \Omega_n(U)$ and $h \in \Omega_p(U)$. Then:*

(a) $g \wedge f = (-1)^{mn} f \wedge g$.

(b) $(f \wedge g) \wedge h = f \wedge (g \wedge h)$.

(c) *If f and g are of class C^k then $f \wedge g$ is also of class C^k.*

PROOF. (a) follows from Proposition 18.2. (b) follows from Proposition 18.3. To show (c) we observe that $f \wedge g = T \circ S$ where

$$S : x \in U \to (f(x), g(x)) \in \mathcal{L}^a(^mE) \times \mathcal{L}^a(^nE)$$

and

$$T : (A,B) \in \mathcal{L}^a(^mE) \times \mathcal{L}^a(^nE) \to A \wedge B \in \mathcal{L}^a(^{m+n}E).$$

Since f and g are of class C^k the mapping S is of class C^k by Exercise 14.A. The mapping T is bilinear and continuous, and hence is of class C^∞ by Exercise 14.B. Hence $f \wedge g = T \circ S$

is of class C^∞ by Theorem 14.9.

From now on we shall be primarily interested in those differential forms which are differentiable, and to simplity matters we shall restrict our attention to C^∞ differential forms. We want to associate with each differential form $f \in C_m^\infty(U;F) = C^\infty(U;\mathcal{L}^a({}^mE;F))$ a differential form $df \in C_{m+1}^\infty(U;F) = C^\infty(U;\mathcal{L}^a({}^{m+1}E;F))$. Note that for each $x \in U$ we have that

$$Df(x) \in \mathcal{L}(E;\mathcal{L}^a({}^mE;F)),$$

and therefore, with the notation of the preceding section, we have that

$$Df(x) \in \mathcal{L}^{a1m}({}^{m+1}E;F) \subset \mathcal{L}({}^{m+1}E;F).$$

Thus we are ready to give the following definition.

19.4. DEFINITION. Let U be an open subset of E. Given a differential form $f \in C_m^\infty(U;F)$, its *exterior differential* $df \in C_{m+1}^\infty(U;F)$ is defined by

$$df(x) = (m + 1)[Df(x)]^a = (m + 1)\ [Df(x)]^{a1m}$$

for every $x \in U$. From the second expression for $df(x)$ we see that

$$df(x)(t_1,\ldots,t_{m+1}) = \sum_{j=1}^{m+1} (-1)^{j-1} Df(x)(t_j)(t_1,\ldots,\hat{t}_j,\ldots,t_{m+1})$$

where the hat over t_j means that t_j is omitted.

The mapping $d : C_m^\infty(U;F) \to C_{m+1}^\infty(U;F)$ is clearly linear. To establish other properties of this operator we need an auxiliary lemma. But first we give a definition.

19.5. DEFINITION. We shall denote by $\mathcal{L}^{amo}({}^{m+n}E;F)$ the subspace of all $A \in \mathcal{L}({}^{m+n}E;F)$ which are alternating in the first m variables. Likewise, we shall denote by $\mathcal{L}^{aon}({}^{m+n}E;F)$ the subspace of all $A \in \mathcal{L}({}^{m+n}E;F)$ which are alternating in the last n variables.

19.6. **LEMMA.** *For each* $A \in \mathcal{L}({}^{m+n}E;F)$ *let* $A^{amo} \in \mathcal{L}^{amo}({}^{m+n}E;F)$ *and* $A^{aon} \in \mathcal{L}^{aon}({}^{m+n}E;F)$ *be defined by*

$$A^{amo}(x_1,\ldots,x_{m+n})$$

$$= \frac{1}{m!} \sum_{\sigma \in S_m} (-1)^{\sigma} A(x_{\sigma(1)},\ldots,x_{\sigma(m)},x_{m+1},\ldots,x_n)$$

and

$$A^{aon}(x_1,\ldots,x_{m+n})$$

$$= \frac{1}{n!} \sum_{\sigma \in S_n} (-1)^{\sigma} A(x_1,\ldots,x_m,x_{\sigma(m+1)},\ldots,x_{\sigma(m+n)}).$$

Then:

(a) *The mapping* $A \to A^{amo}$ *is a continuous projection from* $\mathcal{L}({}^{m+n}E;F)$ *onto* $\mathcal{L}^{amo}({}^{m+n}E;F)$. *Likewise, the mapping* $A \to A^{aon}$ *is a continuous projection from* $\mathcal{L}({}^{m+n}E;F)$ *onto* $\mathcal{L}^{aon}({}^{m+n}E;F)$.

(b) $A^a = 0$ *for every* $A \in \mathcal{L}({}^{m+n}E;F)$ *such that* $A^{amo} = 0$ *or* $A^{aon} = 0$.

(c) $(A^{amo})^a = (A^{aon})^a = A^a$ *for every* $A \in \mathcal{L}({}^{m+n}E;F)$.

PROOF. (a) is clear. To prove (b) and (c) it suffices to repeat the proofs of parts (a) and (b) in Lemma 18.4.

19.7. PROPOSITION. *Let* U *be an open subset of* E. *Then for every* $f \in C_m^{\infty}(U;F)$ *and* $g \in C_n^{\infty}(U;F)$ *we have the formula*

$$d(f \wedge g) = df \wedge g + (-1)^m f \wedge dg.$$

PROOF. By Proposition 19.3, $f \wedge g \in C_{m+n}^{\infty}(U;F)$, and using Exercise 13.D we get that

$$D(f \wedge g)(x)(t) = Df(x)(t) \wedge g(x) + f(x) \wedge Dg(x)(t)$$

$$= Df(x)(t) \wedge g(x) + (-1)^{mn} Dg(x)(t) \wedge f(x)$$

for every $x \in U$ and $t \in E$. If we fix $x \in U$ then

$$D(f \wedge g)(x) = A + (-1)^{mn} B$$

where

$$A(t_1, \ldots, t_{m+n+1}) = [Df(x)(t_1) \wedge g(x)](t_2, \ldots, t_{m+n+1})$$

and

$$B(t_1, \ldots, t_{m+n+1}) = [Dg(x)(t_1) \wedge f(x)](t_2, \ldots, t_{m+n+1}).$$

Now

$$A(t_1, \ldots, t_{m+n+1}) = \frac{(m+n)!}{m!\,n!} [Df(x)(t_1) \otimes g(x)]^a (t_2, \ldots, t_{m+n+1})$$

$$= \frac{(m+n)!}{m!\,n!} [Df(x) \otimes g(x)]^{a0,m+n} (t_1, \ldots, t_{m+n+1})$$

Thus

$$A = \frac{(m+n)!}{m!\,n!} [Df(x) \otimes g(x)]^{a0,m+n}$$

and using Lemmas 19.6 and 18.4 we get that

$$A^a = \frac{(m+n)!}{m!\,n!} [Df(x) \otimes g(x)]^a = \frac{(m+n)!}{m!\,n!} [Df(x))^a \otimes g(x)]^a$$

$$= \frac{(m+n)!}{m!\,n!(m+1)} [df(x) \otimes g(x)]^a = (m+n+1)^{-1} df(x) \wedge g(x).$$

Similarly we get that

$$B^a = (m+n+1)^{-1} dg(x) \wedge f(x)$$

and therefore

$$d(f \wedge g)(x) = (m + n + 1)(A^a + (-1)^{mn} B^a)$$

$$= df(x) \wedge g(x) + (-1)^{mn} dg(x) \wedge f(x)$$

$$= df(x) \wedge d(x) + (-1)^m f(x) \wedge dg(x),$$

as we wanted.

19.8. **PROPOSITION.** *If* U *is an open subset of* E *then* $d^2 f = d(df) = 0$ *for every* $f \in C_m^\infty(U;F)$.

PROOF. Since $df(x) = (m + 1)[Df(x)]^a$, an application of the Chain Rule shows that

$$D(df)(x)(t) = (m + 1)[D(Df)(x)(t)]^a$$

for every $x \in U$ and $t \in E$, and therefore

$$D(df)(x) = (m + 1)[D^2 f(x)]^{a0,m+1}$$

for every $x \in U$. Thus, using Lemma 19.6 we get that

$$d^2 f(x) = (m + 2)[D(df)(x)]^a$$

$$= (m + 2)(m + 1)[D^2 f(x)]^a.$$

Now,

$$2[D^2 f(x)]^{a20}(t_1, t_2, t_3, \ldots, t_{m+2})$$

$$= D^2 f(x)(t_1, t_2, t_3, \ldots, t_{m+2}) - D^2 f(x)(t_2, t_1, t_3, \ldots, t_{m+2}) = 0$$

since, by Theorem 14.3, $D^2 f(x) \in \mathcal{L}^s({}^2E; \mathcal{L}^a({}^mE;F))$. Thus $[D^2 f(x)]^a = 0$, by Lemma 19.6, and therefore $d^2 f(x) = 0$, as asserted.

To complete this section we introduce an important operation for differential forms.

19.9. **DEFINITION.** Let E, F, G be real Banach spaces, let $U \subset E$

and $V \subset F$ be open sets, and let $\varphi \in C^{\infty}(U;F)$ with $\varphi(U) \subset V$. Then for each $f \in \Omega_m(V;G)$ let $\varphi^* f \in \Omega_m(U;G)$ be defined by

$$\varphi^* f(x)(t_1,\ldots,t_m) = f(\varphi(x))(D\varphi(x)(t_1),\ldots,D\varphi(x)(t_m))$$

for all $x \in U$ and $t_1,\ldots,t_m \in E$.

The mapping $\varphi^* : \Omega_m(V;G) \to \Omega_m(U;G)$ is clearly linear. Other properties of this operator are given next.

19.10. **PROPOSITION.** *Let* $U \subset E$ *and* $V \subset F$ *be open sets, and let* $\varphi \in C^{\infty}(U;F)$ *with* $\varphi(U) \subset V$. *Then for every* $f \in \Omega_m(V)$ *and* $g \in \Omega_n(V)$ *we have that*

$$\varphi^*(f \wedge g) = (\varphi^* f) \wedge (\varphi^* g).$$

PROOF. For every $x \in U$ and $t_1,\ldots,t_{m+n} \in E$ we have that

$$\varphi^*(f \wedge g)(x)(t_1,\ldots,t_{m+n})$$

$$= (f \wedge g)(\varphi(x))(D\varphi(x)(t_1),\ldots,D\varphi(x)(t_{m+n}))$$

$$= \sum_{\sigma \in S_{mn}} (-1)^{\sigma} f(\varphi(x))(D\varphi(x)(t_{\sigma(1)}),\ldots,D\varphi(x)(t_{\sigma(m)})) \cdot$$
$$g(\varphi(x))(D\varphi(x)(t_{\sigma(m+1)}),\ldots,D\varphi(x)(t_{\sigma(m+n)}))$$

$$= \sum_{\sigma \in S_{mn}} (-1)^{\sigma} (\varphi^* f)(x)(t_{\sigma(1)},\ldots,t_{\sigma(m)})(\varphi^* g)(x)(t_{\sigma(m+1)},\ldots,t_{\sigma(m+n)})$$

$$= ((\varphi^* f) \wedge (\varphi^* g))(x)(t_1,\ldots,t_{m+n}).$$

19.11. **PROPOSITION.** *Let* E, F, G, H *be real Banach spaces, let* $U \subset E$, $V \subset F$ *and* $W \subset G$ *be open sets, and let* $\varphi \in C^{\infty}(U;F)$ *and* $\psi \in C^{\infty}(V;G)$ *with* $\varphi(U) \subset V$ *and* $\psi(V) \subset W$. *Then*

$$(\psi \circ \varphi)^* f = \varphi^*(\psi^* f)$$

for every $f \in \Omega_m(W;H)$.

PROOF. For every $x \in U$ and $t_1,\dots,t_m \in E$ we have that

$$(\psi \circ \varphi)^*f(x)(t_1,\dots,t_m)$$

$$= f(\psi \circ \varphi(x))(D(\psi \circ \varphi)(x)(t_1),\dots,D(\psi \circ \varphi)(x)(t_m))$$

$$= f(\psi(\varphi(x)))(D\psi(\varphi(x))D\varphi(x)(t_1)),\dots,D\psi(\varphi(x))(D\varphi(x)(t_m)))$$

$$= \psi^*f(\varphi(x))(D\varphi(x)(t_1),\dots,D\varphi(x)(t_m))$$

$$= \varphi^*(\psi^*f)(x)(t_1,\dots,t_m).$$

19.12. **PROPOSITION.** *Let E, F, G be real Banach spaces, let $U \subset E$ and $V \subset F$ be open sets, and let $\varphi \in C^\infty(U;F)$ with $\varphi(U) \subset V$. Then:*

(a) φ^* *maps* $C_m^\infty(V;G)$ *into* $C_m^\infty(U;G)$.

(b) $d(\varphi^*f) = \varphi^*(df)$ *for every* $f \in C_m^\infty(V;G)$.

PROOF. (a) If $f \in C_m^\infty(V;G)$ then $\varphi^*f = T \circ S$, where

$$S : U \to \mathcal{L}^a({}^mF;G) \times [\mathcal{L}(E;F)]^m$$

and

$$T : \mathcal{L}^a({}^mF;G) \times [\mathcal{L}(E;F)]^m \to \mathcal{L}^a({}^mE;G)$$

are defined by

$$S(x) = (f \circ \varphi(x), D\varphi(x),\dots,D\varphi(x))$$

for every $x \in U$ and

$$T(A,B_1,\dots,B_m)(t_1,\dots,t_m) = A(B_1(t_1),\dots,B_m(t_m))$$

for every $(A,B_1,\dots,B_m) \in \mathcal{L}^a({}^mF;G) \times [\mathcal{L}(E;F)]^m$ and $t_1,\dots,t_m \in E$. Clearly S is of class C^∞, and so is T, since it is $(m+1)$-linear and continuous. Hence φ^*f is also of class C^∞, as asserted.

(b) We have that

$$d(\varphi^*f)(x)(t_o,\ldots,t_m) = \sum_{j=0}^{m} (-1)^j D(\varphi^*f)(x)(t_j)(t_o,\ldots,\hat{t}_j,\ldots,t_m).$$

Since $\varphi^*f = T \circ S$ it follows from Exercise 13.D that

$$D(\varphi^*f)(x)(t) = T(D(f\circ\varphi)(x)(t), D\psi(x),\ldots,D\varphi(x)) +$$

$$\sum_{k=1}^{m} T(f\circ\varphi(x), D\varphi(x),\ldots,D^2\varphi(x)(t),\ldots,D\varphi(x)).$$

Hence

$$D(\varphi^*f)(x)(t_j)(t_o,\ldots,\hat{t}_j,\ldots,t_m)$$

$$= D(f\circ\varphi)(x)(t_j)(D\varphi(x)(t_o),\ldots,D\varphi(x)(\hat{t}_j),\ldots,D\varphi(x)(t_m)) +$$

$$\sum_{k=0}^{j-1} f\circ\varphi(x)(D\varphi(x)(t_o),\ldots,D^2\varphi(x)(t_j,t_k),\ldots,D\varphi(x)(\hat{t}_j),\ldots,D\varphi(x)(t_m)) +$$

$$\sum_{k=j+1}^{m} f\circ\varphi(x)(D\varphi(x)(t_o),\ldots,D\varphi(x)(\hat{t}_j),\ldots,D^2\varphi(x)(t_j,t_k),\ldots,D\varphi(x)(t_m)).$$

Since $f\circ\varphi(x)$ is alternating and $D^2\varphi(x)$ is symmetric, it follows that

$$d(\varphi^*f)(x)(t_o,\ldots,t_m)$$

$$= \sum_{j=0}^{m} (-1)^j Df(\varphi(x))(D\varphi(x)(t_j))(D\varphi(x)(t_o),\ldots,D\varphi(x)(\hat{t}_j),\ldots,D\varphi(x)(t_m))$$

$$+ \sum_{j\neq k} a_{jk}$$

where $a_{jk} = -a_{kj}$ whenever $j \neq k$. Hence $\sum_{j\neq k} a_{jk} = 0$ and therefore

$$d(\varphi^*f)(x)(t_o,\ldots,t_m) = df(\varphi(x))(D\varphi(x)(t_o),\ldots,D\varphi(x)(t_m))$$

$$= \varphi^*(df)(x)(t_o,\ldots,t_m),$$

completing the proof.

EXERCISES

19.A. Let U be an open subset of $\mathbb{R}^n$, let $x_1,\dots,x_n$ denote the coordinate functionals of $\mathbb{R}^n$, and let $1 \leq m \leq n$. Show that each $f \in \Omega_m(U;F)$ can be uniquely represented as a sum

$$f = \Sigma f_{j_1\dots j_m} dx_{j_1} \wedge \dots \wedge dx_{j_m}$$

where $f_{j_1\dots j_m} : U \to F$ and the summation is taken over all multi-indices $(j_1,\dots,j_m)$ such that $1 \leq j_1 < \dots < j_m \leq n$. This representation will be often written in the abbreviated form $f = \sum_J f_J dx_J$. Show that the differential form f is of class C^k if and only if each of the mappings $f_J = f_{j_1\dots j_m}$ if of class C^k.

19.B. Let U be an open subset of $\mathbb{R}^n$, and let $x_1,\dots,x_n$ denote the coordinate functionals of $\mathbb{R}^n$.

(a) Show that if $f \in C^\infty(U;F)$ then

$$df = \sum_{j=1}^{n} \frac{\partial f}{\partial x_j} dx_j .$$

(b) Show that if $f = \sum_J f_J dx_J \in C_m^\infty(U;F)$ then

$$df = \sum_J df_J \wedge dx_J = \sum_J \sum_{k=1}^{n} \frac{\partial f_J}{\partial x_k} dx_k \wedge dx_J .$$

20. THE POINCARE LEMMA

This section is devoted to the proof of a single theorem, a partial converse to Proposition 19.8. As in the preceding section, the letters E and F represent real Banach spaces.

20.1. DEFINITION. Let U be an open subset of E. A differential form $g \in C^\infty_m(U;F)$ is said to be *closed* if $dg = 0$, and is said to be *exact* if there is a differential form $f \in C^\infty_{m-1}(U;F)$ such that $df = g$.

Proposition 19.8 asserts that every exact differential form is closed. We shall presently prove a partial converse when the open set U satisfies a certain condition.

20.2. DEFINITION. A subset A of E is said to be *star-shaped* with respect to a point $a \in A$ if $(1-\theta)a + \theta x \in A$ for every $x \in A$ and $\theta \in [0,1]$.

Now we can prove the following theorem, which is known as the *Poincaré Lemma*.

20.3. THEOREM. *Let U be an open subset of E which is star-shaped with respect to one of its points. If $m \geq 1$ then the equation $df = g$ has a solution $f \in C^\infty_{m-1}(U;F)$ for each $g \in C^\infty_m(U;F)$ such that $dg = 0$. In other words, every closed form in $C^\infty_m(U;F)$ is exact.*

PROOF. Without loss of generality we may assume that U is star-shaped with respect to the origin. For each $m \geq 1$ we shall define a mapping $K : C^\infty_m(U;F) \to C^\infty_{m-1}(U;F)$ such that

$$d(Kg) + K(dg) = g$$

for every $g \in C^\infty_m(U;F)$. The conclusion of the theorem follows at once from this. Now, given $g \in C^\infty_m(U;F)$ we define $Kg \in \Omega_{m-1}(U;F)$ by

$$Kg(x) = \int_0^1 \theta^{m-1} g(\theta x)x d\theta$$

for every $x \in U$. Or more explicitly

$$Kg(x)(t_1,\ldots,t_{m-1}) = \int_0^1 \theta^{m-1} g(\theta x)(x,t_1,\ldots,t_{m-1})d\theta$$

for every $x \in U$ and $t_1,\ldots,t_{m-1} \in E$. If we set $\varphi(x,\theta) = \theta^{m-1}g(\theta x)x$

then $\varphi(x,\theta)$ is a C^∞ function of x in U for each θ in $[0,1]$, and each of the mappings $(x,\theta) \to D_x^k\varphi(x,\theta)$ is continuous on $U \times [0,1]$. By repeated applications of Proposition 13.14 we conclude that Kg is a C^∞ differential form and

$$D(Kg)(x) = \int_0^1 D_x[\theta^{m-1}g(\theta x)x]\,d\theta$$

for every $x \in U$. Whence

$$D(Kg)(x)(t) = \int_0^1 \theta^{m-1}[Dg(\theta x)(\theta t)x + g(\theta x)t]\,d\theta$$

$$= \int_0^1 \theta^m Dg(\theta x)tx\,d\theta + \int_0^1 \theta^{m-1} g(\theta x)t\,d\theta$$

for every $x \in U$ and $t \in E$. Thus

$$d(Kg)(x)(t_1,\ldots,t_m) = \sum_{j=1}^m (-1)^{j-1} D(Kg)(x)(t_j,t_1,\ldots,\hat{t}_j,\ldots,t_m)$$

$$= \sum_{j=1}^m (-1)^{j-1}\int_0^1 \theta^m Dg(\theta x)(t_j,x,t_1,\ldots,\hat{t}_j,\ldots,t_m)\,d\theta +$$

$$\sum_{j=1}^m (-1)^{j-1}\int_0^1 \theta^{m-1} g(\theta x)(t_j,t_1,\ldots,\hat{t}_j,\ldots,t_m)\,d\theta$$

$$= \sum_{j=1}^n (-1)^{j-1}\int_0^1 \theta^m Dg(\theta x)(t_j,x,t_1,\ldots,\hat{t}_j,\ldots,t_m)\,d\theta +$$

$$m\int_0^1 \theta^{m-1} g(\theta x)(t_1,\ldots,t_m)\,d\theta.$$

On the other hand,

$$K(dg)(x)(t_1,\ldots,t_m) = \int_0^1 \theta^m\, dg(\theta x)(x,t_1,\ldots,t_m)\,d\theta$$

$$= \int_0^1 \theta^m Dg(\theta x)(x,t_1,\ldots,t_m)\,d\theta +$$

$$\sum_{j=1}^m (-1)^j \int_0^1 \theta^m Dg(\theta x)(t_j,x,t_1,\ldots,\hat{t}_j,\ldots,t_m)\,d\theta.$$

Thus we get that

$$d(Kg)(x)(t_1,\ldots,t_m) + K(dg)(x)(t_1,\ldots,t_m)$$

$$= m\int_0^1 \theta^{m-1} g(\theta x)(t_1,\ldots,t_m)d\theta + \int_0^1 \theta^m Dg(\theta x)(x,t_1,\ldots,t_m)d\theta.$$

Or more simply

$$d(Kg)(x) + K(dg)(x) = \int_0^1 [m\theta^{m-1} g(\theta x) + \theta^m Dg(\theta x)]d\theta$$

$$= \int_0^1 \frac{d}{d\theta}[\theta^m g(\theta x)]d\theta = g(x)$$

and the proof is complete.

EXERCISES

20.A. Let E, F, G be real Banach spaces, let $U \subset E$ and $V \subset F$ be open sets, and let $\varphi \in C^\infty(U;F)$ be an injective mapping such that $\varphi(U) = V$ and $\varphi^{-1} \in C^\infty(V;E)$. If every closed form in $C_m^\infty(U;G)$ is exact, show that the same is true for $C_m^\infty(V;G)$.

20.B. Let $U = \mathbb{R}^2 \setminus \{0\}$. Show that the differential form $g \in C_1^\infty(U)$ defined by

$$g = \frac{-y}{x^2+y^2}dx + \frac{x}{x^2+y^2}dy$$

is closed but is not exact.

21. THE $\bar\partial$ OPERATOR

In this section we introduce complex differential forms and define the $\bar\partial$ operator. Throughout this section the letters E and F will represent complex Banach spaces.

If U is an open subset of E then the space $\Omega_m(U;F)$ makes sense, since we may regard E and F as real Banach spaces.

21.1. DEFINITION. If U is an open subset of E then we shall denote by $\Omega_{pq}(U;F)$ the subspace of all $f \in \Omega_{p+q}(U;F)$ such that $f(x) \in \mathcal{L}^a({}^{pq}E;F)$ for every $x \in U$. Likewise, we shall denote by $C^k_{pq}(U;F)$ the subspace of all $f \in C^k_{p+q}(U;F)$ such that $f(x) \in \mathcal{L}^a({}^{pq}E;F)$.

21.2. PROPOSITION. *If U is an open subset of E then:*

(a) *$\Omega_m(U;F)$ is the algebraic direct sum of the subspaces $\Omega_{pq}(U;F)$ with $p+q=m$.*

(b) *$C^k_m(U;F)$ is the algebraic direct sum of the subspaces $C^k_{pq}(U;F)$ with $p+q=m$.*

PROOF. By Theorem 18.8 there are continuous projections

$$u_q : \mathcal{L}^a({}^mE_{\mathbb{R}};F_{\mathbb{R}}) \to \mathcal{L}^a({}^{m-q,q}E;F)$$

such that $u_o + \ldots + u_m = identity$. This induces projections

$$\tilde{u}_q : f \in \Omega_m(U;F) \to u_q \circ f \in \Omega_{m-q,q}(U;F)$$

such that $\tilde{u}_o + \ldots + \tilde{u}_m = identity$. If $f \in \Omega_m(U;F)$ is of class C^k then it is clear that each $u_q \circ f$ is also of class C^k.

If $f \in C^\infty_m(U;F)$ and $x \in U$ then it follows from Proposition 13.15 that $Df(x)$ can be decomposed in the form $Df(x) = D'f(x) + D''f(x)$, where $D'f(x) \in \mathcal{L}({}^{10}E;\mathcal{L}^a({}^mE_{\mathbb{R}};F_{\mathbb{R}})$ and $D''f(x) \in \mathcal{L}({}^{01}E;\mathcal{L}^a({}^mE_{\mathbb{R}};F_{\mathbb{R}})$. Since in particular

$$D'f(x), D''f(x) \in \mathcal{L}^{a1m}({}^{m+1}E_{\mathbb{R}};F_{\mathbb{R}}) \subset \mathcal{L}({}^{m+1}E_{\mathbb{R}};F_{\mathbb{R}}),$$

we are ready to give the following definition.

21.3. DEFINITION. Let U be an open subset of E. For each f in $C^\infty_m(U;F)$ let ∂f and $\bar{\partial} f$ in $C^\infty_{m+1}(U;F)$ be defined by

$$\partial f(x) = (m+1)[D'f(x)]^a = (m+1)[D'f(x)]^{a1m}$$

and

$$\bar{\partial} f(x) = (m+1)[D''f(x)]^a = (m+1)[D''f(x)]^{a1m}$$

for every $x \in U$. In other words,

$$\partial f(x)(t_1,\ldots,t_{m+1}) = \sum_{j=1}^{m+1} (-1)^{j-1} D'f(x)(t_j)(t_1,\ldots,\hat{t}_j,\ldots,t_{m+1})$$

and

$$\bar{\partial} f(x)(t_1,\ldots,t_{m+1}) = \sum_{j=1}^{m+1} (-1)^{j-1} D''f(x)(t_j)(t_1,\ldots,\hat{t}_j,\ldots,t_{m+1}).$$

21.4. **PROPOSITION.** *If* U *is an open subset of* E *then:*

(a) *Both mappings* $\partial : C_m^\infty(U;F) \to C_{m+1}^\infty(U;F)$ *and* $\bar{\partial} : C_m^\infty(U;F) \to C_{m+1}^\infty(U;F)$ *are lienar.*

(b) $df = \partial f + \bar{\partial} f$ *for every* $f \in C_m^\infty(U;F)$.

(c) ∂ *maps* $C_{pq}^\infty(U;F)$ *into* $C_{p+1,q}^\infty(U;F)$. $\bar{\partial}$ *maps* $C_{pq}^\infty(U;F)$ *into* $C_{p,q+1}^\infty(U;F)$.

(d) $\partial^2 f = 0$, $\bar{\partial}^2 f = 0$ *and* $\partial\bar{\partial} f + \bar{\partial}\partial f = 0$ *for every* $f \in C_m^\infty(U;F)$.

(e) *If* $f \in C_m^\infty(U)$ *and* $g \in C_n^\infty(U)$ *then*

$$\partial(f \wedge g) = \partial f \wedge g + (-1)^m f \wedge \partial g$$

$$\bar{\partial}(f \wedge g) = \bar{\partial} f \wedge g + (-1)^m f \wedge \bar{\partial} g.$$

PROOF. (a), (b) and (c) are straighforward consequences of the definitions. In view of Proposition 21.2 it is sufficient to show (d) for each $f \in C_{pq}^\infty(U;F)$. Then by (b) and Proposition 19.8 we have that

$$0 = d^2 f = \partial^2 f + (\partial\bar{\partial} f + \bar{\partial}\partial f) + \bar{\partial}^2 f.$$

Now, by (c) we have that $\partial^2 f \in C^\infty_{p+2,q}(U;F)$, $\partial\bar{\partial}f + \bar{\partial}\partial f \in C^\infty_{p+1,q+1}(U;F)$ and $\bar{\partial}^2 f \in C^\infty_{p,q+2}(U;F)$. Hence $\partial^2 f = 0$, $\partial\bar{\partial}f + \bar{\partial}\partial f = 0$ and $\bar{\partial}^2 f = 0$. Finally, it is sufficient to prove (e) for $f \in C^\infty_{pq}(U)$ and $g \in C^\infty_{rs}(U)$. Then by Proposition 19.7 we have that

$$d(f \wedge g) = df \wedge g + (-1)^{p+q} f \wedge dg.$$

Then using (b) we get that

$$[\partial(f \wedge g) - \partial f \wedge g - (-1)^{p+q} f \wedge \partial g]$$

$$\pm [\bar{\partial}(f \wedge g) - \bar{\partial} f \wedge g - (-1)^{p+1} f \wedge \bar{\partial} g] = 0.$$

By (c) the first term belongs to $C^\infty_{p+r+1,q+s}(U)$, whereas the second term belongs to $C^\infty_{p+r,q+s+1}(U)$. The desired conclusion follows.

21.5. PROPOSITION. *Let E, F, G be complex Banach spaces, let $U \subset E$ and $V \subset F$ be open sets, and let $\varphi \in \mathcal{H}(U;F)$ with $\varphi(U) \subset V$. Then:*

(a) *φ^* maps $\Omega_{pq}(V;G)$ into $\Omega_{pq}(U;G)$.*

(b) *$\partial(\varphi^* f) = \varphi^*(\partial f)$ and $\bar{\partial}(\varphi^* f) = \varphi^*(\bar{\partial} f)$ for every $f \in C^\infty_m(V;G)$.*

PROOF. (a) If $f \in \Omega_{pq}(V;G)$ then

$$(\varphi^* f)(x)(t_1,\ldots,t_{p+q}) = f(\varphi(x))(D\varphi(x)(t_1),\ldots,D\varphi(x)(t_{p+q})).$$

Since the mapping $D\varphi(x)$ is $\mathbb{C}$-linear, it follows that $\varphi^* f \in \Omega_{pq}(U;G)$.

(b) If suffices to show (b) for $f \in C^\infty_{pq}(V;G)$. By Propositions 19.12 and 21.4

$$0 = d(\varphi^* f) - \varphi^*(df) = [\partial(\varphi^* f) - \varphi^*(\partial f)] + [\bar{\partial}(\varphi^* f) - \varphi^*(\bar{\partial} f)].$$

By Proposition 21.4, $\partial(\varphi^*f) - \varphi^*(\partial f) \in C^\infty_{p+1,q}(U;G)$ whereas $\overline{\partial}(\varphi^*f) - \varphi^*(\overline{\partial}f) \in C^\infty_{p,q+1}(U;G)$. The desired conclusion follows.

If $f \in C^\infty(U;F)$ is a C^∞ mapping then $\overline{\partial}f = D''f$ and it follows from Corollary 13.17 that $f \in \mathcal{H}(U;F)$ if and only if $\overline{\partial}f = 0$ on U. This generalizes to differential forms as follows.

21.6. **PROPOSITION.** *Let U be an open subset of E and let $f \in C^\infty_{p0}(U;F)$. Then $\overline{\partial}f = 0$ on U if and only $f \in \mathcal{H}(U;\mathcal{L}^a({}^pE;F))$.*

PROOF. In view of Corollary 13.17 it suffices to prove that $\overline{\partial}f(x) = 0$ if and only $D''f(x) = 0$. Since $D''f(x) = 0$ clearly implies that $\overline{\partial}f(x) = 0$, we have to prove the converse. If we assume $\overline{\partial}f(x) = 0$ then for arbitrary vectors $t_o,\dots,t_p \in E$ and numbers $\lambda_o,\dots,\lambda_p \in \mathbb{C}$ we have that

$$(21.1)\quad 0 = \overline{\partial}f(x)(\lambda_o t_o,\dots,\lambda_p t_p)$$

$$= \sum_{k=0}^{p} (-1)^k D''f(x)(\lambda_k t_k)(\lambda_o t_o,\dots,\widehat{\lambda_k t_k},\dots,\lambda_p t_p)$$

$$= \sum_{k=0}^{p} (-1)^k \lambda_o \dots \overline{\lambda}_k \dots \lambda_p D''f(x)(t_k)(t_o,\dots,\hat{t}_k,\dots,t_p).$$

Choose $p+1$ distinct numbers $\theta_o,\dots,\theta_p$ in $(0,\pi)$ and set $\alpha_k = e^{i\theta_k}$, $\beta_k = \alpha_k/\overline{\alpha}_k = e^{2i\theta_k}$ and $\zeta_k = \alpha_o \dots \overline{\alpha}_k \dots \alpha_p$ for $k = 0,\dots,p$. Then the complex numbers $\beta_o,\dots,\beta_p$ are all distinct, and so are the numbers $\zeta_o,\dots,\zeta_p$. By applying (21.1) with $\lambda_o = \alpha_o^j,\dots,\lambda_p = \alpha_p^j$ we get a system of equations

$$0 = \sum_{j=0}^{p} (-1)^k \zeta_k^j D''f(x)(t_k)(t_o,\dots,\hat{t}_k,\dots,t_p) \qquad (j = 0,\dots,p)$$

which has a nonzero Vandermonde determinant. In particular it follows that $D''f(x)(t_o)(t_1,\dots,t_p) = 0$, completing the proof.

EXERCISES

21.A. Let U be an open subset of $\mathbb{C}^n$ and let $z_1,\dots,z_n$ denote the complex coordinate functionals of $\mathbb{C}^n$. Show that if $0 \le p \le n$, $0 \le q \le n$ and $p + 1 \ge 1$ then each $f \in \Omega_{pq}(U;F)$ can by uniquely represented as a sum

$$f = \Sigma\, f_{j_1 \dots j_p k_1 \dots k_q}\, dz_{j_1} \wedge \dots \wedge dz_{j_p} \wedge d\bar{z}_{k_1} \wedge \dots \wedge d\bar{z}_{k_q}$$

where $f_{j_1 \dots j_p k_1 \dots k_q} : U \to F$ and the summation is taken over all multi-indices $(j_1,\dots,j_p,k_1,\dots,k_q)$ such that $1 \le j_1 < \dots < j_p \le n$ and $1 \le k_1 < \dots < k_q \le n$. This representation will be often written in the abbreviated form $f = \sum_{J,K} f_{JK} dz_J \wedge d\bar{z}_K$.

Show that the differential form f is of class C^k if and only if each of the mappings f_{JK} is of class C^k.

21.B. Let U be an open subset of $\mathbb{C}^n$ and let $z_1,\dots,z_n$ denote the complex coordinate functionals of $\mathbb{C}^n$.

(a) Show that if $f \in C^\infty(U;F)$ then

$$\partial f = \sum_{j=1}^{n} \frac{\partial f}{\partial z_j}\, dz_j \quad \text{and} \quad \bar{\partial} f = \sum_{j=1}^{n} \frac{\partial f}{\partial \bar{z}_j}\, d\bar{z}_j .$$

(b) Show that if $f = \sum_{J,K} f_{JK} dz_J \wedge d\bar{z}_K \in C^\infty_{pq}(E;F)$ then

$$\partial f = \sum_{J,K} \partial f_{JK} \wedge dz_J \wedge d\bar{z}_K = \sum_{J,K} \sum_{r=1}^{n} \frac{\partial f_{JK}}{\partial z_r}\, dz_r \wedge dz_J \wedge d\bar{z}_K$$

and

$$\bar{\partial} f = \sum_{J,K} \bar{\partial} f_{JK} \wedge dz_J \wedge d\bar{z}_K = \sum_{J,K} \sum_{r=1}^{n} \frac{\partial f_{JK}}{\partial \bar{z}_r}\, d\bar{z}_r \wedge dz_J \wedge d\bar{z}_K .$$

21.C. Let U be an open subset of $\mathbb{C}^n$ and let $f = \sum_J f_J\, dz_J \in C^\infty_{p0}(U;F)$. Show that $\bar{\partial} f = 0$ if and only if $f_J \in \mathcal{H}(U;F)$ for every J.

22. DIFFERENTIAL FORMS WITH BOUNDED SUPPORT

In this section we begin the study of the equation $\bar{\partial} f = g$, which constitutes one of the central problems treated in this book. As in the preceding section the letters E and F will represent complex Banach spaces. To begin with, we give a definition which parallels Definition 20.1.

22.1. DEFINITION. Let U be an open subset of E. A differential form $g \in C^{\infty}_{pq}(U;F)$ is said to be *$\bar{\partial}$-closed* if $\bar{\partial} g = 0$, and is said to be *$\bar{\partial}$-exact* if there is a differential form $f \in C^{\infty}_{p,q-1}(U;F)$ such that $\bar{\partial} f = g$.

By Proposition 21.4 every $\bar{\partial}$-exact form is $\bar{\partial}$-closed and we would like to know whether the converse is true. In this section we solve this problem in the simplest possible case, that is, when the given form g has bounded support. To solve this problem we shall need an integral formula for differentiable functions which generalizes the Cauchy integral formula.

22.2. THEOREM. *Let U be an open subset of E, and let $f \in C^{\infty}(U;F)$. Let $a \in U$, $b \in E$ and $r > 0$ be such that $a + \zeta b \in U$ for all $\zeta \in \bar{\Delta}(0;r)$. Then for each $\lambda \in \Delta(0;r)$ we have the following generalized Cauchy integral formula*

$$2\pi i f(a+\lambda b) = \int_{|\zeta|=r} f(a+\zeta b)\frac{d\zeta}{\zeta-\lambda} + \int_{|\zeta|\le r} D''f(a+\zeta b)(b)\frac{d\zeta \wedge d\bar{\zeta}}{\zeta-\lambda}.$$

To prove this theorem we need the following lemma.

22.3. LEMMA. *Let U be an open set in $\mathbb{C}$, let $\varphi \in C^{\infty}(U)$ and let $\bar{\Delta}(a;r) \subset U$. Then for each $\lambda \in \Delta(a;r)$ we have the formula*

$$2\pi i \varphi(\lambda) = \int_{|\zeta|=r} \varphi(\zeta)\frac{d\zeta}{\zeta-\lambda} + \int_{|\zeta|\le r} \frac{\partial\varphi(\zeta)}{\partial\bar{\zeta}}\frac{d\zeta \wedge d\bar{\zeta}}{\zeta-\lambda}$$

PROOF. Set $D = \Delta(a;r)$, fix $\lambda \in D$ and set $D_\rho = \{\zeta \in D : |\zeta - \lambda| > \rho\}$, where $0 < \rho < d_D(\lambda)$. By Stokes' theorem in the plane we have that

$$\int_{\partial D_\rho} \frac{\varphi(\zeta)}{\zeta - \lambda}\, d\zeta = \int_{D_\rho} d\left(\frac{\varphi(\zeta)}{\zeta - \lambda}\, d\zeta\right).$$

Using Exercise 21.B we get that

$$d\left(\frac{\varphi(\zeta)}{\zeta - \lambda}\, d\zeta\right) = \frac{\partial}{\partial \bar\zeta}\left(\frac{\varphi(\zeta)}{\zeta - \lambda}\right) d\bar\zeta \wedge d\zeta = -\frac{\partial \varphi(\zeta)}{\partial \bar\zeta}\, \frac{d\zeta \wedge d\bar\zeta}{\zeta - \lambda},$$

and therefore

$$(22.1)\quad \int_{\partial D} \frac{\varphi(\zeta)}{\zeta-\lambda}\, d\zeta - \int_{|\zeta-\lambda|=\rho} \frac{\varphi(\zeta)}{\zeta-\lambda}\, d\zeta = -\int_{D_\rho} \frac{\partial \varphi(\zeta)}{\partial \bar\zeta}\, \frac{d\zeta \wedge d\bar\zeta}{\zeta - \lambda}.$$

Since $(\zeta - \lambda)^{-1} d\zeta \wedge d\bar\zeta$ defines a finite measure on each bounded subset of the plane, the Dominated Convergence Theorem guarantees that

$$(22.2)\quad \lim_{\rho \to 0} \int_{D_\rho} \frac{\partial \varphi(\zeta)}{\partial \bar\zeta}\, \frac{d\zeta \wedge d\bar\zeta}{\zeta - \lambda} = \int_D \frac{\partial \varphi(\zeta)}{\partial \bar\zeta}\, \frac{d\zeta \wedge d\bar\zeta}{\zeta - \lambda}.$$

On the other hand,

$$\int_{|\zeta-\lambda|=\rho} \frac{\varphi(\zeta)}{\zeta - \lambda}\, d\zeta - 2\pi i\, \varphi(\lambda) = \int_{|\zeta-\lambda|=\rho} [\varphi(\zeta) - \varphi(\lambda)]\, \frac{d\zeta}{\zeta - \lambda}$$

and since φ is continuous, it follows that

$$(22.3)\quad \lim_{\rho \to 0} \int_{|\zeta-\lambda|=\rho} \frac{\varphi(\zeta)}{\zeta - \lambda}\, d\zeta = 2\pi i\, \varphi(\lambda).$$

If we let $\rho \to 0$ in (22.1) and use (22.2) and (22.3), then the desired conclusion follows.

PROOF OF THEOREM 22.2. Let $\psi \in F'$ and consider the function

$\varphi(\zeta) = \psi \circ f(a + \zeta b)$, which is defined and is of class C^{∞} on a neighborhood of the disc $\overline{\Delta}(0;r)$. Then, by the Chain Rule, for each $\eta \in \mathbb{C}$ we have that

$$D\varphi(\zeta)(\eta) = \psi[Df(a + \zeta b)(\eta b)]$$

and using Exercise 13.H we get that

$$\eta \frac{\partial \varphi(\zeta)}{\partial \zeta} = D'\varphi(\zeta)(\eta) = \eta\psi[D'f(a + \zeta b)(t)]$$

$$\overline{\eta} \frac{\partial \varphi(\zeta)}{\partial \overline{\zeta}} = D''\varphi(\zeta)(\zeta) = \overline{\eta}\psi[D''f(a + \zeta b)(t)].$$

Hence, after applying Lemma 22.3 to φ, we get that

$$2\pi i\psi \circ f(a + \lambda b) = \int_{|\zeta|=r} \psi \circ f(a + \zeta b)\frac{d\zeta}{\zeta - \lambda} +$$

$$\int_{|\zeta|\leq r} \psi[D''f(a + \zeta b)(b)] \frac{d\zeta \wedge d\overline{\zeta}}{\zeta}$$

for every $\lambda \in \Delta(0;r)$. Since F' separates the points of F, the desired conclusion follows.

22.4. THEOREM. *For each $\overline{\partial}$-closed differential form $g \in C^{\infty}_{p,q+1}(E;F)$, with bounded support, there exists a differential form $f \in C^{\infty}_{pq}(E;F)$ such that $\overline{\partial} f = g$. In other words, every $\overline{\partial}$-closed differential form in $C^{\infty}_{p,q+1}(E;F)$, with bounded support, is $\overline{\partial}$-exact.*

PROOF. The proof is similar to the proof of the Poincaré Lemma. For each differential form $g \in C^{\infty}_{p,q+1}(E;F)$, with bounded support, we shall define a differential form $Kg \in C^{\infty}_{pq}(E;F)$ such that

$$\overline{\partial}(Kg) + K(\overline{\partial} g) = g.$$

The conclusion of the theorem follows at once from this. Now, fix $b \in E$ with $b \neq 0$. Then, given a differential form $g \in C^{\infty}_{p,q+1}(E;F)$, with bounded support, we define a differential form $Kg \in \Omega_{pq}(E;F)$ by

$$Kg(x) = \frac{1}{2\pi i}\int_{\mathbb{C}} g(x+\zeta b)b\,\frac{d\zeta \wedge d\bar{\zeta}}{\zeta}$$

for every $x \in E$. Or more explicitly

$$Kg(x)(t_1,\ldots,t_{p+q}) = \frac{1}{2\pi i}\int_{\mathbb{C}} g(x+\zeta b)(b,t_1,\ldots,t_{p+q})\,\frac{d\zeta \wedge d\bar{\zeta}}{\zeta}$$

for all $x, t_1,\ldots,t_{p+q} \in E$. Since g has bounded support, for each bounded open set U in E we can find $r > 0$ such that $g(x+\zeta b) = 0$ for all $x \in U$ and all $\zeta \in \mathbb{C}$ with $|\zeta| \geq r$. Hence for each $x \in U$ we can write

$$Kg(x) = \frac{1}{2\pi i}\int_{|\zeta|\leq r} g(x+\zeta b)b\,\frac{d\zeta \wedge d\bar{\zeta}}{\zeta}$$

Thus we may apply Proposition 13.14 to conclude that Kg is of class C^{∞} on U and

$$D''(Kg)(x)(t) = \frac{1}{2\pi i}\int_{|\zeta|\leq r} D''g(x+\zeta b)(t)b\,\frac{d\zeta \wedge d\bar{\zeta}}{\zeta}$$

for all $x \in U$ and $t \in E$. Thus for all $x \in U$ and $t_1,\ldots,t_{p+q+1} \in E$ we have that

$$\bar{\partial}(Kg)(x)(t_1,\ldots,t_{p+q+1})$$

$$= \sum_{j=1}^{p+q+1}(-1)^{j-1}D''(Kg)(x)(t_j)(t_1,\ldots,\hat{t}_j,\ldots,t_{p+q+1})$$

$$= \sum_{j=1}^{p+q+1}\frac{(-1)^{j-1}}{2\pi i}\int_{|\zeta|\leq r} D''g(x+\zeta b)(t_j)(b,t_1,\ldots,\hat{t}_j,\ldots,t_{p+q+1})\,\frac{d\zeta \wedge d\bar{\zeta}}{\zeta}.$$

On the other hand,

$$K(\bar{\partial}g)(x)(t_1,\ldots,t_{p+q+1})$$

$$= \frac{1}{2\pi i}\int_{|\zeta|\le r} (\bar{\partial}g)(x+\zeta b)(b,t_1,\ldots,t_{p+q+1})\frac{d\zeta\wedge d\bar{\zeta}}{\zeta}$$

$$= \frac{1}{2\pi i}\int_{|\zeta|\le r} D''g(x+\zeta b)(b)(t_1,\ldots,t_{p+q+1})\frac{d\zeta\wedge d\bar{\zeta}}{\zeta} +$$

$$\sum_{j=1}^{p+q+1}\frac{(-1)^j}{2\pi i}\int_{|\zeta|\le r} D''g(x+\zeta b)(t_j)(b,t_1,\ldots,\hat{t}_j,\ldots,t_{p+q+1})\frac{d\zeta\wedge d\bar{\zeta}}{\zeta}.$$

Thus we get that

$$\bar{\partial}(Kg)(x)(t_1,\ldots,t_{p+q+1}) + K(\bar{\partial}g)(x)(t_1,\ldots,t_{p+q+1})$$

$$= \frac{1}{2\pi i}\int_{|\zeta|\le r} D''g(x+\zeta b)(b)(t_1,\ldots,t_{p+q+1})\frac{d\zeta\wedge d\bar{\zeta}}{\zeta},$$

and using Theorem 22.2 we conclude that

$$\bar{\partial}(Kg)(x) + K(\bar{\partial}g)(x)$$

$$= \frac{1}{2\pi i}\int_{|\zeta|\le r} D''g(x+\zeta b)(b)\frac{d\zeta\wedge d\bar{\zeta}}{\zeta} = g(x).$$

Thus we have shown that $Kg \in C^\infty_{pq}(U;F)$ and $\bar{\partial}(Kg) + K(\bar{\partial}g) = g$ on U. Since U was an arbitrary bounded open set in E, we conclude that $Kg \in C^\infty_{pq}(E;F)$ and $\bar{\partial}(Kg) + K(\bar{\partial}g) = g$ on E. This completes the proof of the theorem.

22.5. THEOREM. *If* $\dim E \ge 2$ *then for each* $\bar{\partial}$*-closed differential form* $g \in C^\infty_{p1}(E;F)$, *with bounded suport, there exists a unique differential form* $f \in C^\infty_{p0}(E;F)$, *with bounded support, such that* $\bar{\partial}f = g$. *The differential form* f *is holomorphic on*

$E \setminus \operatorname{supp} g$, and is identically zero on the unbounded component of $E \setminus \operatorname{supp} g$.

PROOF. Let $g \in C^\infty_{p1}(E;F)$ be a $\bar\partial$-closed differential form with bounded support, and let $f = Kg \in C^\infty_{p0}(E;F)$ be the differential form defined in the proof of the preceding theorem. We shall prove that f has bounded support. Let M be a closed subspace of E such that $E = M \oplus \mathbb{C}b$. Then each $x \in E$ can be uniquely represented as a sum $x = Tx + \eta(x)b$, where $T \in \mathcal{L}(E;M)$ and $\eta \in E'$. Then $f(x)$ can be represented in the form

$$f(x) = \frac{1}{2\pi i} \int_{\mathbb{C}} g(Tx + \eta(x)b + \zeta b)b \, \frac{d\zeta \wedge d\bar\zeta}{\zeta}$$

or in the equivalent form

$$f(x) = \frac{1}{2\pi i} \int_{\mathbb{C}} g(Tx + \zeta b)b \, \frac{d\zeta \wedge d\bar\zeta}{\zeta - \eta(x)} \tag{22.4}$$

since the Lebesgue measure is translation-invariant. Since E is topologically isomorphic to the product $M \times \mathbb{C}$, there exists $c > 0$ such that

$$\| y + \zeta b \| \geq c \max \{ \|y\|, |\zeta| \} \geq c \|y\|$$

for all $y \in M$ and $\zeta \in \mathbb{C}$. Since g has bounded support we can then find $\rho > 0$ such that $g(y + \zeta b) = 0$ for every $\zeta \in \mathbb{C}$ and every $y \in M$ with $\|y\| > \rho$. Then it follows from (22.4) that $f(x) = 0$ for every $x \in E$ with $\|Tx\| > \rho$, that is, f is identically zero on an unbounded open subset of E. On the other hand, $\bar\partial f = g = 0$ on $E \setminus \operatorname{supp} g$, and Proposition 21.6 guarantees that f is holomorphic on $E \setminus \operatorname{supp} g$. Then it follows from the Identity Principle that f is identically zero on the unbounded component of $E \setminus \operatorname{supp} g$, and in particular f has bounded support, as asserted.

To establish uniqueness, let $f_1, f_2 \in C^\infty_{p0}(E;F)$ be two differential forms with bounded support such that $\bar\partial f_1 = \bar\partial f_2 = g$. then $f_1 - f_2$ is holomorphic on E, and $f_1 - f_2$ is identically zero outside a bounded set. By the Identity Principle, $f_1 - f_2$ is identically zero on E.

As an important application of Theorem 22.5 we prove the following *extension theorem of Hartogs*.

22.6. THEOREM. *Let E be a separable Hilbert space with $\dim E \geq 2$. Let U be an open subset of E, and let A be a closed, bounded subset of E, contained in U, and such that $U \setminus A$ is connected. Then for each $f \in \mathcal{H}(U \setminus A;F)$ there exists $g \in \mathcal{H}(U;F)$ such that $g = f$ on $U \setminus A$.*

PROOF. Since E is normal, we can find an open set V such that $A \subset V \subset \overline{V} \subset U$, where the closure is taken in E. And since A is bounded, we may assume that V is bounded too. By applying Corollary 15.5 to the closed sets A and $E \setminus V$ we can find a function $\varphi \in C^{\infty}(E)$ such that $0 \leq \varphi \leq 1$ on E, $\varphi = 1$ on a neighborhood of A, and $supp\ \varphi \subset V$. Let $h \in C^{\infty}(U;F)$ be defined by $h = (1 - \varphi)f$ on $U \setminus A$ and $h = 0$ on a neighborhood of A. We want to find a function $u \in C^{\infty}(U;F)$ such that $h - u \in \mathcal{H}(U;F)$. Thus we should find a function $u \in C^{\infty}(U;F)$ such that $\overline{\partial} u = \overline{\partial} h$ on U. If we define $v \in C^{\infty}_{01}(E;F)$ by $v = \overline{\partial} h$ on U and $v = 0$ on $E \setminus supp\ \varphi$, then $\overline{\partial} v = 0$ on E and $supp\ v \subset supp\ \varphi \subset V$, so that v has bounded support. By Theorem 22.5 there is a unique function $u \in C^{\infty}(E;F)$ with bounded support such that $\overline{\partial} u = v$ on E. Hence $\overline{\partial} u = \overline{\partial} h$ on U and $g = h - u \in \mathcal{H}(U;F)$. To complete the proof we shall show that $g = f$ on $U \setminus A$. By Theorem 22.5, $u = 0$ on W, the unbounded component of $E \setminus supp\ \varphi$. Since $h = f$ on $U \setminus supp\ \varphi$, we see that $g = f$ on $W \cap (U \setminus supp\ \varphi)$, and this is a nonvoid open subset of $U \setminus A$. Since $U \setminus A$ is connected we conclude that $g = f$ on $U \setminus A$ and the proof is complete.

23. THE $\overline{\partial}$ EQUATION IN POLYDISCS

In this section we give a proof of the Dolbeault Lemma, which is a complex version of the Poincaré Lemma. As in the preceding section the letters E and F represent complex

Banach spaces.

23.1. THEOREM. *If* $0 < r < R \leq \infty$ *then for each* $\bar\partial$ *closed differential form* $g \in C^{\infty}_{p,q+1}(\Delta^n(a;R);F)$ *there exists a differential form* $f \in C^{\infty}_{pq}(\Delta^n(a;r);F)$ *such that* $\bar\partial f = g$ *on* $\Delta^n(a;r)$.

Before proving this theorem we show the following lemma.

23.2. LEMMA. *Let* $E = M \oplus \mathbb{C}b \oplus N$ *be a direct sum decomposition of* E. *Let* $U \subset M$ *and* $V \subset N$ *be open sets. Let* $0 < r < R \leq \infty$ *and let* $W_r = U + \Delta(0;r)b + V$ *and* $W_R = U + \Delta(0;R)b + V$. *Let* $g \in C^{\infty}_{p,q+1}(W_R;F)$ *be a differential form such that*

$$\bar\partial g(x)(t_1, \ldots, t_{p+q+2}) = 0 \tag{23.1}$$

for all $x \in W_R$ *and* $t_j \in E$, *and*

$$g(x)(s, t_1, \ldots, t_{p+q}) = 0 \tag{23.2}$$

for all $x \in W_R$, $s \in M$ *and* $t_j \in E$. *Then there exists a differential form* $f \in C^{\infty}_{pq}(W_r;F)$ *such that*

$$\bar\partial f(x)(s, t_1, \ldots, t_{p+q}) = g(x)(s, t_1, \ldots, t_{p+q}) \tag{23.3}$$

for all $x \in W_r$, $s \in M \oplus \mathbb{C}b$ *and* $t_j \in E$.

PROOF. Every $x \in E$ can be uniquely written as a sum $x = Sx + \eta(x)b + Tx$, where $S \in \mathcal{L}(E;M)$, $T \in \mathcal{L}(E;N)$ and $\eta \in E'$. Without loss of generality we may assume that $R < \infty$. Let $\varphi \in C^{\infty}(\mathbb{C})$ such that $\varphi = 1$ on $\Delta(0;r)$ and $supp\ \varphi \subset \Delta(0;R)$. Define $f \in \Omega_{pq}(W_r;F)$ by

$$f(x) = \frac{1}{2\pi i} \int_{\mathbb{C}} \varphi(\eta(x) + \zeta) g(x + \zeta b) b \frac{d\zeta \wedge d\bar\zeta}{\zeta},$$

where the integrand is defined to be zero when $x + \zeta b \notin W_R$. Note that if $x + \zeta b \notin W_R$ then $|\eta(x) + \zeta| \geq R$ and therefore $\varphi(\eta(x) + \zeta) = 0$. Since the integrand vanishes outside the disc

$|\zeta| \leq R + r$ for every $x \in W_r$ we may apply Proposition 13.14 to conclude that f is of class C^∞ and

$$D''f(x)(t) = \frac{1}{2\pi i}\int_{\mathbb{C}} D''_x\,[\varphi(\eta(x) + \zeta)g(x + \zeta b)b](t)\ \frac{d\zeta \wedge d\bar{\zeta}}{\zeta}$$

$$= \frac{1}{2\pi i}\int_{\mathbb{C}} D''\,\varphi(\eta(x) + \zeta)(\eta(t))g(x + \zeta b)b\ \frac{d\zeta \wedge d\bar{\zeta}}{\zeta} +$$

$$\frac{1}{2\pi i}\int_{\mathbb{C}} \varphi(\eta(x) + \zeta)D''g(x + \zeta b)(t)b\ \frac{d\zeta \wedge d\bar{\zeta}}{\zeta}$$

for every $x \in W_r$ and $t \in E$. Hence for all $x \in W_r$ and $t_j \in E$ we get that

$$\bar{\partial}f(x)(t_o,\ldots,t_{p+q}) = \sum_{j=0}^{p+q}(-1)^j D''f(x)(t_j)(t_o,\ldots,\hat{t}_j,\ldots,t_{p+q})$$

$$= \sum_{j=0}^{p+q}\frac{(-1)^j}{2\pi i}\int_{\mathbb{C}} D''\varphi(\eta(x) + \zeta)(\eta(t_j))g(x + \zeta b)(b,t_o,\ldots,\hat{t}_j,\ldots,t_{p+q})\frac{d\zeta \wedge d\bar{\zeta}}{\zeta}$$

$$+ \sum_{j=0}^{p+q}\frac{(-1)^j}{2\pi i}\int_{\mathbb{C}} \varphi(\eta(x) + \zeta)D''g(x + \zeta b)(t_j)(b,t_o,\ldots,\hat{t}_j,\ldots,t_{p+q})\frac{d\zeta \wedge d\bar{\zeta}}{\zeta}.$$

Note that the last written sum is nothing but

$$\frac{1}{2\pi i}\int_{\mathbb{C}} \varphi(\eta(x) + \zeta)D''g(x + \zeta b)(b)(t_o,\ldots,t_{p+q})\ \frac{d\zeta \wedge d\bar{\zeta}}{\zeta}$$

$$- \frac{1}{2\pi i}\int_{\mathbb{C}} \varphi(\eta(x) + \zeta)\bar{\partial}g(x + \zeta b)(b,t_o,\ldots,t_{p+q})\ \frac{d\zeta \wedge d\bar{\zeta}}{\zeta},$$

and the last written term vanishes by hypothesis (23.1), since $\varphi(\eta(x) + \zeta) = 0$ whenever $x + \zeta b \notin W_R$. Thus

$$\bar{\partial}f(x)(t_o,\ldots,t_{p+q})$$

$$= \sum_{j=0}^{p+q}\frac{(-1)^j}{2\pi i}\int_{\mathbb{C}} D''\varphi(\eta(x) + \zeta)(\eta(t_j))g(x + \zeta b)(b,t_o,\ldots,\hat{t}_j,\ldots,t_{p+q})\ \frac{d\zeta \wedge d\bar{\zeta}}{\zeta}$$

$$+ \frac{1}{2\pi i} \int_{\mathbb{C}} \varphi(\eta(x) + \zeta) D''g(x + \zeta b)(b)(t_0, \ldots, t_{p+q}) \frac{d\zeta \wedge d\bar{\zeta}}{\zeta}$$

If $t_0 = s$ lies in M then by hypothesis (23.2) all the terms in the last written expression for $\bar{\partial} f(x)(t_0, \ldots, t_{p+q})$ vanish, and therefore

$$(23.4) \qquad \bar{\partial} f(x)(s, t_1, \ldots, t_{p+q}) = 0 = g(x)(s, t_1, \ldots, t_{p+q})$$

for all $x \in W_r$, $s \in M$ and $t_j \in E$. On the other hand, the preceding expression for $\bar{\partial} f(x)(t_0, \ldots, t_{p+q})$ can be rewritten in the form

$$\bar{\partial} f(x)(t_0, \ldots, t_{p+q})$$

$$= \sum_{j=1}^{p+q} \frac{(-1)^j}{2\pi i} \int_{\mathbb{C}} D''\varphi(\eta(x) + \zeta)(\eta(t_j))g(x + \zeta b)(b, t_0, \ldots, \hat{t}_j, \ldots, t_{p+q}) \frac{d\zeta \wedge d\bar{\zeta}}{\zeta}$$

$$+ \frac{1}{2\pi i} \int_{\mathbb{C}} D''\varphi(\eta(x) + \zeta)(\eta(t_j))g(x + \zeta b)(b, t_1, \ldots, t_{p+q}) \frac{d\zeta \wedge d\bar{\zeta}}{\zeta}$$

$$+ \frac{1}{2\pi i} \int_{\mathbb{C}} \varphi(\eta(x) + \zeta) D''g(x + \zeta b)(b)(t_0, \ldots, t_{p+q}) \frac{d\zeta \wedge d\bar{\zeta}}{\zeta} .$$

If $t_0 = b$ then all the terms in the first sum vanish and it follows that

$$\bar{\partial} f(x)(b, t_1, \ldots, t_{p+q})$$

$$= \frac{1}{2\pi i} \int_{\mathbb{C}} D''_x [\varphi(\eta(x) + \zeta) g(x + \zeta b) b](b)(t_1, \ldots, t_{p+q}) \frac{d\zeta \wedge d\bar{\zeta}}{\zeta} .$$

Hence an application of Theorem (22.2) shows that

$$(23.5) \qquad \bar{\partial} f(x)(b, t_1, \ldots, t_{p+q}) = \varphi(\eta(x)) g(x)(b, t_1, \ldots, t_{p+q})$$

$$= g(x)(b, t_1, \ldots, t_{p+q})$$

for all $x \in W_r$ and $t_j \in E$. Since (23.3) follows at once from

(23.4) and (23.5), the proof of the lemma is complete.

PROOF OF THEOREM 23.1. Without loss of generality we may assume that $a = 0$, so that $g \in C^{\infty}_{p,q+1}(\Delta^n(0;R);F)$, for each $k = 1,\ldots,n$ set

$$D_k = \{z \in \mathbb{C}^n : |z_j| < r \text{ for } j \leq k \text{ and } |z_j| < R \text{ for } j < k\}.$$

For each $k = 1,\ldots,n$ we identify $\mathbb{C}^k$ with the subspace of $\mathbb{C}^n$ generated by the first k vectors of the canonical basis. An application of Lemma 23.2 with $M = \{0\}$ yields a differential form $f_1 \in C^{\infty}_{pq}(D_1;F)$ such that

$$\bar{\partial} f_1(x)(s,t_1,\ldots,t_{p+q}) = g(x)(s,t_1,\ldots,t_{p+q})$$

for all $x \in D_1$, $s \in \mathbb{C}^1$ and $t_j \in \mathbb{C}^n$. By repeated applications of Lemma 23.2 we can find differential forms $f_2,\ldots,f_n$ such that $f_k \in C^{\infty}_{pq}(D_k;F)$ and

$$\bar{\partial} f_k(x)(s,t_1,\ldots,t_{p+q}) = (g(x) - \sum_{j=1}^{k-1} \bar{\partial} f_j(x))(s,t_1,\ldots,t_{p+q})$$

for all $x \in D_k$, $s \in \mathbb{C}^k$ and $t_j \in \mathbb{C}^n$. Then $f = f_1 + \ldots + f_n$ is the required differential form.

If g is a $\bar{\partial}$-closed F-valued differential form of bidegree $(p,q+1)$ on a neighborhood of a compact polydisc $\bar{\Delta}^n(a;r)$ then Theorem 23.1 guarantees the existence of an F-valued differential form f of bidegree (p,q) such that $\bar{\partial} f = g$ on a neighborhood of $\bar{\Delta}^n(a;r)$. Note that we may assume that f is defined on all of $\mathbb{C}^n$, for otherwise it suffices to multiply f by a function $\varphi \in C^{\infty}(\mathbb{C}^n)$ such that φ is identically one on a neighborhood of $\bar{\Delta}^n(a;r)$, and the support of φ is contained in the domain of definition of f. These remarks will be repeatedly used in the sequel.

Next we improve Theorem 23.1 as follows.

23.3. THEOREM. *If* $0 < R \leq \infty$ *then for each* $\bar{\partial}$ *closed differential form* $g \in C^{\infty}_{p,q+1}(\Delta^n(a;R);F)$ *there eixsts a differential form* $f \in C^{\infty}_{pq}(\Delta^n(a;R);F)$ *such that* $\bar{\partial} f = g$.

PROOF. Let $(r_j)_{j=1}^{\infty}$ be a strictly increasing sequence of positive numbers tending to R. Consider separately the following two cases.

(a) First assume $q \geq 1$. In this case we shall inductively find a sequence $(f_j)_{j=1}^{\infty}$ in $C^{\infty}_{pq}(\mathbb{C}^n;F)$ such that $\bar{\partial} f_j = g$ on a neighborhood of $\bar{\Delta}^n(a;r_j)$ for every $j \geq 1$, and $f_j = f_{j-1}$ on $\Delta^n(a;r_{j-1})$ for every $j \geq 2$. The existence of f_1 is guaranteed by Theorem 23.1 and the preceding remarks. Let $j \geq 2$ and assume that $f_1,\ldots,f_{j-1}$ have already been found. By Theorem 23.1 there exists $u \in C^{\infty}_{pq}(\mathbb{C}^n;F)$ such that $\bar{\partial} u = g$ on a neighborhood of $\bar{\Delta}^n(a;r_j)$. But then $\bar{\partial}(u - f_{j-1}) = 0$ on a neighborhood of $\bar{\Delta}^n(a;r_{j-1})$, and another application of Theorem 23.1 yields a differential form $v \in C^{\infty}_{p,q+1}(\mathbb{C}^n;F)$ such that $\bar{\partial} v = u - f_{j-1}$ on a neighborhood of $\bar{\Delta}^n(a;r_{j-1})$. Then the differential form $f = u - \bar{\partial} v \in C^{\infty}_{pq}(\mathbb{C}^n;F)$ has the required properties.

Once the existence of the sequence (f_j) has been established, we define $f \in \Omega_{pq}(\Delta^n(a;R);F)$ by $f = f_j$ on $\Delta^n(a;r_j)$ for each $j \in \mathbb{N}$. Then it is clear that $f \in C^{\infty}_{pq}(\Delta^n(a;R);F)$ and $\bar{\partial} f = g$ on $\Delta^n(a;R)$.

(b) Next consider the case $g = 0$. In this case we shall inductively find a sequence $(f_j)_{j=1}^{\infty}$ in $C^{\infty}_{p0}(\mathbb{C}^n;F)$ such that $\bar{\partial} f_j = g$ on a neighborhood of $\bar{\Delta}^n(a;r_j)$ and $\| f_j(x) - f_{j-1}(x) \| \leq 2^{-j}$ for every $x \in \bar{\Delta}^n(a;r_{j-1})$ and $j \geq 2$. As before, the existence of f_1 is guaranteed by Theorem 23.1. Let $j \geq 2$ and assume that $f_1,\ldots,f_{j-1}$ have already been found. By Theorem 23.1 there exists $u \in C^{\infty}_{p0}(\mathbb{C}^n;F)$ such that $\bar{\partial} u = g$ on a neighborhood of $\bar{\Delta}^n(a;r_{j-1})$. But then $\bar{\partial}(u - f_{j-1}) = 0$ on a neighborhood V of $\bar{\Delta}^n(a;r_{j-1})$, and Proposition 21.6 guarantees that

$u - f_{j-1} \in \mathcal{H}(V;\mathcal{L}^a({}^p\mathbb{C}^n;F))$. Hence we can find a polynomial $P \in P(\mathbb{C}^n;\mathcal{L}^a({}^p\mathbb{C}^n;F))$ such that $\|u(x) - f_{j-1}(x) - P(x)\| \leq 2^{-j}$ for every $x \in \bar{\Delta}^n(a;r_{j-1})$. Then the differential form $f = u - P \in C^\infty_{p0}(\mathbb{C}^n;F)$ has the required properties.

Thus the existence of the sequence (f_j) has been established. In particular we see that the differential form $f_j - f_{j-1}$ belongs to $\mathcal{H}(\Delta^n(a;r_{j-1}); \mathcal{L}^a({}^p\mathbb{C}^n;F))$ for every $j \geq 2$ and the series $\sum_{j=2}^{\infty}(f_j - f_{j-1})$ converges uniformly on each $\Delta^n(a;r_k)$. Hence the series $\sum_{j=k+1}^{\infty}(f_j - f_{j-1})$ belongs to $\mathcal{H}(\Delta^n(a;r_k);\mathcal{L}^a({}^p\mathbb{C}^n;F))$ for each $k \in \mathbb{N}$. Define $f \in \Omega_{pq}(\Delta^n(a;R);F)$ by

$$f = f_1 + \sum_{j=2}^{\infty}(f_j - f_{j-1}).$$

Since for each $k \in \mathbb{N}$ we can write

$$f = f_k + \sum_{j=k+1}^{\infty}(f_j - f_{j-1}),$$

we see that $f \in C^\infty_{pq}(\Delta^n(a;r_k);F)$ and $\bar{\partial} f = \bar{\partial} f_k = g$ on $\Delta^n(a;r_k)$ for each $k \in \mathbb{N}$. The desired conclusion follows.

To end this section we show how Theorem 23.3 can be used to solve the so called *first Cousin problem* on a polydisc.

23.4. THEOREM. *Let $(U_i)_{i\in I}$ be an open cover of a polydisc $\Delta^n(a;R)$, where $0 < R \leq \infty$. Suppose there are functions $f_{ij} \in \mathcal{H}(U_i \cap U_j;F)$ $(i, j \in I)$ such that*

(23.6) $$f_{ij} + f_{ji} = 0 \quad \text{on} \quad U_i \cap U_j$$

and

(23.7) $$f_{ij} + f_{jk} + f_{ki} = 0 \quad \text{on} \quad U_i \cap U_j \cap U_k.$$

Then there are functions $f_i \in \mathcal{H}(U_i;F)$ $(i \in I)$ *such that*

(23.8) $$f_i - f_j = f_{ij} \quad on \quad U_i \cap U_j .$$

PROOF. First we shall find functions $u_i \in C^\infty(U_i;F)$ such that

(23.9) $$u_i - u_j = f_{ij} \quad \text{on} \quad U_i \cap U_j .$$

To achieve this let (φ_i) be a C^∞ partition of unity on $\Delta^n(a;R)$, subordinated to the cover (U_i). If we define

$$u_i = \sum_k f_{ik}\varphi_k \quad \text{on} \quad U_i ,$$

where $f_{ik}\varphi_k$ is defined to be zero on $U_i \setminus U_k$, then u_i is well defined and belongs to $C^\infty(U_i;F)$. Furthermore, it follows from (23.6) that

$$u_i - u_j = \sum_k (f_{ik} - f_{jk})\varphi_k = \sum_k f_{ij}\varphi_k = f_{ij}$$

on $U_i \cap U_j$, as asserted. To complete the proof of the theorem it suffices to find a function $v \in C^\infty(\Delta^n(a;R);F)$ such that $f_i = u_i + v$ is holomorphic on U_i for every i. Thus we want to find a function $v \in C^\infty(\Delta^n(a;R);F)$ such that $\bar\partial v = -\bar\partial u_i$ on U_i for every i. Since f_{ij} is holomorphic on $U_i \cap U_j$, (23.9) implies that $\bar\partial u_i = \bar\partial u_j$ on $U_i \cap U_j$ for all i and j. Hence we may define $w \in C^\infty_{01}(\Delta^n(a;R);F)$ by $w = \bar\partial u_i$ on U_i for every i. Then it follows from Theorem 23.3 that the equation $\bar\partial v = -w$ has a solution $v \in C^\infty(\Delta^n(a;R);F)$ and the proof of the theorem is complete.

EXERCISES

23.A. Let (U_i) be an open cover of $\mathbb{C}$. Suppose that for each i there is a function f_i meromorphic on U_i such that $f_i - f_j$ is holomorphic on $U_i \cap U_j$ for all i and j. Using Theorem 23.4 find a function f meromorphic on $\mathbb{C}$ such that $f - f_i$

is holomorphic on U_i for every i.

23.B. Let (a_i) be a sequence of distinct points in $\mathbb{C}$ without cluster points. Let $(P_i(z))$ be a corresponding sequence of non-zero polynomials in z without constant term. Using Exercise 23.A find a function f meromorphic on $\mathbb{C}$, whose poles are precisely the points a_i, and such that $P_i[(z - a_i)^{-1}]$ is the singular part of f at a_i. This is the *Mittag-Leffler Theorem.*

NOTES AND COMMENTS

Our presentation of alternating multilinear forms and real differential forms in Sections 18, 19 and 20, follows essentially the book of H. Cartan [3]. Our presentation of complex differential forms in Section 21 follows essentially an article of E. Ligocka [2]. The results in Section 22 on existence of solutions of the equation $\overline{\partial} f = g$, when g has bounded support, are straightforward generalizations of results of L. Hörmander [3], in the case of $\mathbb{C}^n$, and E. Ligocka [2], in the case of Banach spaces. The proof of Hartogs' Theorem 22.6, based on Theorem 22.5, is due to L. Hörmander [3], following an idea of L. Ehrenpreis [1]. Finally, Theorems 23.1 and 23.3 are due to P. Dolbeault [1].

CHAPTER VI

POLYNOMIALLY CONVEX DOMAINS

24. POLYNOMIALLY CONVEX COMPACT SETS IN $\mathbb{C}^n$

This section is devoted to the study of polynomially convex compact sets in $\mathbb{C}^n$. We show that the $\bar{\partial}$ equation has a solution on a neighborhood of each polynomially convex compact set. We also prove a theorem on polynomial approximation known as the Oka-Weil theorem. The letters E and F will always represent complex Banach spaces.

If g is a $\bar{\partial}$-closed differential form of bidegree $(p,q+1)$ on a neighborhood of a compact polydisc $\bar{\Delta}^n(a;r)$ then Theorem 23.1 guarantees the existence of a differential form f of bidegree (p,q) such that $\bar{\partial}f = g$ on some neighborhood of $\bar{\Delta}^n(a;r)$. In this section we extend that result to a larger class of compact sets, namely the polynomially convex compact sets in $\mathbb{C}^n$. To begin with we define polynomially convex compact sets in Banach spaces.

24.1. DEFINITION. (a) The *$P(E)$-hull* or *polynomially convex hull* of a set $A \subset E$ is defined by

$$\hat{A}_{P(E)} = \{x \in E : |P(x)| \leq \sup_A |P| \quad \text{for all} \quad P \in P(E)\}.$$

(b) A compact set $K \subset E$ is said to be *polynomially convex* if $\hat{K}_{P(E)} = K$.

The following examples of polynomially convex compact sets can be readily verified by the reader. To prove (a) use Proposition 11.1.

24.2. EXAMPLES. (a) Every convex compact subset of E is polynomially convex.

(b) If K is a polynomially convex compact subset of E and $P_i \in \mathcal{P}(E)$ for each $i \in I$ then the compact set

$$L = \{x \in K : |P_i(x)| \leq 1 \quad \text{for each} \quad i \in I\}$$

is polynomially convex as well.

(c) In particular, if D is a compact polydisc in $\mathbb{C}^n$, I is a finite set, and $P_i \in \mathcal{P}(\mathbb{C}^n)$ for each $i \in I$, then the compact set

$$L = \{x \in D : |P_i(x)| \leq 1 \quad \text{for each} \quad i \in I\}$$

is polynomially convex. These polynomially convex compact sets are called *compact polynomial polyhedra*.

Next we introduce the notion of germ of a function on a compact set, a notion that serves to describe the behavior of a function on sufficiently small neighborhoods of a compact set.

24.3. DEFINITION. If K is a compact subset of E then we shall denote by $\mathcal{H}(K;F)$ the set of all equivalence classes of F-valued functions which are holomorphic on some open neighborhood of K, under the equivalence relation $f \sim g$ if $f = g$ on some neighborhood of K. The members of $\mathcal{H}(K;F)$ are called *germs of F-valued holomorphic functions on K*.

If U is an open neighborhood of K then there is a natural mapping $\mathcal{H}(U;F) \to \mathcal{H}(K;F)$ which associates with each function $f \in \mathcal{H}(U;F)$ the corresponding germ. The set $\mathcal{H}(K;F)$ is in a natural way a vector space, and the canonical mapping $\mathcal{H}(U;F) \to \mathcal{H}(K;F)$ is linear for each open neighborhood U of K.

One should always remember that the elements of $\mathcal{H}(K;F)$ are equivalence classes of functions. However, it is often convenient to regard these elements as being functions by identifying each equivalence class with a suitable representative, and we shall frequently do so.

The preceding construction can be carried out equally well within the realm of C^∞ functions instead of holomorphic functions. Thus we have the following definition.

24.4. DEFINITION. If K is a compact subset of E then we shall denote by $C^\infty(K;F)$ the set of all equivalence classes of F-valued functions which are of class C^∞ on some open neighborhood of K, under the equivalence relation $f \sim g$ if $f = g$ on some neighborhood of K. The members of $C^\infty(K;F)$ are called *germs of F-valued C^∞ functions on K.*

Clearly all the remarks concerning $\mathcal{H}(K;F)$ apply equally well to $C^\infty(K;F)$. In particular $C^\infty(K;F)$ is a vector space.

24.5. DEFINITION. If K is a compact subset of E then we shall denote by $C^\infty_{pq}(K;F)$ the vector space

$$C^\infty_{pq}(K;F) = C^\infty(K; \mathcal{L}^a(^{pq}E;F)).$$

The members of $C^\infty_{pq}(K;F)$ are called *germs of F-valued C^∞ differential forms of bidegree (p,q) on K.*

Thus to say that $\bar{\partial} f = g$ for $f \in C^\infty_{pq}(K;F)$ and $g \in C^\infty_{p,q+1}(K;F)$ means that $\bar{\partial} f = g$ on some neighborhood of K.

24.6. DEFINITION. A compact subset K of E is said to have the *F-Cousin property* if for each $g \in C^\infty_{p,q+1}(K;F)$ such that $\bar{\partial} g = 0$ there exists $f \in C^\infty_{pq}(K;F)$ such that $\bar{\partial} f = g$.

Thus Theorem 23.1 asserts that each compact polydisc in $\mathbb{C}^n$ has the F-Cousin property for each Banach space F. We shall soon extend that result to all the polynomially convex compact sets in $\mathbb{C}^n$. First we show that each polynomially convex compat set in $\mathbb{C}^n$ can be approximated by compact polynomial polyhedra.

24.7. LEMMA. *Let K be a polynomially convex compact set in $\mathbb{C}^n$ and let U be an open neighborhood of K. Then there is a compact polynomial polyhedron L such that $K \subset L \subset U$.*

PROOF. Let D be any compact polydisc containing K. If $D \subset U$ then if suffices to take $L = D$. If $D \not\subset U$ then for each point $a \in D \setminus U$ there is a polynomial $P \in P(\mathbb{C}^n)$ such that $\sup_K |P| < 1 < |P(a)|$. Since the set $D \setminus U$ is compact we can find polynomials $P_1, \ldots, P_m \in P(\mathbb{C}^n)$ such that $\sup_K |P_j| < 1$ for $j = 1, \ldots, m$ and

$$D \setminus U \subset \bigcup_{j=1}^{m} \{x \in E : |P_j(x)| > 1\}.$$

Thus it suffices to take

$$L = \{x \in D : |P_j(x)| \leq 1 \quad \text{for} \quad j = 1, \ldots, m\}.$$

By Lemma 24.7 we may concentrate our attention on compact polynomial polyhedra.

24.8. THEOREM. *Let*

$$L = \{x \in D : |P_j(x)| \leq 1 \quad for \quad j = 1, \ldots, m\}$$

be a compact polynomial polyhedron in $\mathbb{C}^n$, and let μ denote the mapping

$$\mu : x \in \mathbb{C}^n \to (x, P_1(x), \ldots, P_m(x)) \in \mathbb{C}^{n+m}.$$

Then:

(a) *L has the F-Cousin property for each Banach space F.*

(b) *For each $f \in C^{\infty}_{pq}(L;F)$ such that $\bar{\partial} f = 0$ there exists $f_1 \in C^{\infty}_{pq}(D \times \bar{\Delta}^m;F)$ such that $\bar{\partial} f_1 = 0$ and $\mu^* f_1 = f$.*

The key to the proof of Theorem 24.8 is the following Lemma.

24.9. LEMMA. *Let K be a compact subset of $\mathbb{C}^n$, let $P \in P(\mathbb{C}^n)$ and let*

$$K_P = \{x \in K : |P(x)| \leq 1\}.$$

Let μ denote the mapping

$$\mu : x \in \mathbb{C}^n \to (x, P(x)) \in \mathbb{C}^{n+1}$$

and assume that $K \times \overline{\Delta}$ has the F-Cousin property. Then:

(a) *K_P has the F-Cousin property.*

(b) *For each $f \in C^\infty_{pq}(K_P;F)$ with $\overline{\partial} f = 0$ there exists $f_1 \in C^\infty_{pq}(K \times \overline{\Delta};F)$ such that $\overline{\partial} f_1 = 0$ and $\mu^* f_1 = f$.*

PROOF. First we prove (b). Let V be an open neighborhood of K_P such that $f \in C^\infty_{pq}(V;F)$ and $\overline{\partial} f = 0$ on V. If $\pi : \mathbb{C}^{n+1} \to \mathbb{C}^n$ denotes the canonical projection then $\pi^* f \in C^\infty_{pq}(\pi^{-1}(V);F)$ and $\overline{\partial}(\pi^* f) = \pi^*(\overline{\partial} f) = 0$ on $\pi^{-1}(V)$. Note that $\pi \circ \mu$ is the identity on $\mathbb{C}^n$ and hence $\mu(K_P) \subset \pi^{-1}(V)$. Moreover $K_P = \mu^{-1}(K \times \overline{\Delta})$ and

$$\mu(K_P) = \{(x,\zeta) \in K \times \overline{\Delta} : P(x) = \zeta\}.$$

Choose $\varphi \in C^\infty(\mathbb{C}^{n+1})$ such that $\varphi = 1$ on a neighborhood of $\mu(K_P)$ and $supp\ \varphi \subset \pi^{-1}(V)$. Define $f_1 \in C^\infty_{pq}(K \times \overline{\Delta};F)$ by

$$f_1 = \varphi\, \pi^* f - Qu,$$

where $Q \in P(\mathbb{C}^{n+1})$ is defined by $Q(x,\zeta) = P(x) - \zeta$ and $u \in C^\infty_{pq}(K \times \overline{\Delta};F)$ will be determined so that $\overline{\partial} f_1 = 0$ (the product $\varphi\, \pi^* f$ is understood to be zero outside $\pi^{-1}(V)$, a remark that will not be repeated again in similiar situations). The condition $\overline{\partial} f_1 = 0$ reduces to the equation $\overline{\partial} u = v$, where

$$v = \frac{1}{Q}\, \overline{\partial} \varphi \wedge \pi^* f.$$

Since $\overline{\partial}\varphi$ vanishes on a neighborhood of $\mu(K_P)$ we see that v is a well defined member of $C^{\infty}_{p,q+1}(K \times \overline{\Delta};F)$ satisfying the condition $\overline{\partial}v = 0$. Since by hypothesis $K \times \overline{\Delta}$ has the F-Cousin property the equation $\overline{\partial}u = v$ has a solution, as desired. It is clear that $f_1 = \pi^*f$ on a neighborhood of $\mu(K_P)$ and whence it follows that $\mu^*f_1 = \mu^*\pi^*f = f$ on a neighborhood of K_P. This shows (b).

To show (a) let $g \in C^{\infty}_{p,q+1}(K_P;F)$ with $\overline{\partial}g = 0$. By (b) there exists $g_1 \in C^{\infty}_{p,q+1}(K \times \overline{\Delta};F)$ such that $\overline{\partial}g_1 = 0$ and $\mu^*g_1 = g$. Since by hypothesis $K \times \overline{\Delta}$ has the F-Cousin property there exists $f_1 \in C^{\infty}_{pq}(K \times \overline{\Delta};F)$ such that $\overline{\partial}f_1 = g_1$. Set $f = \mu^*f_1 \in C^{\infty}_{pq}(K_P;F)$. Then $\overline{\partial}f = \mu^*(\overline{\partial}f_1) = \mu^*g_1 = g$ and the proof is complete.

PROOF OF THEOREM 24.8. We proceed by induction on the number m of polynomials involved. If $m = 0$ then part (b) is obvious and part (a) is true by Theorem 23.1. Let $m \geq 1$ and assume the theorem is true for compact polynomial polyhedra involving less than m polynomials. If we set

$$K = \{x \in D : |P_j(x)| \leq 1 \quad \text{for} \quad j = 2,\dots,m\}$$

then

$$K \times \overline{\Delta} = \{(x,\zeta) \in D \times \overline{\Delta} : |P_j(x)| \leq 1 \quad \text{for} \quad j = 2,\dots,m\}$$

and

$$L = K_{P_1} = \{x \in K : |P_1(x)| \leq 1\}.$$

Then $K \times \overline{\Delta}$ has the F-Cousin property by the induction hypothesis, and L has the F-Cousin property by Lemma 24.9. This shows (a). To show (b) consider the mappings

$$\mu_1 : x \in \mathbb{C}^n \to (x, P_1(x)) \in \mathbb{C}^{n+1}$$

and

$$\mu_2 : (x,\zeta) \in \mathbb{C}^{n+1} \to (x,\zeta,P_2(x),\dots,P_m(x)) \in \mathbb{C}^{n+m}$$

so that $\mu_2 \circ \mu_1 = \mu$. Let $f \in C^\infty_{pq}(L;F)$ with $\bar\partial f = 0$. By Lemma 24.9 there exists $f_1 \in C^\infty_{pq}(K \times \bar\Delta;F)$ such that $\bar\partial f_1 = 0$ and $\mu_1^* f_1 = f$. By the induction hypothesis there exists $f_2 \in C^\infty_{pq}(D \times \bar\Delta^m;F)$ such that $\bar\partial f_2 = 0$ and $\mu_2^* f_2 = f_1$. Hence $\mu^* f_2 = \mu_1^*(\mu_2^* f_2) = \mu_1^* f_1 = f$ and the proof is complete.

Theorem 24.8 has many consequences.

24.10. THEOREM. *Each polynomially convex compact subset of* $\mathbb{C}^n$ *has the F-Cousin property for each Banach space F.*

PROOF. Apply Lemma 24.7 and Theorem 24.8.

24.11. THEOREM. *Let*

$$L = \{x \in D : |P_j(x)| \leq 1 \quad \text{for} \quad j = 1,\dots,m\}$$

be a compact polynomial polyhedron in $\mathbb{C}^n$. *Then for each* $f \in \mathcal{H}(L;F)$ *there exists* $f_1 \in \mathcal{H}(D \times \bar\Delta^m;F)$ *such that*

$$f(x) = f_1(x, P_1(x), \dots, P_m(x))$$

for all x *in some neighborhood of* L.

PROOF. This is Theorem 24.8 (b) with $p = q = 0$.

The following theorem on polynomial approximation is known as the *Oka-Weil Theorem.*

24.12. THEOREM. *Let* K *be a polynomially convex compact subset of* $\mathbb{C}^n$. *Then for each* $f \in \mathcal{H}(K;F)$ *there is a sequence of polynomials* $P_j \in P(\mathbb{C}^n;F)$ *which converges to* f *uniformly on* K.

PROOF. Let U be an open neighborhood of K such that $f \in \mathcal{H}(U;F)$. By Lemma 24.7 there is a compact polynomial polyhedron

$$L = \{x \in D : |Q_j(x)| \leq 1 \quad \text{for} \quad j = 1,\dots,m\}$$

such that $K \subset L \subset U$. By Theorem 24.11 there exists $f_1 \in \mathcal{H}(D \times \overline{\Delta}^m;F)$ such that

$$f(x) = f_1(x, Q_1(x), \ldots, Q_m(x))$$

for all x in some neighborhood of L. There is a sequence of polynomials $R_j \in P(C^{n+m};F)$ which converges to f_1 uniformly on the polydisc $D \times \overline{\Delta}^m$, namely the partial sums of the Taylor series of f_1 at the center of the polydisc. If we define $P_j \in P(\mathbb{C}^n;F)$ by

$$P_j(x) = R_j(x, Q_1(x), \ldots, Q_m(x))$$

then it is clear that (P_j) converges to f uniformly on L, as we wanted.

EXERCISES

24.A. Show that $\hat{K}_{P(E)} = \hat{K}_{\mathcal{H}(E)}$ for each compact subset K of E.

24.B. Show that if A is a balanced subset of E then the set $\hat{A}_{P(E)}$ is balanced as well.

24.C. Show that if M is a complemented subspace of E then $\hat{A}_{P(M)} = \hat{A}_{P(E)}$ for each subset A of M.

24.D. Show that each finite subset of E is polynomially convex. More generally, show that if $(a_m)_{m=1}^{\infty}$ is a sequence in E which converges to a point a, then the compact set $K = \{a_m : m \in \mathbb{N}\} \cup \{a\}$ is polynomially convex.

24.E. Let K be a connected compact subset of $\mathbb{C}^n$. Using the Oka-Weil Theorem 24.12 show that the set $\hat{K}_{P(\mathbb{C}^n)}$ is connected as well.

24.F. Let K be a polynomially convex compact subset of $\mathbb{C}^n$.

Using the Oka-Weil Theorem 24.12 show that each open and closed subset of K is polynomially convex as well.

24.G. Using the classical theorem of Runge on polynomial approximation for holomorphic functions of one complex variable show that a compact set K in $\mathbb{C}$ is polynomially convex if and only if the open set $\mathbb{C} \setminus K$ is connected.

25. POLYNOMIALLY CONVEX DOMAINS IN $\mathbb{C}^n$

This section is devoted to the study of polynomially convex domains in $\mathbb{C}^n$. Most to the results are straightforward consequences of the results from the preceding section. To begin with we define polynomially convex domains in Banach spaces, keeping in mind the definitions and results from Section 11.

25.1. DEFINITION. Let U be an open subset of E.

(a) U is said to be *polynomially convex* if $\hat{K}_{P(E)} \cap U$ is compact for each compact set $K \subset U$.

(b) U is said to be *strongly polynomially convex* if $\hat{K}_{P(E)} \subset U$ for each compact set $K \subset U$.

One can readily see that U is polynomially convex if and only if $d_U(\hat{K}_{P(E)} \cap U) > 0$ for each compact set $K \subset U$. The following examples of polynomially convex domains can be easily verified by the reader.

25.2. EXAMPLES. (a) Every convex open set in E is strongly polynomially convex.

(b) For each $P \in P(E)$ the open set $U = \{x \in E : |P(x)| < 1\}$ is strongly polynomially convex.

(c) The intersection of two (strongly) polynomially convex open subsets of E is also a (strongly) polynomially convex open subset of E.

(d) If U is a (strongly) polynomially convex open subset

of E then $U \cap M$ is a (strongly) polynomially convex open subset of M for each closed subspace M of E.

Clearly every strongly polynomially convex open subset of E is polynomially convex. The next proposition shows that in $\mathbb{C}^n$ the converse is true.

25.3. PROPOSITION. *An open subset U of $\mathbb{C}^n$ is polynomially convex if and only if U is strongly polynomially convex.*

PROOF. To prove the nontrivial implication let U be a polynomially convex open subset of $\mathbb{C}^n$ and let K be a compact subset of U. If $\hat{K}_{P(\mathbb{C}^n)} \not\subset U$ then we can write $\hat{K}_{P(\mathbb{C}^n)} = A \cup B$, where $A = \hat{K}_{P(\mathbb{C}^n)} \cap U$ and $B = \hat{K}_{P(\mathbb{C}^n)} \setminus U$. Then A and B are two disjoint compact sets. If we define $f = 0$ on a neighborhood of A and $f = 1$ on a neighborhood of B then $f \in \mathcal{H}(\hat{K}_{P(\mathbb{C}^n)})$ and an application of the Oka-Weil Theorem 24.12 yields a polynomial $P \in P(\mathbb{C}^n)$ such that $|P - f| < ½$ on $\hat{K}_{P(\mathbb{C}^n)}$. Since $K \subset A$ we conclude that $|P(x)| < ½$ for every $x \in K$ whereas $|P(b)| > ½$ for every $b \in B$. This is impossible, for $B \subset \hat{K}_{P(\mathbb{C}^n)}$.

25.4. THEOREM. *Let U be a polynomially convex open subset of $\mathbb{C}^n$. Then for each $f \in \mathcal{H}(U;F)$ there is a sequence (P_j) in $P(\mathbb{C}^n;F)$ which converges to f uniformly on each compact subset of U.*

PROOF. By Proposition 25.3 the open set U is strongly polynomially convex. Hence we can find a sequence (K_j) of polynomially convex compact sets such that $U = \bigcup_{j=1}^{\infty} K_j$ and $K_j \subset \overset{\circ}{K}_{j+1}$ for each j. By the Oka-Weil Theorem 24.12 for each j we can find $P_j \in P(\mathbb{C}^n;F)$ such that $\| P_j - f \| \leq 1/j$ on K_j. Whence it follows that (P_j) converges to f uniformly on each compact subset of U.

25.5. THEOREM. *Let U be a polynomially convex open subset of*

$\mathbb{C}^n$. *Then for each* $g \in C^{\infty}_{p,q+1}(U;F)$ *such that* $\bar{\partial} g = 0$ *there exists* $f \in C^{\infty}_{pq}(U;F)$ *such that* $\bar{\partial} f = g$.

After writting U as the union of an increasing sequence of polynomially convex compact sets, the proof of Theorem 25.5 is almost a repetition of the proof of Theorem 23.3, but using Theorem 24.10 instead of Theorem 23.1. The details are left to the reader as an exercise.

We also leave as an exercise the analogue of Theorem 23.4 for polynomially convex domains in $\mathbb{C}^n$.

EXERCISES

25.A. Let V be an open subset of F, let $T \in \mathcal{L}(E;F)$ and let $U = T^{-1}(V)$. Show that if V is (strongly) polynomially convex then U is (strongly) polynomially convex as well.

25.B. Show that if U_i is a (strongly) polynomially convex open subset of a Banach space E_i for $i = 1,2$ then $U_1 \times U_2$ is a (strongly) polynomially convex open subset of $E_1 \times E_2$.

25.C. Show that if U is the union of an increasing sequence of strongly polynomially convex open subsets of E then U is strongly polynomially convex as well.

25.D. Show that if U is a strongly polynomially convex open subset of E then $d_U(\hat{K}_{P(E)}) = d_U(K)$ for each compact set $K \subset U$.

25.E. Show that if U_i is a strongly polynomially convex open subset of E for each $i \in I$ then the open set $U = \operatorname{int} \bigcap_{i \in I} U_i$ is strongly polynomially convex as well.

25.F. Show that an open set U in $\mathbb{C}^n$ is polynomially convex if and only if U is holomorphically convex and $P(\mathbb{C}^n)$ is dense in $(\mathcal{H}(U), \tau_c)$.

25.G. Show that if U is a polynomially convex open subset of

$\mathbb{C}^n$ then the union of each collection of components of U is polynomially convex as well.

26. SCHAUDER BASES

The most important results from Sections 24 and 25 were restricted to $\mathbb{C}^n$ and the corresponding proofs do not seem to apply in the case of arbitrary Banach spaces. This section is devoted to the study of Banach spaces with a Schauder basis, and in Section 28 we shall extend some of the results from Sections 24 and 25 to the case of Banach spaces with a Schauder basis.

26.1. DEFINITION. A sequence $(e_n)_{n=1}^{\infty}$ in E is said to be a *Schauder basis* if each $x \in E$ has a unique series representation of the form $x = \sum_{n=1}^{\infty} \xi_n(x)e_n$, where $\xi_n(x) \in \mathbb{K}$ for every n. A Schauder basis (e_n) is said to be *normalized* if $\|e_n\| = 1$ for every n. A Schauder basis (e_n) is said to be *monotone* if $\|\sum_{j=1}^{n} \xi_j(x)e_j\| \leq \|x\|$ for all $x \in E$ and $n \in \mathbb{N}$.

If E has a Schauder basis (e_n) then the coordinate functionals $\xi_n : x \in E \to \xi_n(x) \in \mathbb{K}$ are clearly linear, and so are the mappings $T_n : x \in E \to \sum_{j=1}^{n} \xi_j(x)e_j \in E$. If E_n denotes the subspace generated by $e_1,\ldots,e_n$ then T_n is a projection from E onto E_n.

If (e_n) is a monotone Schauder basis then $\|T_n x\| \leq \|x\|$ and $|\xi_n(x)|\,\|e_n\| \leq 2\|x\|$ for all $x \in E$ and $n \in \mathbb{N}$. The following theorem tells us that every Schauder basis is monotone with respect to some equivalent norm. In particular the mappings ξ_n and T_n are always continuous.

26.2. THEOREM. *Let E be a Banach space with a Schauder basis. Then E is canonically topologically isomorphic to a Banach space* *of sequences whose unit vectors form a monotone Schauder basis.*

PROOF. Let $(e_n)_{n=1}^{\infty}$ be a Schauder basis for E and let F be the vector space of all scalar sequences $y = (\eta_n)_{n=1}^{\infty}$ such that the series $\sum_{n=1}^{\infty} \eta_n e_n$ converges in E. We claim that F is a Banach space with respect to the norm $\| y \| = \sup_n \| \sum_{j=1}^{n} \eta_j e_j \|$. Indeed, let $(y^p)_{p=1}^{\infty}$ be a Cauchy sequence in F and set $y^p = (\eta_n^p)_{n=1}^{\infty}$ for each p. One can readily see that

$$|\eta_n^p - \eta_n^q| \, \|e_n\| \leq 2 \, \| y^p - y^q \|$$

and hence $(\eta_n^p)_{p=1}^{\infty}$ is a Cauchy sequence in $\mathbb{K}$ for each n. Set $\eta_n = \lim_{p \to \infty} \eta_n^p$ for each n. We shall prove that the sequence $y = (\eta_n)_{n=1}^{\infty}$ belongs to F and that (y^p) converges to y in F. Indeed, since (y^p) is a Cauchy sequence in F, given $\varepsilon > 0$ we can find p_o such that

$$\| y^p - y^q \| = \sup_n \| \sum_{j=1}^{n} (\eta_j^p - \eta_j^q) e_j \| \leq \varepsilon$$

for all $p, q \geq p_o$. Hence it follows that

$$\| \sum_{j=1}^{n} (\eta_j^p - \eta_j) e_j \| \leq \varepsilon \tag{26.1}$$

for all $p \geq p_o$ and all $n \in \mathbb{N}$. Fix $p \geq p_o$ and choose n_o such that

$$\| \sum_{j=1}^{m} \eta_j^p e_j - \sum_{j=1}^{n} \eta_j^p e_j \| \leq \varepsilon \tag{26.2}$$

for all $m, n > n_o$. Then it follows from (26.1) and (26.2) that

$$\| \sum_{j=1}^{m} \eta_j e_j - \sum_{j=1}^{n} \eta_j e_j \| \leq \| \sum_{j=1}^{m} (\eta_j - \eta_j^p) e_j \|$$
$$+ \| \sum_{j=1}^{m} \eta_j^p e_j - \sum_{j=1}^{n} \eta_j^p e_j \| + \| \sum_{j=1}^{n} (\eta_j^p - \eta_j) e_j \| \leq 3\varepsilon$$

for all $m, n \geq n_o$. Hence the series $\sum_{n=1}^{\infty} \eta_n e_n$ converges in E and therefore $y = (\eta_n)$ belongs to F. Then it follows from (26.1) that (y^p) converges to y, and thus F is complete, as asserted.

Next we show that the unit vectors form a monotone Schauder basis in F. Indeed, let f_n denote the nth unit vector in F, and let $y = (\eta_n) \in F$. Then

$$\| y - \sum_{j=1}^{n} \eta_j f_j \| = \sup_{m > n} \| \sum_{n+1}^{m} \eta_j e_j \|$$

and the last written expression tends to zero when $n \to \infty$. This shows that $y = \sum_{j=1}^{\infty} \eta_j f_j$. To show that this series representation is unique it suffices to show that if $\sum_{j=1}^{\infty} \zeta_j f_j = 0$ then $\zeta_j = 0$ for every j. Now, if $\sum_{j=1}^{\infty} \zeta_j f_j = 0$ then for each $\varepsilon > 0$ there exists n_o such that $\| \sum_{j=1}^{n} \zeta_j f_j \| \leq \varepsilon$ for every $n \geq n_o$. Hence it follows that $\| \sum_{j=1}^{m} \zeta_j e_j \| \leq \varepsilon$ for every $m \in \mathbb{N}$ and therefore $|\zeta_m| \, \|e_m\| \leq 2\varepsilon$ for every $m \in \mathbb{N}$. Since $\varepsilon > 0$ was arbitrary we conclude that $\zeta_m = 0$ for every $m \in \mathbb{N}$, as we wanted. Thus (f_n) is a Schauder basis for F, and since

$$\| \sum_{j=1}^{n} \eta_j f_j \| = \sup_{m \leq n} \| \sum_{j=1}^{m} \eta_j e_j \| \leq \| y \|$$

for all $y = (\eta_j) \in F$ and $n \in \mathbb{N}$, we conclude that (f_n) is a monotone Schauder basis, as asserted.

If $(\xi_n)_{n=1}^{\infty}$ is the sequence of coordinate functionals associated with the basis $(e_n)_{n=1}^{\infty}$ then we define $A : E \to F$ by $Ax = (\xi_n(x))_{n=1}^{\infty}$ for every $x \in E$. Clearly A is a vector space isomorphism from E onto F and

$$\|Ax\| = \sup_{n} \| \sum_{j=1}^{n} \xi_j(x) e_j \| \geq \| x \|$$

for every $x \in E$. By the Banach Open Mapping Theorem there is a constant $c \geq 1$ such that $\|Ax\| \leq c \|x\|$ for every $x \in E$. The proof of the theorem is now complete.

26.3. COROLLARY. *Let E be a Banach space with a Schauder basis, and let (T_n) denote the sequence of canonical projections. Then:*

(a) *There is a constant $c \geq 1$ such that $\|T_n x\| \leq c \|x\|$ for all $x \in E$ and $n \in \mathbb{N}$.*

(b) *The sequence (T_n) converges to the identity uniformly on each compact subset of E.*

We remark that we did not give the shortest possible proof of Theorem 26.2, but the proof given shows more than has been stated, and this will be exploited in the next section.

Next we give a useful criterion to recognize a Schauder basis.

26.4. THEOREM. *A sequence (e_n) of nonzero vectors in E is a Schauder basis if and only if the following two conditions hold.*

(a) *The closed vector subspace generated by (e_n) is all of E.*

(b) *There is a constant $c \geq 1$ such that for each scalar sequence (λ_j) and positive integers $m < n$ we have that*

$$\|\sum_{j=1}^{m} \lambda_j e_j\| \leq c \|\sum_{j=1}^{n} \lambda_j e_j\|.$$

PROOF. If (e_n) is a Schauder basis then (a) is obvious and (b) follows from Corollary 26.3. Conversely assume that (a) and (b) hold. Let M be the vector subspace of all $x \in E$ which admit a series representation of the form $x = \sum_{j=1}^{\infty} \lambda_j e_j$. We shall prove that M is closed in E. Indeed, let (x^p) be a sequence in M which converges to some $x \in E$. Write $x^p = \sum_{j=1}^{\infty} \lambda_j^p e_j$ for each $p \in \mathbb{N}$. Given $\varepsilon > 0$ we can find $p_o \in \mathbb{N}$ such that

$\|x^p - x^q\| \leq \varepsilon$ for all $p, q \geq p_o$. We claim that

$$\|\sum_{j=1}^{m} (\lambda_j^p - \lambda_j^q)e_j\| \leq 2c\varepsilon \tag{26.3}$$

for all $p, q \geq p_o$ and all $m \in \mathbb{N}$. Indeed, if we fix $p, q \geq p_o$ and choose $n_o \in \mathbb{N}$ such that

$$\|\sum_{j=1}^{n} (\lambda_j^p - \lambda_j^q)e_j - (x^p - x^q)\| \leq \varepsilon$$

for every $n \geq n_o$ then $\|\sum_{j=1}^{n} (\lambda_j^p - \lambda_j^q)e_j\| \leq 2\varepsilon$ for every $n \geq n_o$ and (26.3) follows from condition (b). It follows from (26.3) that $|\lambda_m^p - \lambda_m^q| \|e_m\| \leq 4c\varepsilon$ for all $p, q \geq p_o$ and all $m \in \mathbb{N}$. Thus the limit $\lambda_m = \lim_{p \to \infty} \lambda_m^p$ exists for each $m \in \mathbb{N}$ and it follows from (26.3) that

$$\|\sum_{j=1}^{m} (\lambda_j^p - \lambda_j)e_j\| \leq 2c\varepsilon \tag{26.4}$$

for all $p \geq p_o$ and $m \in \mathbb{N}$. Since

$$\lim_{m,n\to\infty} \|\sum_{j=1}^{m} \lambda_j^p e_j - \sum_{j=1}^{n} \lambda_j^p e_j\| = 0$$

for every p, we can easily get that

$$\lim_{m,n\to\infty} \|\sum_{j=1}^{m} \lambda_j e_j - \sum_{j=1}^{n} \lambda_j e_j\| = 0$$

and hence the series $\sum_{j=1}^{\infty} \lambda_j e_j$ converges. Then it follows from (26.4) that $\|x^p - \sum_{j=1}^{\infty} \lambda_j e_j\| \leq 2c\varepsilon$ for all $p \geq p_o$, proving that (x^p) converges to $\sum_{j=1}^{\infty} \lambda_j e_j$. Hence $x = \sum_{j=1}^{\infty} \lambda_j e_j$ and M is closed in E, as asserted. In view of condition (a) we conclude that $M = E$ and thus each $x \in E$ can be represented in the form $x = \sum_{j=1}^{\infty} \lambda_j e_j$.

To show that this series representation is unique it suffices to prove that if $\sum_{j=1}^{\infty} \lambda_j e_j = 0$ then $\lambda_j = 0$ for every j. Now, if $\sum_{j=1}^{\infty} \lambda_j e_j = 0$ then for each $\varepsilon > 0$ there exists n_o such that $\| \sum_{j=1}^{n} \lambda_j e_j \| \leq \varepsilon$ for every $n \geq n_o$. Then it follows from condition (b) that $\| \sum_{j=1}^{m} \lambda_j e_j \| \leq c\,\varepsilon$ for every $m \in \mathbb{N}$, and therefore $|\lambda_m| \, \|e_m\| \leq 2c\varepsilon$ for every $m \in \mathbb{N}$. Since $\varepsilon > 0$ was arbitrary we conclude that $\lambda_m = 0$ for every m and the proof of the theorem is complete.

26.5. **EXAMPLES.** The unit vectors form a monotone normalized Schauder basis in each of the spaces c_o and ℓ^p, where $1 \leq p < \infty$. In the exercises at the end of this section we give examples of Schauder bases for the spaces c, $C[0,1]$ and $L^p[0,1]$, where $1 \leq p < \infty$. It is clear that ℓ^∞ cannot have a Schauder basis, since each Banach space with a Schauder basis must be separable.

S. Banach [1] posed the problem whether every separable Banach space has a Schauder basis. This problem, known as the *basis problem*, remained open for a long time and was solved in the negative by P. Enflo [1].

EXERCISES

26.A. Let E be a Banach space with a Schauder basis, and let M be a finite dimensional subspace of E. Show that each basis for M is part of a Schauder basis for E.

26.B. (a) Show that each sequence $(\gamma_n) \in c$ can be uniquely written as a sum $(\gamma_n) = (\alpha_n) + (\beta_n)$, where (α_n) is a constant sequence and (β_n) is a sequence belonging to c_o.

(b) Use (a) to find a Schauder basis for c.

26.C. The *Haar system* $(\varphi_n)_{n=1}^{\infty} \subset L^p[0,1]$ is defined by $\varphi_n(x) \equiv 1$ and

$$\varphi_{2^j+k}(x) = \begin{cases} 1 & \text{if } x \in [(2k-2)2^{-j-1}, (2k-1)2^{-j-1}] \\ -1 & \text{if } x \in ((2k-1)2^{-j-1}, (2k)2^{-j-1}] \\ 0 & \text{otherwise} \end{cases}$$

for $j \geq 0$ and $1 \leq k \leq 2^j$.

(a) Show that the vector subspace generated by the Haar system contains the characteristic functions of all the intervals of the form $[(k-1)2^{-j}, k2^{-j}]$. Conclude that the Haar system satisfies the first condition in Theorem 26.4.

(b) Show that the Haar system satisfies the second condition in Theorem 26.4 with $c = 1$. Conclude that the Haar system is a Schauder basis for $L^p[0,1]$ whenever $1 \leq p < \infty$.

26.D. If $(\varphi_n)_{n=1}^{\infty}$ is the Haar system then the *Schauder system* $(\psi_n)_{n=1}^{\infty} \subset C[0,1]$ is defined by $\psi_1(x) \equiv 1$ and $\psi_n(x) = \int_0^x \varphi_{n-1}(t)dt$ for every $n \geq 2$.

(a) Show that the vector subspace generated by the Schauder system contains all the continuous piecewise linear functions on $[0,1]$ with vertices at the points of the form $k2^{-j}$. Conclude that the Schauder system satisfies the first condition in Theorem 26.4.

(b) Show that the Schauder system satisfies the second condition in Theorem 26.4 with $c = 1$. Conclude that the Schauder system is a Schauder basis for $C[0,1]$.

27. THE APPROXIMATION PROPERTY

This section is devoted to the study of Banach spaces with the approximation property or with the bounded approximation property. In the next section we shall extend some of the results from Sections 24 and 25 to the case of Banach spaces with the approximation property or with the bounded approximation property.

27.1. DEFINITION. E is said to have the *approximation property* if for each compact set $K \subset E$ and $\varepsilon > 0$ there is a finite rank operator $T \in \mathcal{L}(E;E)$ such that $\| Tx - x \| \leq \varepsilon$ for every $x \in K$.

We recall that an operator $T \in \mathcal{L}(E;F)$ is said to have *finite rank* if its image $T(E)$ is finite dimensional. One can readily see that an operator $T \in \mathcal{L}(E;F)$ has finite rank if and only if there exists $\varphi_1,\ldots,\varphi_n \in E'$ and $b_1,\ldots,b_n \in F$ such that $Tx = \sum_{j=1}^{n} \varphi_j(x) b_j$ for every $x \in E$.

27.2. PROPOSITION. *For a Banach space E the following conditions are equivalent:*

(a) *E has the approximation property.*

(b) *Each $T \in \mathcal{L}(E;E)$ can be uniformly approximated on compact sets by operators of finite rank.*

(c) *For each Banach space F each $T \in \mathcal{L}(E;F)$ can be uniformly approximated on compact sets by operators of finite rank.*

(d) *For each Banach space F each $T \in \mathcal{L}(F;E)$ can be uniformly approximated on compact sets by operators of finite rank.*

PROOF. (a) ⇒ (c): Let $T \in \mathcal{L}(E;F)$, $T \neq 0$, let K be a compact subset of E and let $\varepsilon > 0$. By (a) there is a finite rank operator $S \in \mathcal{L}(E;E)$ such that $\| Sx - x \| \leq \varepsilon / \| T \|$ for every $x \in K$. Hence $\| T \circ Sx - Tx \| \leq \varepsilon$ for every $x \in K$. Since $T \circ S$ has finite rank, (c) has been proved.

(a) ⇒ (d): Let $T \in \mathcal{L}(F;E)$, let K be a compact subset of F, and let $\varepsilon > 0$. By (a) there is a finite rank operator $S \in \mathcal{L}(E;E)$ such that $\| Sx - x \| \leq \varepsilon$ for every $x \in T(K)$. Hence

$\| S \circ Ty - Ty \| \leq \varepsilon$ for every $y \in K$. Since $S \circ T$ has finite rank, (d) has been proved.

Since the implications (c) $\Rightarrow$ (b), (d) $\Rightarrow$ (b) and (b) $\Rightarrow$ (a) are obvious, the proof of the proposition is complete.

Next we introduce two stronger versions of the approximation property.

27.3. DEFINITION. (a) E is said to have the *metric approximation property* if for each compact set $K \subset E$ and $\varepsilon > 0$ there is a finite rank operator $T \in \mathcal{L}(E;E)$ such that $\| T \| \leq 1$ and $\| Tx - x \| \leq \varepsilon$ for every $x \in K$.

(b) E is said to have the *bounded approximation property* if there exists a constant $c \geq 1$ such that for each compact set $K \subset E$ and $\varepsilon > 0$ there is a finite rank operator $T \in \mathcal{L}(E;E)$ such that $\| T \| \leq c$ and $\| Tx - x \| < \varepsilon$ for every $x \in K$.

27.4. THEOREM. *For a separable Banach space E the following conditions are equivalent.*

(a) *E has the bounded approximation property.*

(b) *There is a sequence of finite rank operators $T_n \in \mathcal{L}(E;E)$ such that $\lim T_n x = x$ for every $x \in E$.*

(c) *There are a sequence (a_n) in E and a sequence (φ_n) in E' such that $\|a_n\| = 1$ for every $n \in \mathbb{N}$ and $\sum_{n=1}^{\infty} \varphi_n(x) a_n = x$ for every $x \in E$.*

(d) *E is topologically isomorphic to a complemented subspace of a Banach space with a monotone Schauder basis.*

Before proving this theorem we give two preparatory results.

27.5. PROPOSITION. *Let E be a Banach space of dimension n. Then there are n vectors $e_1, \ldots, e_n$ of norm one in E and n vectors $\xi_1, \ldots, \xi_n$ for norm one in E' such that $\xi_j(e_k) = \delta_{jk}$*

for all $j, k = 1,\ldots,n$.

PROOF. Fix a basis in E and let $A \in \mathcal{L}(^nE)$ by defined by $A(x_1,\ldots,x_n) = \det(a_{jk})$ where $a_{1k},\ldots,a_{nk}$ are the coordinates of x_k with respect to the fixed basis. Let $(e_1,\ldots,e_n)$ be an n-tuple of vectors of norm one in E where A attains its norm. If we define $\xi_1,\ldots,\xi_n \in E'$ by

$$\xi_j(x) = \frac{A(e_1,\ldots,e_{j-1}, x, e_{j+1},\ldots,e_n}{A(e_1,\ldots,e_n)}$$

then the n-tuples $(e_1,\ldots,e_n)$ and $(\xi_1,\ldots,\xi_n)$ have the required properties.

27.6. LEMMA. *Let E be a Banach space of dimension n. Then there are operators $B_1,\ldots,B_{n^2} \in \mathcal{L}(E;E)$, of rank one, such that $\sum_{j=1}^{n^2} B_j(x) = x$ for every $x \in E$ and $\| \sum_{j=1}^{k} B_j \| \leq 2$ for every $k = 1,\ldots,n^2$.*

PROOF. By Proposition 27.5 there are operators $A_1,\ldots,A_n \in \mathcal{L}(E;E)$, of rank one, such that $\sum_{j=1}^{n} A_j x = x$ for every $x \in E$ and $\|A_j\| = 1$ for every $j = 1,\ldots,n$. Indeed, if $(e_1,\ldots,e_n)$ and $(\xi_1,\ldots,\xi_n)$ are the n-tuples given by Proposition 27.5 then it suffices to define $A_j x = \xi_j(x)e_j$ for every $x \in E$ and $j = 1,\ldots,n$.

Then we define $B_j = n^{-1}A_s$ if $j = rn + s$, where $0 \leq r \leq n - 1$ and $1 \leq s \leq n$. Then

$$\sum_{j=1}^{rn+s} B_j x = \sum_{j=1}^{rn} B_j x + \sum_{rn+1}^{rn+s} B_j x$$

$$= \frac{r}{n} \sum_{t=1}^{n} A_t x + \frac{1}{n} \sum_{t=1}^{s} A_t x$$

$$= \frac{r}{n} x + \frac{1}{n} \sum_{t=1}^{s} A_t x$$

and if follows that $\sum_{j=1}^{n^2} B_j x = x$ and $\| \sum_{j=1}^{rn+s} B_j x \| \leq 2 \| x \|$, as we wanted.

PROOF OF THEOREM 27.4. (a) ⇒ (b): Let E be a separable Banach space with the bounded approximation property. Let (x_j) be a sequence which is dense in E. Then there are a constant $c \geq 1$ and a sequence of finite rank operators $T_n \in \mathcal{L}(E;E)$ such that $\|T_n\| \leq c$ and $\|T_n x_j - x_j\| \leq 1/n$ for $j = 1,\ldots,n$. Let $x \in E$ and $\varepsilon > 0$ be given. First choose j such that $\|x_j - x\| \leq \varepsilon$ and next choose $n_o \geq max\,\{j, 1/\varepsilon\}$. Then for $n \geq n_o$ we have that

$$\|T_n x - x\| \leq \|T_n x - T_n x_j\| + \|T_n x_j - x_j\| + \|x_j - x\| \leq (c+2)\varepsilon$$

and hence $\|T_n x - x\|$ tends to zero.

(b) ⇒ (c): Suppose there is a sequence of finite rank operators $T_n \in \mathcal{L}(E;E)$ such that $lim\, T_n x = x$ for every $x \in E$. If we define $S_1 = T_1$ and $S_n = T_n - T_{n-1}$ for $n \geq 2$ then each S_n is a finite rank operator and $\sum_{n=1}^{\infty} S_n x = x$ for every $x \in E$. Set $M_n = S_n(E)$ and $m_n = (dim\, M_n)^2$ for every $n \in \mathbb{N}$. By Lemma 27.6 for each $n \in \mathbb{N}$ there are operators $B_{n1},\ldots,B_{nm_n} \in \mathcal{L}(M_n;M_n)$, of rank one, such that $\sum_{j=1}^{m_n} B_{nj} x = x$ for every $x \in M_n$ and $\| \sum_{j=1}^{k} B_{nj} \| \leq 2$ for every $k = 1,\ldots,m_n$. Set $m_o = 0$ and define $A_k = B_{ns} \circ S_n$ if $k = m_o + \ldots + m_{n-1} + s$, where $n \in \mathbb{N}$ and $1 \leq s \leq m_n$. Then

$$\sum_{j=1}^{k} A_j = \sum_{r=1}^{n-1} \sum_{t=1}^{m_r} B_{rt} \circ S_r + \sum_{t=1}^{s} B_{rt} \circ S_n = \sum_{r=1}^{n-1} S_r + \left(\sum_{t=1}^{s} B_{rt}\right) \circ S_n .$$

Whence

$$\| \sum_{j=1}^{k} A_j x - x \| \leq \| \sum_{r=1}^{n-1} S_r x - x \| + 2 \| S_n x \|$$

and the last written expression tends to zero when $n \to \infty$. Thus we have found a sequence of operators $A_n \in \mathcal{L}(E;E)$, of rank one, such that $\sum_{n=1}^{\infty} A_n x = x$ for every $x \in E$. Then for each $n \in \mathbb{N}$ we can find $a_n \in A_n(E)$ and $\varphi_n \in E'$ such that $\|a_n\| = 1$ and $A_n x = \varphi_n(x) a_n$ for every $x \in E$. Thus (c) has been proved.

(c) $\Rightarrow$ (d): Suppose there are a sequence (a_n) in E and a sequence (φ_n) in E' such that $\|a_n\| = 1$ for every $n \in \mathbb{N}$ and $\sum_{n=1}^{\infty} \varphi_n(x) a_n = x$ for every $x \in E$. Let F be the vector space of all scalar sequences $y = (\eta_n)$ such that the series $\sum_{n=1}^{\infty} \eta_n a_n$ converges in E, and endow F with the norm $\|y\| = \sup_n \|\sum_{j=1}^{n} \eta_j a_j\|$. Then the proof of Theorem 26.2 shows that F is a Banach space whose unit vectors form a monotone Schauder basis. Define $A: E \to F$ by $Ax = (\varphi_n(x))$. Since $\sum_{n=1}^{\infty} \varphi_n(x) a_n = x$ for every $x \in E$ the Principle of Uniform Boundedness guarantees the existence of a constant $C > 0$ such that $\|\sum_{j=1}^{n} \varphi_j(x) a_j\| \leq C \|x\|$ for all $x \in E$ and $n \in \mathbb{N}$. Hence $\|Ax\| \leq C\|x\|$ for every $x \in E$ and $A \in \mathcal{L}(E;F)$. Next define $B: F \to E$ by $By = \sum_{n=1}^{\infty} \eta_n a_n$. Then $\|By\| \leq \sup_n \|\sum_{j=1}^{n} \eta_j a_j\| = \|y\|$ for every $y \in F$ and therefore $B \in \mathcal{L}(F;E)$. It is also clear that $BAx = x$ for every $x \in E$ and $(AB)(AB)y = ABy$ for every $y \in F$. Hence A is a topological isomorphism between E and $A(E)$ and AB is a continuous projection from F onto $A(E)$. Thus (d) has been proved.

(d) $\Rightarrow$ (a): Since every Banach space with a Schauder basis has the bounded approximation property, the desired conclusion follows from Exercise 27.A. The proof of the theorem is now complete.

Clearly every Banach space with a Schauder basis has the

bounded approximation property, and every Banach space with a monotone Schauder basis has the metric approximation property. Next we give more interesting examples.

27.7. EXAMPLE. If X is a compact Hausdorff space then $C(X)$ has the metric approximation property.

PROOF. Let K be a compact subset of $C(X)$ and let $\varepsilon > 0$. Since K is compact we can find functions $f_1, \dots, f_m \in K$ such that $K \subset \bigcup_{j=1}^{m} B(f_j;\varepsilon)$. For each $a \in X$ consider the open set

$$U_a = \{x \in X : |f_j(x) - f_j(a)| \leq \varepsilon \quad \text{for} \quad j = 1,\dots,m\}.$$

Since X is compact we can find points $a_1,\dots,a_n \in X$ such that $X = \bigcup_{k=1}^{n} U_{a_k}$. By Exercise 15.G we can find nonnegative functions $\varphi_1,\dots,\varphi_n \in C(X)$ such that $supp\ \varphi_k \subset U_{a_k}$ for $k = 1,\dots,n$ and $\varphi_1(x) + \dots + \varphi_n(x) = 1$ for every $x \in X$. For each $f \in C(X)$ define $Tf \in C(X)$ by $(Tf)(x) = \sum_{k=1}^{n} f(a_k)\varphi_k(x)$ for every $x \in X$. Then clearly T is a finite rank operator and $\|Tf\| \leq \|f\|$ for every $f \in C(X)$. It is also clear that

$$|(Tf_j)(x) - f_j(x)| \leq \sum_{k=1}^{n} |f_j(a_k) - f_j(x)|\ \varphi_k(x) \leq \varepsilon$$

for every $x \in X$ and $j = 1,\dots,n$. Since each $f \in K$ lies in $B(f_j;\varepsilon)$ for some $j = 1,\dots,m$ we conclude that

$$\|Tf - f\| \leq \|Tf - Tf_j\| + \|Tf_j - f_j\| + \|f_j - f\| \leq 3\varepsilon$$

for every $f \in K$.

27.8. EXAMPLE. If X is a completely regular Hausdorff space then the Banach space $C_b(X)$ of all bounded continuous functions on X has the metric approximation property.

PROOF. If βX denotes the Stone-Cech compactification of X

then each $f \in C_b(X)$ has a unique extension $\beta f \in C(\beta X)$. Clearly the mapping $f \in C_b(X) \to \beta f \in C(\beta X)$ is a vector space isomorphism and an isometry. Then the desired conclusion follows from the preceding example.

27.9. EXAMPLE. ℓ^∞ has the metric approximation property.

PROOF. ℓ^∞ can be identified with $C_b(\mathbb{N})$ if $\mathbb{N}$ is endowed with the discrete topology.

A. Grothendieck [1] posed the problem whether every Banach space has the approximation property. This problem, known as the *approximation problem*, was solved in the negative by P. Enflo [1].

EXERCISES

27.A. Show that if E has the (bounded) approximation property then each complemented subspace of E has the (bounded) approximation property as well.

27.B. Show that if each closed separable subspace of E has the approximation property then E has the approximation property as well.

27.C. Let (X, Σ, μ) be a measure space and let $f_1, \ldots, f_m$ be m simple functions such that $\mu\{x \in X : f_j(X) \neq 0\} < \infty$ for $j = 1,\ldots,m$.

(a) Show the existence of n disjoint sets $A_1,\ldots,A_n \in \Sigma$ such that $0 < \mu(A_k) < \infty$ for each k, each f_j is constant on each A_k and each f_j is identically zero on $X \setminus \bigcup_{k=1}^{n} A_k$.

(b) For each p with $1 \leq p < \infty$ and each $f \in L^p(X, \Sigma, \mu)$ define a simple function $Tf \in L^p(X, \Sigma, \mu)$ by

$$Tf = \sum_{k=1}^{n} \left(\int_{A_k} f d\mu/\mu(A_k)\right)\chi_{A_k}.$$

Show that T is a finite rank linear operator on $L^p(X, \Sigma, \mu)$ with $\|T\| \leq 1$. Furthermore, show that $Tf_j = f_j$ on A_k for every $j = 1,\ldots,m$ and $k = 1,\ldots,n$.

(c) Show that $L^p(X, \Sigma, \mu)$ has the metic approximation property for each p with $1 \leq p < \infty$.

28. POLYNOMIAL APPROXIMATION IN BANACH SPACES

In this section we present Banach space versions of the results on polynomial approximation established in Sections 24 and 25. The letters E and F will always represent complex Banach spaces.

Let us recall that $\mathcal{P}_f(E;F)$ denotes the vector space of all continuous polynomials of finite type from E into F. This space was introduced in Exercise 2.K. With this notation we have the following result, which extends Theorem 25.4.

28.1. THEOREM. *Let E be a Banach space with the approximation property and let U be a polynomially convex open subset of E. Then for each $f \in \mathcal{H}(U;F)$ and each compact set $K \subset U$ there is a sequence of polynomials $P_j \in \mathcal{P}_f(E;F)$ which converges to f uniformly on K.*

PROOF. Let $\varepsilon > 0$ be given. Since f is continuous and K is compact we can find $\delta > 0$ such that $K + B(0;\delta) \subset U$ and $\|f(y) - f(x)\| < \varepsilon$ for all $x \in K$ and $y \in B(x;\delta)$. Since E has the approximation property there is a finite rank operator $T \in \mathcal{L}(E;E)$ such that $\|Tx - x\| < \delta$ for every $x \in K$. Whence it follows that $T(K) \subset U$ and $\|f \circ T(x) - f(x)\| < \varepsilon$ for every $x \in K$. By Example 25.2 the intersection $U \cap T(E)$ is a polynomially convex open set in $T(E)$. Since $T(E)$ is finite dimensional, Theorem 25.4 guarantees the existence of a polynomial $P \in \mathcal{P}(T(E);F)$, such that $\|P(y) - f(y)\| \leq \varepsilon$ for every $y \in T(K)$. Since $T(E)$ is finite dimensional, Exercise 2.K guarantees that P is of finite type, that is, P is of the form $P = \sum_j c_j \varphi_j^{m_j}$ where $c_j \in F$ and $\varphi_j \in (T(E))'$. Whence if follows that

$P \circ T = \sum_j c_j (\varphi_j \circ T)^{m_j} \in P_f(E;F)$. Furthermore, we have that

$$\|P \circ T(x) - f(x)\| \leq \|P \circ T(x) - f \circ T(x)\| + \|f \circ T(x) - f(x)\| \leq 2\varepsilon$$

for every $x \in K$, and the proof is complete.

Next we want to extend the Oka-Weil Theorem 24.12 to the case of Banach spaces. The key to the proof is the following theorem.

28.2. THEOREM. *Let K be a polynomially convex compact subset of E and let U be an open neighborhood of K. Then there is a strongly polynomially convex open set V such that $K \subset V \subset U$.*

PROOF. Since the closed, convex hull $\overline{co}(K)$ of K is compact, the proof of Lemma 24.7 shows the existence of polynomials $P_1,\ldots,P_m \in P(E)$ such that $|P_j| < 1$ on K for $j = 1,\ldots,m$ and

$$(28.1) \qquad \{x \in \overline{co}(K) : |P_j(x)| \leq 1 \text{ for } j = 1,\ldots,m\} \subset U.$$

We claim there exists $\delta > 0$ such that

$$(28.2) \quad (\overline{co}(K) + B(0;\delta)) \cap \{x \in E : |P_j(x)| \leq 1 \text{ for } j = 1,\ldots,m\} \subset U.$$

Otherwise there is a sequence (a_n) such that

$$a_n \in (\overline{co}(K) + B(0;1/n)) \cap \{x \in E : |P_j(x)| \leq 1 \text{ for } j = 1,\ldots,m\} \setminus U$$

for every n. Choose for each n an element $b_n \in \overline{co}(K)$ such that $\|a_n - b_n\| < 1/n$. Since $\overline{co}(K)$ is compact the sequence (b_n) has a subsequence (b_{n_k}) which converges to a point $b \in \overline{co}(K)$. But then the corresponding subsequence (a_{n_k}) of (a_n) also converges to b. Whence it follows that

$$b \in \{x \in \overline{co}(K) : |P_j(x)| \leq 1 \quad \text{for} \quad j = 1,\ldots,m\} \setminus U,$$

contradicting (28.1). This shows the existence of $\delta > 0$ satisfying

(28.2). If we define

$$V = (\overline{co}(K) + B(0;\delta)) \cap \{x \in E : |P_j(x)| < 1 \quad \text{for} \quad j = 1,\ldots,m\},$$

then $K \subset V \subset U$ and it follows from Example 25.2 that V is strongly polynomially convex.

Now it is easy to extend the Oka-Weil Theorem 24.12 to the case of Banach spaces.

28.3. THEOREM. *Let E be a Banach space with the approximation property and let K be a polynomially convex compact subset of E. Then for each $f \in \mathcal{H}(K;F)$ there is a sequence of polynomials $P_j \in \mathcal{P}_f(E;F)$ which converges to f uniformly on K.*

PROOF. By Theorem 28.2 we may assume that $f \in \mathcal{H}(U;F)$, where U is a polynomially convex open subset of E containing K. Then it suffices to apply Theorem 28.1.

28.4. PROPOSITION. *Let E be a Banach space with the approximation property. Then an open subset U of E is polynomially convex if and only if U is strongly polynomially convex.*

PROOF. In view of Theorem 28.3 the proof of Proposition 25.3 applies.

In the case of separable Banach spaces with the bounded approximation property the conclusion of Theorem 28.1 can be sharpened as follows.

28.5. THEOREM. *Let E be a separable Banach space with the bounded approximation property and let U be a polynomially convex open subset of E. Then for each $f \in \mathcal{H}(U;F)$ there is a sequence of polynomials $P_j \in \mathcal{P}_f(E;F)$ which converges to f uniformly on each compact subset of U.*

PROOF. (a) Let us assume first that E has a monotone Schauder basis. Let (e_n) denote the basis, let (T_n) denote the sequence of canonical projections and let $E_n = T_n(E)$ for $n = 1,2,\ldots$.

Finally let E_∞ denote the dense subspace $E_\infty = \bigcup_{n=1}^{\infty} E_n$. Since E_∞ is dense in E and since U is a Lindelof space we can easily find a sequence (a_j) in E_∞ and a sequence (r_j) of positive numbers such that

$$U = \bigcup_{j=1}^{\infty} B(a_j;r_j) \quad \text{and} \quad B(a_j;2r_j) \subset U$$

for every j. If we set

$$A_k = \bigcup_{j=1}^{k} B(a_j;r_j) \quad \text{and} \quad s_k = \min\{r_1,\ldots,r_k\}$$

for every k, then (A_k) is an increasing sequence of bounded open sets such that

$$(28.3) \qquad U = \bigcup_{k=1}^{\infty} A_k \quad \text{and} \quad A_k + B(0;s_k) \subset U$$

for every k, let (n_k) be an increasing sequence of positive integers such that $a_j \in E_{n_k}$ for $j = 1,\ldots,k$. Hence it is clear that

$$T_n(B(a_j;r_j)) = B(a_j;r_j) \cap E_n \quad \text{for} \quad j \leq k \quad \text{and} \quad n \geq n_k$$

and therefore

$$(28.4) \qquad T_n(A_k) = A_k \cap E_n \quad \text{for} \quad n \geq n_k .$$

It follows from (28.3) and (28.4) that $T_{n_k}(A_k)$ is a relatively compact subset of $U \cap E_{n_k}$ for every k. Since $U \cap E_{n_k}$ is a polynomially convex open subset of the finite dimensional subspace E_{n_k}, Theorem 25.4 guarantees the existence of a polynomial $P_k \in P(E_{n_k};F)$ such that $\|P_k(y) - f(y)\| \leq 1/k$ for every $y \in T_{n_k}(A_k)$. Thus $P_k \circ T_{n_k} \in P_f(E;F)$ and to complete the proof we show that $(P_k \circ T_{n_k})$ converges to f uniformly on each compact subset of U. Let K be a compact subset of U and let $\varepsilon > 0$. Then first choose $\delta > 0$ such that $K + B(0;\delta) \subset U$ and $\|f(y) - f(x)\| < \varepsilon$

for all $x \in K$ and $y \in B(x;\delta)$. Next choose k_o such that $1/k_o < \varepsilon$, $K \subset A_{k_o}$ and $\|T_{n_k} x - x\| < \delta$ for all $x \in K$ and $k \geq k_o$. Then for every $x \in K$ and $k \geq k_o$ we get that

$$\|P_k \circ T_{n_k}(x) - f(x)\| \leq \|P_k \circ T_{n_k}(x) - f \circ T_{n_k}(x)\| + \|f \circ T_{n_k}(x) - f(x)\| \leq 2\varepsilon.$$

This completes the proof when E has a monotone Schauder basis.

(b) Consider next the general case, in which E is a separable Banach space with the bounded approximation property. By Theorem 27.4 we may assume that E is a complemented subspace of a Banach space G which has a monotone Schauder basis. Let $\pi \in \mathcal{L}(G;E)$ be a continuous projection from G onto E. If U is a polynomially convex open subset of E then $\pi^{-1}(U)$ is a polynomially convex open subset of G by Exercise 25.A. If $f \in \mathcal{H}(U;F)$ then $f \circ \pi \in \mathcal{H}(\pi^{-1}(U);F)$ and by part (a) there is a sequence of polynomials $Q_j \in \mathcal{P}_f(G;F)$ which converges to $f \circ \pi$ uniformly on each compact subset of $\pi^{-1}(U)$. If we set $P_j = Q_j | E$ then $P_j \in \mathcal{P}_f(E;F)$ and (P_j) converges to f uniformly on each compact subset of U. The proof of the theorem is now complete.

By combining Theorems 28.2 and 28.5 we can sharpen the conclusion of Theorem 28.3 as follows.

28.6. THEOREM. *Let E be a separable Banach space with the bounded approximation property and let K be a polynomially convex compact subset of E. Then for each $f \in \mathcal{H}(K;F)$ there are an open set U containing K and a sequence of polynomials $P_j \in \mathcal{P}_f(E;F)$ such that $f \in \mathcal{H}(U;F)$ and (P_j) converges to f uniformly on each compact subset of U.*

To conclude this section we present a first attempt to solve Problem 11.6. In Section 45 we shall present a more satisfactory solution.

28.7. THEOREM. *Let E be a separable Banach space with the bounded approximation property. Then every polynomially convex open set in E is a domain of existence.*

PROOF. (a) Assume first that E has a monotone Schauder basis and let U be a polynomially convex open set in E. In the proof of Theorem 28.5 we found an increasing sequence of bounded open sets A_k such that

$$U = \bigcup_{k=1}^{\infty} A_k \quad \text{and} \quad A_k + B(0;s_k) \subset U$$

for every k. If we can prove that

$$(\hat{A}_k)_{P(E)} + B(0;s_k) \subset U \tag{28.5}$$

for every k then Theorem 11.4 will guarantee that U is a domain of existence. Now, by (28.4) we have that $T_n(A_k) \subset A_k \cap E_n$ for $n \geq n_k$ and this immediately implies that

$$T_n((\hat{A}_k)_{P(E)}) \subset (A_k \cap E_n)\hat{}_{P(E_n)} \quad \text{for} \quad n \geq n_k. \tag{28.6}$$

Fix $\varepsilon > 0$ small. Since $A_k + B(0;s_k) \subset U$ we get that $\bar{A}_k + B(0;s_k - \varepsilon) \subset U$ and therefore

$$\bar{A}_k \cap E_n + B(0;s_k - \varepsilon) \cap E_n \subset U \cap E_n$$

for all k and n. Now, $U \cap E_n$ is a domain of holomorply in E_n, since $U \cap E_n$ is polynomially convex and E_n is finite dimensional. Since $\bar{A}_k \cap E_n$ is a compact set it follows from Theorem 11.4 that

$$(A_k \cap E_n)\hat{}_{\mathcal{H}(U \cap E_n)} + B(0;s_k - \varepsilon) \cap E_n \subset U \cap E_n.$$

Since $P(E_n)$ is dense in $(\mathcal{H}(U \cap E_n), \tau_c)$, by Theorem 25.4, we see that $(A_k \cap E_n)\hat{}_{\mathcal{H}(U \cap E_n)} = (A_k \cap E_n)\hat{}_{P(E_n)}$. Whence we get that

$$(A_k \cap E_n)\hat{}_{P(E_n)} + B(0;s_k - \varepsilon) \cap E_n \subset U$$

and therefore

$$(A_k \cap E_n)\hat{}_{P(E_n)} + B(0;s_k - 2\varepsilon) \cap E_n \subset U_\varepsilon \tag{28.7}$$

where $U_\varepsilon = \{x \in U : d_U(x) \geq \varepsilon\}$. Let $x \in (\hat{A}_k)_{P(E)}$ and $y \in B(0;s_k - 2\varepsilon)$. Then it follows from (28.6) and (28.7) that

$$T_n x + T_n y \in (A_k \cap E_n)\hat{}_{P(E_n)} + B(0;s_k - 2\varepsilon) \cap E_n \subset U_\varepsilon$$

for every $n \geq n_k$. Letting $n \to \infty$ we get that $x + y \in U_\varepsilon \subset U$, proving that

$$(\hat{A}_k)_{P(E)} + B(0;s_k - 2\varepsilon) \subset U.$$

Since $\varepsilon > 0$ was arbitrarily small (28.5) follows. This completes the proof when E has a monotone Schauder basis.

(b) In the general case we may assume that E is a complemented subspace of a Banach space G which has a monotone Schauder basis. Let π be a continuous projection from G onto E and let U be a polynomially convex open subset of E. Then $\pi^{-1}(U)$ is a polynomially convex open subset of G and hence a domain of existence, by part (a). But then $U = \pi^{-1}(U) \cap E$ is a domain of existence in E by Proposition 11.7. The proof of the theorem is now complete.

EXERCISES

28.A. An open subset U of E is said to be *Runge* if $P(E)$ is dense in $(\mathcal{H}(U), \tau_c)$. Show that an open subset U of E is polynomially convex if U is holomorphically convex and Runge.

28.B. An open subset U of E is said to be *finitely polynomially convex* (resp. *finitely Runge*) if $U \cap M$ is polynomially convex (resp. Runge) for each finite dimensional subspace M of E. Consider the following conditions on an open subset U of E.

(a) U is polynomially convex.

(b) U is finitely polynomially convex.

(c) U is finitely Runge.

(d) U is Runge.

Show that the implications (a) $\Rightarrow$ (b) and (b) $\Rightarrow$ (c) are always true. Show that (c) $\Rightarrow$ (d) if E has the approximation property. Finally show that if E has the approximation property and U is a holomorphically convex open subset of E then the conditions (a), (b), (c) and (d) are equivalent.

28.C. Let E be a Banach space with the approximation property and let U be a (finitely) polynomially convex open subset of E. Show that the union of each collection of connected components of U is (finitely) polynomially convex as well.

28.D. Let E be a Banach space with the approximation property and let K be a compact subset of E. Show that each connected component of $\hat{K}_{P(E)}$ contains points of K.

28.E. Let E be a Banach space with the approximation property and let K be a polynomially convex compact subset of E. Show that the union of each collection of connected components of K is also polynomially convex.

28.F. Let E be a separable Banach space with the bounded approximation property. Show that an open subset U of E is polynomially convex if and only if U is finitely polynomially convex.

28.G. Show that if U is a balanced polynomially convex open set in E then $\hat{K}_{P(E)} \subset U$ for each compact set $K \subset U$.

28.H. Let E be a Banach space with the approximation property, let U be a polynomially convex open subset of E and let $h : (\mathcal{H}(U), \tau_c) \to \mathbb{C}$ be a continuous algebra homomorphism.

(a) Show the existence of a compact set $K \subset U$ such that $|h(f)| \leq \sup_K |f|$ for every $f \in \mathcal{H}(U)$.

(b) Using the Mackey-Arens Theorem show the existence of a unique point $a \in E$ such that $h(P) = P(a)$ for each $P \in P_f(E)$.

(c) Using Theorem 28.1 twice show that $a \in U$ and $h(f) = f(a)$ for every $f \in \mathcal{H}(U)$.

NOTES AND COMMENTS

Theorems 24.11 and 24.12 are due to K. Oka [1]. Theorem 24.12 extends an earlier result of A. Weil [1]. Our presentation in Sections 24 and 25 follows essentially the book of L. Hörmander [3].

Schauder bases were introduced by J. Schauder [1], [2], who gave the examples of Schauder bases in $L^p[0,1]$ and $C[0,1]$ which appear in Exercises 26.C and 26.D. Theorem 26.2 is due to S. Banach [1], whereas Theorem 26.4 has been taken from the book of J. Lindenstrauss and L. Tzafriri [1].

The approximation property was introduced by A. Grothendieck [1]. Theorem 27.4 is due to A. Pelczynski [1]. Proposition 27.5, due to H. Auerbach, can be found in the book of S. Banach [1].

R. Aron and M. Schottenloher [1] obtained Theorem 28.1 and the result in Exercise 28.B, thus extending earlier results of S. Dineen [3] and P. Noverraz [2]. Theorem 28.2 is due to E. Ligocka [1]. Theorem 28.3, due to M. Schottenloher [5], extends an earlier result of P. Noverraz [2]. The simple proof given here is due to J. Mujica [3]. Theorem 28.5 is due to C. Matyszczyk [1], whereas Theorem 28.7 is essentially due to S. Dinnen and A. Hirschowitz [1]. The result in Exercise 28.H, due to J. Mujica [1], extends an earlier result of J. M. Isidro [1].

CHAPTER VII

COMMUTATIVE BANACH ALGEBRAS

29. BANACH ALGEBRAS

There is a close interplay between the theory of holomorphic functions and the theory of commutative Banach algebras. In this section we present some basic facts about Banach algebras, not necessarily commutative.

29.1. DEFINITION. A is said to be a *Banach algebra* if A is an algebra (always assumed complex and with unit element $e \neq 0$) and a Banach space such that

(a) $\|xy\| \leq \|x\| \|y\|$ for all $x, y \in A$;

(b) $\|e\| = 1$.

A Banach algebra A is said to be *commutative* if $xy = yx$ for all $x, y \in A$.

29.2. EXAMPLES. (a) Let X be a compact Hausdorff space. If multiplication is defined pointwise then $C(X)$ is a commutative Banach algebra.

(b) Let K be a compact subset of a complex Banach space and let $P(K)$ denote the closure of $P(E)$ in $C(K)$. Then $P(K)$ is a commutative Banach algebra, as a closed subalgebra of $C(K)$.

(c) Let E be a complex Banach space. If multiplication is defined as composition then $\mathcal{L}(E;E)$ is a Banach algebra which is not commutative unless E has dimension one.

The next theorem gives some basic properties of the set of

invertible elements of a Banach algebra.

29.3. THEOREM. *Let A be a Banach algebra. Then:*

(a) *For each $x \in A$ with $\|x\| < 1$ the element $e-x$ is invertible and $(e-x)^{-1} = \sum_{m=0}^{\infty} x^m$.*

(b) *The set U of all invertible elements of A is open.*

(c) *The mapping $x \in U \to x^{-1} \in A$ is holomorphic.*

PROOF. (a) Let $x \in A$ with $\|x\| < 1$. Since $\sum_{m=0}^{\infty} \|x^m\| \leq \sum_{m=0}^{\infty} \|x\|^m < \infty$, the series $\sum_{m=0}^{\infty} x^m$ converges absolutely. If we set $S_n = e + x + x^2 + \ldots + x^n$ for every n then $(e-x)S_n = S_n(e-x) = e - x^{n+1}$ and after letting $n \to \infty$ we get that $(e-x)\sum_{m=0}^{\infty} x^m = (\sum_{m=0}^{\infty} x^m)(e-x) = e$, as we wanted.

(b) If x is invertible then it follows from (a) that $x + t = x(e + x^{-1}t)$ is invertible for every $t \in A$ such that $\|t\| < 1/\|x^{-1}\|$. This shows (b). But it follows also from (a) that

$$(x+t)^{-1} = (e + x^{-1}t)^{-1}x^{-1} = \sum_{m=0}^{\infty} (-x^{-1}t)^m x^{-1}.$$

and the last written series converges uniformly for $\|t\| < r$ whenever $0 < r < 1/\|x^{-1}\|$. Since the mapping $P_m : A \to A$ defined by $P_m(t) = (-x^{-1}t)^m x^{-1}$ clearly belongs to $P(^mA;A)$ we conclude that the mapping $x \in U \to x^{-1} \in A$ is holomorphic and the proof of the theorem is complete.

Next we define the spectrum of an element of a Banach algebra and establish its basic properties.

29.4. DEFINITION. Let A be a Banach algebra and let $x \in A$. The *spectrum* of x, to be denoted by $\sigma(x)$, is the set of all

$\lambda \in \mathbb{C}$ such that $x - \lambda e$ is not invertible. The *spectral radius* of x is the number

$$\rho(x) = \sup\{|\lambda| : \lambda \in \sigma(x)\}.$$

Since $x - \lambda e = -\lambda(e - x/\lambda)$, it follows from Theorem 29.3 that $x - \lambda e$ is invertible for each $\lambda \in \mathbb{C}$ such that $|\lambda| > \|x\|$. Hence $\rho(x) \leq \|x\|$ for each $x \in A$.

29.5. **THEOREM.** *Let A be a Banach algebra and let $x \in A$. Then:*

(a) *The set $\sigma(x)$ is compact and nonempty.*

(b) *$\lambda^m \in \sigma(x^m)$ for each $\lambda \in \sigma(x)$ and $m \in \mathbb{N}$.*

(c) *$\rho(x) = \lim_{m\to\infty} \|x^m\|^{1/m}$. This is the spectal radius formula.*

PROOF. (a) Since $\rho(x) \leq \|x\|$ the set $\sigma(x)$ is bounded. By Theorem 29.3 the set of noninvertible elements of A is closed. Since the mapping $\lambda \in \mathbb{C} \to x - \lambda e \in A$ is continuous, the set $\sigma(x)$ is also closed, and therefore compact. Suppose that $\sigma(x)$ is empty and consider the mapping $f : \mathbb{C} \to A$ defined by $f(\lambda) = (x - \lambda e)^{-1}$. Then $f \in \mathcal{H}(\mathbb{C};A)$ by Theorem 29.3. Furthermore, $-\lambda f(\lambda) = (e - x/\lambda)^{-1} \to e$ when $|\lambda| \to \infty$. Thus $f(\lambda) \to 0$ when $|\lambda| \to \infty$ and in particular f is bounded on $\mathbb{C}$. Hence f is identically zero by Liouville's Theorem 5.10. But this is impossible since $f(0) = x^{-1}$. Thus $\sigma(x)$ must be nonempty.

(b) If $x^m - \lambda^m e$ is invertible then if follows from the factorization formula

$$x^m - \lambda^m e = (x - \lambda e)(x^{m-1} + \lambda x^{m-2} + \ldots + \lambda^{m-2}x + \lambda^{m-1}e)$$

that $x - \lambda e$ is invertible too.

(c) It follows from (b) that $\rho(x)^m \leq \rho(x^m) \leq \|x^m\|$ for every $m \in \mathbb{N}$. Hence

(29.1) $$\rho(x) \leq \liminf_{m\to\infty} \|x^m\|^{1/m}.$$

Next consider the mapping $g(\lambda) = (e - \lambda x)^{-1}$, which is defined and is holomorphic for $|\lambda| < 1/\rho(x)$. Indeed, $g(\lambda)$ is defined and is holomorphic for $|\lambda| < 1/\|x\|$, by Theorem 29.3, and since $e - \lambda x = -\lambda(x - e/\lambda)$, $g(\lambda)$ is defined and is holomorphic for $0 < |\lambda| < 1/\rho(x)$, by the definition of $\rho(x)$. By Theorem 29.3 the Taylor series of g at the origin is $\sum_{m=0}^{\infty} \lambda^m x^m$. Then it follows from the Cauchy-Hadamard Formula that

$$\text{(29.2)} \qquad \limsup_{m\to\infty} \|x^m\|^{1/m} = 1/r_c g(0) \leq \rho(x).$$

From (29.1) and (29.2) we get the desired conclusion.

We conclude this section with the *Gelfand-Mazur Theorem*.

29.6. THEOREM. *Let A be a Banach algebra in which each non-zero element is invertible. Then A is isometrically isomorphic to $\mathbb{C}$.*

PROOF. The set $\sigma(x)$ is nonempty for each $x \in A$ by Theorem 29.5. If λ and μ belong to $\sigma(x)$ then $x - \lambda e$ and $x - \mu e$ are not invertible, hence $x - \lambda e$ and $x - \mu e$ are both equal to zero, and therefore $\lambda = \mu$. Thus for each $x \in A$ the set $\sigma(x)$ consists of a single point $\lambda(x)$ and $x = \lambda(x)e$. Then the mapping $x \in A \to \lambda(x) \in \mathbb{C}$ is obviously an algebra isomorphic and $\|x\| = \|\lambda(x)e\| = |\lambda(x)|$ for every $x \in A$.

30. COMMUTATIVE BANACH ALGEBRAS

This section is devoted to the study of ideals and homomorphisms of commutative Banach algebras.

30.1. DEFINITION. Let A be a commutative Banach algebra. A subset I of A is said to be an *ideal* of A if I is a vector subspace of A and $xa \in I$ for all $x \in I$ and $a \in A$. An ideal $I \neq A$ is said to be a *proper ideal*. A proper ideal which is not contained in any larger proper ideal is said to be a *maximal ideal*.

Let A be a commutative Banach algebra. An ideal I of A is a proper ideal if and only if I contains no invertible elements. A standard application of Zorn's lemma shows that every proper ideal of A is contained in some maximal ideal of A. If I is a closed proper ideal of A then the quotient space A/I is a commutative Banach algebra and the quotient mapping $A \to A/I$ is a continuous algebra homomorphism (we shall always require that an algebra homomorphism map the unit element of the first algebra onto the unit element of the second one). The proofs of these assertions are straightforward and are left to the reader as exercises.

30.2. PROPOSITION. *Let A be a commutative Banach algebra. Then:*

(a) *The closure of each proper ideal of A is a proper ideal.*

(b) *Each maximal ideal of A is closed.*

PROOF. (a) Let I be a proper ideal of A. Then $\overline{I}$ is easily seen to be an ideal. Furthermore, the set U of all invertible elements of A is open and is disjoint from I. Hence U and $\overline{I}$ are also disjoint and therefore $\overline{I} \neq A$.

(b) Let M be a maximal ideal of A. By (a) $\overline{M}$ is a proper ideal and therefore $\overline{M} = M$.

If A is a Banach algebra then we shall denote by $S(A)$ the set of all *complex homomorphisms* of A, that is, the set of all algebra homomorphisms $h : A \to \mathbb{C}$. The set $S(A)$ is called the *spectrum* of A. The complex homomorphisms of A are also called *multiplicative linear functionals*. Recall that, by a previous convention, $h(e) = 1$ for every $h \in S(A)$. This is, of course, just a matter of convenience.

Our next theorem shows that in the case of a commutative Banach algebra A, there is a one-to-one correspondence between the set of complex homomorphisms of A and the set of maximal ideals of A.

30.3. THEOREM. *Let A be a commutative Banach algebra. Then:*

(a) *The kernel of each complex homomorphism of A is a maximal ideal of A.*

(b) *Each maximal ideal of A is the kernel of a unique complex homomorphism of A.*

PROOF. (a) If $h : A \to \mathbb{C}$ is a complex homomorphism then each $x \in A$ can be uniquely decomposed in the form $x = \lambda e + z$, with $\lambda \in \mathbb{C}$ and $z \in \operatorname{Ker} h$. Indeed, $\lambda = h(x)$ and $z = x - h(x)e$. Whence the ideal $\operatorname{Ker} h$ has codimension one and is therefore a maximal ideal.

(b) If M is a maximal ideal of A then A/M is a field and is therefore isomorphic to $\mathbb{C}$ by the Gelfand-Mazur Theorem. Let $\pi : A \to A/M$ be the quotient mapping, let $\varphi : A/M \to \mathbb{C}$ be the isomorphism given by the Gelfand-Mazur Theorem and let $h = \varphi \circ \pi$. Then h is a complex homomorphism of A and $M = \operatorname{Ker} h$. To show uniqueness let h_1 and h_2 be two complex homomorphisms of A such that $\operatorname{Ker} h_1 = \operatorname{Ker} h_2 = M$. From the uniqueness of the decomposition $x = \lambda e + z$, with $\lambda \in \mathbb{C}$ and $z \in \operatorname{Ker} h_1 = \operatorname{Ker} h_2 = M$, we see that $h_1(x) = h_2(x) = \lambda$ for each $x \in A$.

30.4. COROLLARY. *Let A be a commutative Banach algebra and let $x \in A$.*

(a) *x is invertible if and only $h(x) \neq 0$ for each $h \in S(X)$.*

(b) $\sigma(x) = \{h(x) : h \in S(X)\}$.

(c) *$|h(x)| \leq \rho(x) \leq \|x\|$ for each $h \in S(X)$. In particular each $h \in S(X)$ is continuous.*

PROOF. (a) If $xx^{-1} = e$ then $h(x)h(x^{-1}) = h(e) = 1$ and therefore $h(x) \neq 0$ for each $h \in S(X)$. Conversely, if $h(x) \neq 0$ for each $h \in S(X)$ then x is not contained in any maximal ideal and is therefore invertible.

(b) If $\lambda \in \sigma(x)$ then $x - \lambda e$ is not invertible. Then by (a) there exists $h \in S(X)$ such that $h(x - \lambda e) = 0$ and therefore

$\lambda = h(x)$. Conversely, since $x - h(x)e \in Ker\, h$, $x - h(x)e$ is not invertible by (a), and therefore $h(x) \in \sigma(x)$.

(c) This follows at once from (b).

30.5. DEFINITION. Let A be a commutative Banach algebra. The formula $\hat{x}(h) = h(x)$ $(h \in S(A))$ defines a function $\hat{x}: S(A) \to \mathbb{C}$ for each $x \in A$. The function $\hat{x}$ is called the *Gelfand transform* of x. Let $\hat{A}$ denote the algebra of all $\hat{x}$ with $x \in A$. The spectrum $S(A)$ of A will be always endowed with the weakest topology that makes each $\hat{x} \in \hat{A}$ continuous. This topology is called the *Gelfand topology*. Because of the one-to-one correspondence between complex homomorphisms of A and maximal ideals of A, the topological space $S(A)$ is often called the *maximal ideal space of* A.

30.6. THEOREM. *Let A be a commutative Banach algebra. Then*

(a) *$S(A)$ is a compact Hausdorff space.*

(b) *The Gelfand mapping $x \in A \to \hat{x} \in C(S(A))$ is a continuous algebra homomorphism, and $\|\hat{x}\| = \rho(x)$ for every $x \in A$.*

PROOF. By Corollary 30.4(c) $S(A)$ is a subset of the closed unit ball of the dual A' of A. Furthermore, the Gelfand topology on $S(A)$ is precisely the topology induced on $S(A)$ by the weak topology $\sigma(A',A)$ of A'. By Alaoglu's Theorem, the closed unit ball of A' is $\sigma(A',A)$-compact. Since one can readily see that $S(A)$ is $\sigma(A',A)$-closed in A', (a) follows.

The Gelfand mapping $x \in A \to \hat{x} \in C(S(A))$ is clearly an algebra homomorphism. And it follows at once from Corollary 30.4(b) that $\|\hat{x}\| = \rho(x) \leq \|x\|$ for every $x \in A$. This shows (b).

To conclude this section we characterize the spectra of some Banach algebras.

30.7. PROPOSITION. *Let X be a compact Hausdorff space.*

(a) *If $F \subset C(X)$ is a family of functions without common* *zeros then there exist* $f_1,\dots,f_n \in F$ and $g_1,\dots,g_n \in C(X)$ *such that*

$$f_1(x)g_1(x) + \dots + f_n(x)g_n(x) = 1$$

for every $x \in X$.

(b) *For each* $h \in S(C(X))$ *there exists a unique* $a \in X$ *such that* $h(f) = f(a)$ *for every* $f \in C(X)$.

PROOF. (a) For each $a \in X$ there exists $f \in F$ such that $f(a) \neq 0$ and hence $f \neq 0$ on some neighborhood of a. Since X can be covered by finitely many such neighborhoods we can find $f_1,\dots,f_n \in F$ without common zeros. Hence the function

$$f(x) = \sum_{j=1}^{n} f_j(x)\,\overline{f_j(x)} = \sum_{j=1}^{n} |f_j(x)|^2$$

is strictly positive on X. If we define $g_j(x) = \overline{f_j(x)} \,/\, f(x)$ for each $x \in X$ and $j = 1,\dots,n$ then $f_1(x)g_1(x) + \dots + f_n(x)g_n(x) = 1$ for every $x \in X$.

(b) If $h \in S(C(X))$ then $Ker\,h$ is a proper ideal of $C(X)$. Then by (a) there exists $a \in X$ such that $f(a) = 0$ for every $f \in Ker\,h$. Hence $f(a) - h(f) = 0$ for every $f \in C(X)$. Uniqueness of a is clear since continuous functions separate the points of X.

30.8. PROPOSITION. *Let* E *be a complex Banach space with the approximation property and let* K *be a polynomially convex compact subset of* E.

(a) *For each* $h \in S(P(K))$ *there exists a unique* $a \in K$ *such that* $h(f) = f(a)$ *for every* $f \in P(K)$.

(b) *If* $F \subset P(K)$ *is a family of functions without common zeros then there exist* $f_1,\dots,f_n \in F$ *and* $g_1,\dots,g_n \in P(K)$ *such that*

$$f_1(x)g_1(x) + \dots + f_n(x)g_n(x) = 1$$

for every $x \in K$.

PROOF. (a) let $h \in S(\mathcal{P}(K))$. By Corollary 30.4

$$(30.1) \qquad |h(f)| \leq \|f\| = \sup_K |f|$$

for every $f \in \mathcal{P}(K)$. In particular $|h(\varphi)| \leq \sup_K |\varphi|$ for every $\varphi \in E'$ and by the Mackey-Arens Theorem there exists a unique $a \in E$ such that $h(\varphi) = \varphi(a)$ for every $\varphi \in E'$. Then it is clear that

$$(30.2) \qquad h(P) = P(a)$$

for every $P \in \mathcal{P}_f(E)$. Now, by Theorem 28.1 each $P \in \mathcal{P}(E)$ can be uniformly approximated on K by polynomials belonging to $\mathcal{P}_f(E)$. Then it follows from (30.1) and (30.2) that (30.2) is valid for every $P \in \mathcal{P}(E)$. Hence $|P(a)| = |h(P)| \leq \sup_K |P|$ for every $P \in \mathcal{P}(E)$ and therefore $a \in \hat{K}_{\mathcal{P}(E)} = K$. Since $\mathcal{P}(E)$ is dense in $\mathcal{P}(K)$ we conclude that $h(f) = f(a)$ for every $f \in \mathcal{P}(K)$ and (a) has been proved.

(b) If the desired conclusion were not true then the ideal $I(F)$ generated by F would be a proper ideal. Then $I(F)$ would be contained in the kernel of some $h \in S(\mathcal{P}(K))$. Then by (a) all the functions $f \in I(F)$ would vanish at a certain point $a \in K$, a contradiction.

31. THE JOINT SPECTRUM

This section is devoted to the study of the joint spectrum of elements of a commutative Banach algebra. The connection with Section 24 is apparent. The main results in this section are proved with the aid of Oka's Extension Theorem 24.11.

If x is an element of a commutative Banach algebra then the spectrum $\sigma(x)$ of x is the set of all $\lambda \in \mathbb{C}$ such that $x - \lambda e$ generates a proper ideal. This may be generalized as follows:

31.1. DEFINITION. Let $x_1,\ldots,x_n$ be elements of a commutative Banach algebra A. We shall denote by $\sigma(x_1,\ldots,x_n)$ the set of all $\lambda = (\lambda_1,\ldots,\lambda_n) \in \mathbb{C}^n$ such that the elements $x_1 - \lambda_1 e,\ldots,x_n - \lambda_n e$ generate a proper ideal. The set $\sigma(x_1,\ldots,x_n)$ is called the *joint spectrum* of $x_1,\ldots,x_n$.

31.2. PROPOSITION. *Let $x_1,\ldots,x_n$ be elements of a commutative Banach algebra A. Then*

$$\sigma(x_1,\ldots,x_n) = \{(h(x_1),\ldots,h(x_n)) : h \in S(A)\}.$$

PROOF. If $\lambda \in \sigma(x_1,\ldots,x_n)$ then the elements $x_1 - \lambda_1 e,\ldots,x_n - \lambda_n e$ generate a proper ideal and are therefore contained in the kernel of some $h \in S(A)$. Whence it follows that $(\lambda_1,\ldots,\lambda_n) = (h(x_1),\ldots,h(x_n))$, as we wanted.

Conversely, if $\lambda \notin \sigma(x_1,\ldots,x_n)$ then there are elements $y_1,\ldots,y_n \in A$ such that

$$\sum_{j=1}^{n} (x_j - \lambda_j e)y_j = e.$$

Hence

$$\sum_{j=1}^{n} (h(x_j) - \lambda_j)h(y_j) = 1$$

and therefore $(\lambda_1,\ldots,\lambda_n) \neq (h(x_1),\ldots,h(x_n))$ for every $h \in S(A)$, completing the proof.

31.3. THEOREM. *Let A be a commutative Banach algebra which is generated by n elements $x_1,\ldots,x_n$. Then:*

(a) *The mapping*

$$\varphi : h \in S(A) \to (h(x_1),\ldots,h(x_n)) \in \mathbb{C}^n$$

is a homeomorphism from $S(A)$ onto $\sigma(x_1,\ldots,x_n)$.

(b) *A point $\lambda \in \mathbb{C}^n$ belongs to $\sigma(x_1,\ldots,x_n)$ if and only if*

$$|P(\lambda)| \leq \|P(x_1,\ldots,x_n)\|$$

for every $P \in P(\mathbb{C}^n)$.

(c) $\sigma(x_1,\ldots,x_n)$ *is a polynomially convex compact subset of* $\mathbb{C}^n$.

PROOF. (a) The mapping φ is continuous by the definition of the Gelfand topology. φ maps $S(A)$ onto $\sigma(x_1,\ldots,x_n)$ by Proposition 31.2. To show that φ is injective assume

$$(h(x_1),\ldots,h(x_n)) = (h'(x_1),\ldots,h'(x_n))$$

with $h, h' \in S(A)$. whence it follows that

$$h(P(x_1,\ldots,x_n)) = h'(P(x_1,\ldots,x_n))$$

for every $P \in P(\mathbb{C}^n)$. Since the subalgebra

$$M = \{P(x_1,\ldots,x_n) : P \in P(\mathbb{C}^n)\}$$

is dense in A we conclude that $h = h'$. Since $\varphi : S(A) \to \sigma(x_1,\ldots,x_n)$ is continuous and bijective, and $S(A)$ is compact, we conclude that φ is a homeomorphism.

(b) If $\lambda = (h(x_1,\ldots,h(x_n))$ for some $h \in S(A)$ then for each $P \in P(\mathbb{C}^n)$ we have that

$$|P(\lambda)| = |h(P(x_1,\ldots,x_n))| \leq \|P(x_1,\ldots,x_n)\|.$$

Conversely, assume $|P(\lambda)| \leq \|P(x_1,\ldots,x_n)\|$ for every $P \in P(\mathbb{C}^n)$. If we define $h : M \to \mathbb{C}$ by

$$h(P(x_1,\ldots,x_n)) = P(\lambda)$$

for every $P \in P(\mathbb{C}^n)$, then h is a continuous complex homomorphism on M and hence has a unique continuous extension to all of A. Hence $\lambda = (h(x_1),\ldots,h(x_n))$ belongs to $\sigma(x_1,\ldots,x_n)$.

(c) If $\lambda \in [\sigma(x_1,\ldots,x_n)]\hat{}_{P(\mathbb{C}^n)}$ then for each $P \in P(\mathbb{C}^n)$ we have that

$$|P(\lambda)| \leq \sup \{ |P(\zeta)| : \zeta \in \sigma(x_1,\ldots,x_n)\}$$

$$\leq \sup \{ |P(h(x_1),\ldots,h(x_n))| : h \in S(A)\}$$

$$\leq \sup \{ |h(P(x_1,\ldots,x_n))| : h \in S(A)\}$$

$$\leq \|P(x_1,\ldots,x_n)\|.$$

Thus $\lambda \in \sigma(x_1,\ldots,x_n)$ by (b), and $\sigma(x_1,\ldots,x_n)$ is polynomially convex, as asserted.

31.4. THEOREM. *Let $x_1,\ldots,x_n$ be elements of a commutative Banach algebra A, and let f be a function which is holomorphic on an open neighborhood of $\sigma(x_1,\ldots,x_n)$. Then there are elements $x_{n+1},\ldots,x_N \in A$ and a function g holomorphic on an open neighborhood of the polydisc*

$$\{\zeta \in \mathbb{C}^N : |\zeta_j| \leq \|x_j\| \quad \text{for} \quad j = 1,\ldots,N\}$$

such that

$$g \circ (\hat{x}_1,\ldots,\hat{x}_N) = f \circ (\hat{x}_1,\ldots,\hat{x}_n).$$

Before proving this theorem we give two auxiliary lemmas. The first one reduces the proof of the theorem to the case of finitely generated algebras.

If $x_1,\ldots,x_n$ are elements of a commutative Banach algebra A, and B is any closed subalgebra of A containing $x_1,\ldots,x_n$, then we shall denote by $\sigma_B(x_1,\ldots,x_n)$ the joint spectrum of $x_1,\ldots,x_n$ relative to B.

31.5. LEMMA. *Let $x_1,\ldots,x_n$ be elements of a commutative Banach algebra A, and let U be an open neighborhood of $\sigma(x_1,\ldots,x_n)$.*

Then there exist elements $x_{n+1},\ldots,x_{n+m} \in A$ *such that* $\sigma_B(x_1,\ldots,x_n) \subset U$, *where* B *is the closed subalgebra of* A *generated by* $x_1,\ldots,x_{n+m}$.

PROOF. (a) Let $\lambda \in \mathbb{C}^n$, $\lambda \notin \sigma(x_1,\ldots,x_n)$. Then there are $y_1,\ldots,y_n \in A$ such that $\sum_{j=1}^{n} (x_j - \lambda_j e)y_j = e$. Let B_λ be the closed subalgebra of A generated by $x_1,\ldots,x_n,y_1,\ldots,y_n$. Then $\lambda \notin \sigma_{B_\lambda}(x_1,\ldots,x_n)$ and hence there is an open neighborhood V_λ of λ which is disjoint from $\sigma_{B_\lambda}(x_1,\ldots,x_n)$

(b) Let B_o be the closed subalgebra of A generated by $x_1,\ldots,x_n$. If $\sigma_{B_o}(x_1,\ldots,x_n) \subset U$ then there is nothing to prove. If $\sigma_{B_o}(x_1,\ldots,x_n) \not\subset U$ then by applying the argument in (a) to each point λ of the compact set $\sigma_{B_o}(x_1,\ldots,x_n) \setminus U$ we can find open sets $V_1,\ldots,V_k$ and finitely generated closed subalgebras $B_1,\ldots,B_k$ of A, each containing B_o, such that

$$(31.1) \qquad V_j \cap \sigma_{B_j}(x_1,\ldots,x_n) = \phi \qquad \text{for} \quad j = 1,\ldots,k$$

and

$$(31.2) \qquad \sigma_{B_o}(x_1,\ldots,x_n) \setminus U \subset V_1 \cup \ldots \cup V_k .$$

Let B be the closed subalgebra of A generated by $B_1 \cup \ldots \cup B_k$. Then B is finitely generated and contains B_o. From the inclusions $B \supset B_j \supset B_o$ it follows that

$$\sigma_B(x_1,\ldots,x_n) \subset \sigma_{B_j}(x_1,\ldots,x_n) \subset \sigma_{B_o}(x_1,\ldots,x_n)$$

for $j = 1,\ldots,k$. Then it follows from (31.1) and (31.2) that $\sigma_B(x_1,\ldots,x_n) \subset U$, completing the proof.

31.6. **LEMMA.** *Let* B *be a commutative Banach algebra which is generated by* $n+m$ *elements* $x_1,\ldots,x_{n+m}$. *Let* D *denote the polydisc*

$$D = \{\zeta \in \mathbb{C}^{n+m} : |\zeta_j| \leq \|x_j\| \quad \textit{for} \quad j = 1,\ldots,n+m\}$$

and let $\pi : \mathbb{C}^{n+m} \to \mathbb{C}^n$ *denote the projection onto the first* n *coordinates. Then for each open neighborhood* U *of* $\sigma(x_1,\ldots,x_n)$ *in* $\mathbb{C}^n$ *there are polynomials* $P_1,\ldots,P_k \in P(\mathbb{C}^{n+m})$ *such that*

$$\{\zeta \in D : |P_j(\zeta)| \leq \|P_j(x_1,\ldots,x_{n+m})\| \quad \textit{for} \quad j = 1,\ldots,k\} \subset \pi^{-1}(U).$$

PROOF. If $D \subset \pi^{-1}(U)$ then there is nothing to prove. Suppose $D \not\subset \pi^{-1}(U)$ and let $\lambda \in D \setminus \pi^{-1}(U)$. Then $\pi(\lambda) \notin \sigma(x_1,\ldots,x_n)$ and by Theorem 31.3 there exists $P \in P(\mathbb{C}^n)$ such that

$$|P \circ \pi(\lambda)| > \|P(x_1,\ldots,x_n)\| = \|P \circ \pi(x_1,\ldots,x_{n+m})\|.$$

Thus by compactness of $D \setminus \pi^{-1}(U)$ we can find polynomials $P_1,\ldots,P_k \in P(\mathbb{C}^{n+m})$ such that

$$D \setminus \pi^{-1}(U) \subset \bigcup_{j=1}^{k} \{\zeta \in C^{n+m} : |P_j(\zeta)| > \|P(x_1,\ldots,x_{n+m})\|\}.$$

The desired conclusion follows.

PROOF OF THEOREM 31.4. Let $f \in \mathcal{H}(U)$, where U is an open neighborhood of $\sigma(x_1,\ldots,x_n)$. By Lemma 31.5 we can find $x_{n+1},\ldots,x_{n+m} \in A$ such that $\sigma_B(x_1,\ldots,x_n) \subset U$, where B is the closed subalgebra of A generated by $x_1,\ldots,x_{n+m}$. Let D denote the polydisc

$$D = \{\zeta \in \mathbb{C}^{n+m} : |\zeta_j| \leq \|x_j\| \quad \text{for} \quad j = 1,\ldots,n+m\}$$

and let $\pi : \mathbb{C}^{n+m} \to \mathbb{C}^n$ denote the projection onto the first n coordinates. By Lemma 31.6 we can find $P_1,\ldots,P_k \in P(\mathbb{C}^{n+m})$ such that $L \subset \pi^{-1}(U)$, where

$$L = \{\zeta \in D : |P_j(\zeta)| \leq \|P_j(x_1,\ldots,x_{n+m})\| \quad \text{for} \quad j = 1,\ldots,k\}.$$

Set $x_{n+m+j} = P_j(x_1,\ldots,x_{n+m})$ for $j = 1,\ldots,k$. Since the function $f \circ \pi$ is holomorphic on an open neighborhood of L, an application of Oka's Extension Theorem 24.11 shows the existence of a function g, holomorphic on an open neighborhood of the polydisc

$$\{\zeta \in \mathbb{C}^{n+m+k} : |\zeta_j| \leq \|x_j\| \quad \text{for} \quad j = 1,\ldots,n+m+k\},$$

and such that

$$g(\zeta, P_1(\zeta),\ldots,P_k(\zeta)) = f \circ \pi(\zeta) \tag{31.3}$$

for every ζ in a certain neighborhood of L. Note that $\sigma(x_1,\ldots,x_{n+m}) \subset L$ by Theorem 31.3. Hence, by applying (31.3) with $\zeta = (\hat{x}_1(h),\ldots,\hat{x}_{n+m}(h))$ we get that

$$g(\hat{x}_1(h),\ldots,\hat{x}_{n+m+k}(h)) = f(\hat{x}_1(h),\ldots,\hat{x}_n(h))$$

for every $h \in S(A)$. This completes the proof.

31.7. THEOREM. *Let $x_1,\ldots,x_n$ be elements of a commutative Banach algebra A, and let f be a function which is holomorphic on an open neighborhood of $\sigma(x_1,\ldots,x_n)$. Then there exists $y \in A$ such that*

$$\hat{y} = f \circ (\hat{x}_1,\ldots,\hat{x}_n).$$

PROOF. By Theorem 31.4 there are elements $x_{n+1},\ldots,x_N \in A$ and a function g, holomorphic on an open neighborhood of the polydisc

$$\{\zeta \in \mathbb{C}^N : |\zeta_j| \leq \|x_j\| \quad \text{for} \quad j = 1,\ldots,N\},$$

and such that

$$g \circ (\hat{x}_1,\ldots,\hat{x}_N) = f \circ (\hat{x}_1,\ldots,\hat{x}_n).$$

If $\sum_\alpha c_\alpha \zeta^\alpha$ is the Taylor series of g at the origin then

$$\sum_{\alpha} c_{\alpha} \|x_1\|^{\alpha_1} \ldots \|x_N\|^{\alpha_N} < \infty$$

and hence the series $\sum_{\alpha} c_{\alpha} x_1^{\alpha_1} \ldots x_N^{\alpha_N}$ converges to an element $y \in A$. Then for each $h \in S(A)$ we have that

$$h(y) = \sum_{\alpha} c_{\alpha} h(x_1)^{\alpha_1} \ldots h(x_N)^{\alpha_N}$$

$$= g(h(x_1), \ldots, h(x_N))$$

$$= f(h(x_1), \ldots, h(x_n)),$$

completing the proof.

31.8. THEOREM. *Let A be a commutative Banach algebra and assume that $S(A)$ is the union of two disjoint compact sets H and K. Then there exists $y \in A$ such that $\hat{y}(h) = 0$ for every $h \in H$ and $\hat{y}(k) = 0$ for every $k \in K$.*

PROOF. Since H and K are disjoint, for each pair $(h_o, k_o) \in H \times K$ there exists $x \in A$ such that $\hat{x}(h_o) \neq \hat{x}(k_o)$ and hence $\hat{x}(h) \neq \hat{x}(k)$ for all (h,k) in a suitable neighborhood of (h_o, k_o). Since $H \times K$ is compact, it can be covered by finitely many such neighborhoods. This yields elements $x_1, \ldots, x_n$ of A such that for each $(h,k) \in H \times K$ there exists some j $(j = 1, \ldots, n)$ such that $\hat{x}_j(h) \neq \hat{x}_j(k)$. Hence the sets

$$\tilde{H} = \{(\hat{x}_1(h), \ldots, \hat{x}_n(h)) : h \in H\}$$

and

$$\tilde{K} = \{(\hat{x}_1(k), \ldots, \hat{x}_n(k)) : k \in K\}$$

are two disjoint compact subsets of $\mathbb{C}^n$ whose union is the joint spectrum $\sigma(x_1, \ldots, x_n)$ of $x_1, \ldots, x_n$. Then the function f which is equal to zero on an open neighborhood of $\tilde{M}$ and equal to one on an open neighborhood of $\tilde{K}$, is holomorphic on an open neighborhood of $\sigma(x_1, \ldots, x_n)$. By Theorem 31.7 there exists $y \in A$ such that

$$\hat{y}(h) = f(\hat{x}_1(h),\ldots,\hat{x}_n(h))$$

for every $h \in S(A)$. Hence $\hat{y}(h) = 0$ for every $h \in H$ and $\hat{y}(k) = 1$ for every $k \in K$, as we wanted.

32. PROJECTIVE LIMITS OF BANACH ALGEBRAS

Very often we have to deal with topological algebras of functions which are not Banach algebras. A typical example in this book is $(\mathcal{H}(U), \tau_c)$, where U is an open subset of a complex Banach space. In this section we introduce an important class of topological algebras whose study may be reduced to a large extent to the theory of Banach algebras.

32.1. DEFINITIONS. (a) A is said to be a *topological algebra* if A is a complex algebra and a topological vector space in which multiplication is jointly continuous.

(b) A is said to be a *locally multiplicatively convex algebra*, or for short a *locally m-convex algebra*, if A is a topological algebra whose topology is given by a family of seminorms p_i such that

$$(32.1) \qquad p_i(xy) \leq p_i(x)p_i(y) \quad \text{for all} \quad x, y \in A$$

and

$$(32.2) \qquad p_i(e) = 1.$$

32.2. EXAMPLES. (a) Let X be a topological space. If multiplication is defined pointwise then $(C(X), \tau_c)$ is a commutative locally m-convex algebra.

(b) Let U be an open subset of a complex Banach space. Then $(\mathcal{H}(U), \tau_c)$ is a commutative locally m-convex algebra, as a closed subalgebra of $(C(U), \tau_c)$.

If A is a (commutative) topological algebra then we can define the spectrum of an element of A, ideals of A, and the joint spectrum of elements of A, exactly as in the case of (commutative) Banach algebras, for these are purely algebraic notions. But we have to distinguish between the set $S_a(A)$ of all complex homomorphisms of A, called the *algebraic spectrum of* A, and the subset $S(A)$ of all continuous members of $S_a(A)$, called the *spectrum of* A, or more precisely, the *topological spectrum of* A.

Let A be a locally m-convex algebra and let p be a continuous seminorm on A satisfying (32.1) and (32.2). Let (A,p) denote the algebra A, seminormed by p, let A_p denote the normed algebra $(A,p)/p^{-1}(0)$, and let $\pi_p : A \to A_p$ denote the quotient mapping. Then the completion $\hat{A}_p$ of A_p is a Banach algebra and the mapping π_p is a continuous homomorphism from A onto a dense subalgebra of $\hat{A}_p$. Many questions about A can be reduced to questions about the Banach algebras $\hat{A}_p$. To illustrate this point we prove the following proposition.

32.3. PROPOSITION. *If A is a commutative locally m-convex algebra then each closed proper ideal of A is contained in the kernel of a continuous complex homomorphism of A.*

PROOF. Let I be a closed proper ideal of A. Then there exists a continuous seminorm p on A satisfying (32.1) and (32.2), and there exists $\varepsilon > 0$ such that I is disjoint from the open set $U = \{x \in A : p(x - e) < \varepsilon\}$. Hence $\pi_p(I)$ is an ideal of A_p which is dispoint from the ball $\pi_p(U)$. Hence it follows that the closure $\overline{\pi_p(I)}$ of $\pi_p(U)$ in $\hat{A}_p$ is a closed proper ideal of $\hat{A}_p$. Since $\hat{A}_p$ is a Banach algebra, the ideal $\overline{\pi_p(I)}$ is contained in the kernel of some $h_p \in S(\hat{A}_p)$. Hence the ideal I is contained in the kernel of the mapping $h_p \circ \pi_p$, which belongs to $S(A)$.

32.4. COROLLARY. *If A is a commutative locally m-convex algebra then the mapping $h \to \operatorname{Ker} h$ gives a one-to-one correspondence between the continuous complex homomorphisms of A and the closed maximal ideals of A.*

PROOF. If $h \in S(A)$ then the proof of Theorem 30.3 shows that

$Ker\, h$ is a maximal ideal of A, which is closed since h is continuous. Conversely, if M is a closed maximal ideal of A then by Proposition 32.3 there exists $h \in S(A)$ such that $M \subset Ker\, h$. Since M is maximal we conclude that $M = Ker\, h$. And the proof of Theorem 30.3 shows that h is unique, completing the proof.

To study more closely the connection between locally m-convex algebras and Banach algebras we introduced the following definition.

32.5. DEFINITION. Let $(X_i)_{i\in I}$ be a family of topological spaces, indexed by a directed set I. Suppose that for each pair of indices i, j with $i \leq j$ there is a continuous mapping $\pi_{ij} : X_j \to X_i$ with the following properties:

(a) π_{ii} is the identity mapping for every i;

(b) $\pi_{ij} \circ \pi_{jk} = \pi_{ik}$ whenever $i \leq j \leq k$.

Then the collection (X_i, π_{ij}) is said to be a *projection system of topological spaces*. The set

$$X = \{(x_i) \in \Pi X_i : \pi_{ij}(x_j) = x_i \text{ whenever } i \leq j\},$$

endowed with the topology induced by the product ΠX_i, is called the *projective limit* of the topological spaces X_i and is denoted by $proj\, X_i$. The canonical mapping $X \to X_i$ is denoted by π_i.

If each X_i is a topological vector space (resp. a topological algebra) and each $\pi_{ij} : X_j \to X_i$ is linear (resp. an algebra homomorphism) then the collection (X_i, π_{ij}) is said to be a *projective system of topological vector spaces* (resp. a *projective system of topological algebras*). In this case the projective limit $X = proj\, X_i$ is endowed with the vector space structure (resp. algebra structure) induced by the product ΠX_i.

32.6. THEOREM. *Every complete Hausdorff locally m-convex algebra is topologically isomorphic to a projective limit of Banach algebras.*

PROOF. Let A be a complete Hausdorff locally m-convex algebra. If $p_1,\dots,p_n$ are continuous seminorms on A satisfying (32.1) and (32.2) then the seminorm $p = \max\{p_1,\dots,p_n\}$ has the same properties. Thus the topology of A is given by a directed family P of seminorms satisfying (32.1) and (32.2). If for $p \leq q$ we denote by $\pi_{pq} : A_q \to A_p$ the canonical mapping, and by $\hat{\pi}_{pq} : \hat{A}_q \to \hat{A}_p$ its continuous extension, then it is clear that (A_p, π_{pq}) is a projective system of normed algebras, whereas $(\hat{A}_p, \hat{\pi}_{pq})$ is a projective system of Banach algebras. We claim that the projective limit $B = \text{proj}\, A_p$ is a dense subalgebra of the projective limit $\hat{B} = \text{proj}\, \hat{A}_p$. Indeed, let $\hat{y} = (\hat{y}_p) \in \hat{B}$. For each $p \in P$ and $\varepsilon > 0$ we first $x_p \in A_p$ such that $p(x_p - \hat{y}_p) < \varepsilon$, we next choose $x \in A$ such that $\pi_p(x) = x_p$, and then we define $y^{p\varepsilon} = (\pi_q(x))_{q \in P} \in B$. Then one can readily prove that the net $(y^{p\varepsilon})$ converges to $\hat{y}$. Next consider the mapping

$$\pi : x \in A \to (\pi_p(x)) \in \Pi A_p .$$

Clearly π is an algebra homomorphism. π is injective since A is Hausdorff. And clearly π is a homeomorphism between A and its image in ΠA_p. We claim that $\pi(A) = B$. Indeed, given $y = (y_p) \in B$ we choose for each $p \in P$ an $x^p \in A$ such that $\pi_p(x^p) = y_p$. Then (x^p) is easily seen to be a Cauchy net in A, and since A is complete, the net (x^p) converges to some $x \in A$. Then one can readily see that $\pi_p(x) = y_p$ for each p. Thus $\pi(A) = B$ and B is therefore complete. Since we already know that B is dense in $\hat{B}$ we conclude that $\pi(A) = B = \hat{B}$ and the proof of the theorem is complete.

32.7. PROPOSITION. *Let $A = \text{proj}\, A_i$ be a projective limit of topological algebras. Then an element $x = (x_i) \in A$ is invertible in A if and only if x_i is invertible in A_i for every i.*

PROOF. The "only if" part is clear. To show the "if" part let y_i be the inverse of x_i in A_i for each i. It follows from the uniqueness of the inverse that $\pi_{ij}(y_j) = y_i$ whenever $i \leq j$. Hence $y = (y_i)$ belongs to A and y is the inverse of x.

32.8. PROPOSITION. *Let A be a complete, Hausdorff, commutative locally m-convex algebra, and let $x \in A$.*

(a) *x is invertible if and only if $h(x) \neq 0$ for each $h \in S(X)$.*

(b) $\sigma(x) = \{h(x) : h \in S(X)\}$.

PROOF. By Theorem 32.6 A is a projective limit of commutative Banach algebras, say $A = proj\, A_i$. Then it is clear that

$$S(A) = \{h_i \circ \pi_i : h_i \in S(A_i),\ i \in I\}.$$

Hence (a) follows from Proposition 32.7 and Corollary 30.4(a). To show (b), the proof of Corollary 30.4(b) applies.

Let $A = proj\, A_i$ be a projective limit of commutative topological algebras and let I be a principal ideal of A. Then it follows from Proposition 32.7 that $I = A$ if and only if $\pi_i(I)$ generates the improper ideal of A_i for each i. If instead of a principal ideal we consider a finitely generated ideal, then the problem is much more difficult, for if $n \geq 2$ and the equation $\sum_{j=1}^{n} x_j y_j = e$ has a solution $(y_1, \ldots, y_n)$ for a given $(x_1,\ldots,x_n)$, then the solution is not unique. In the case of finitely generated ideals we have the following theorem.

32.9. THEOREM. *Let $A = proj\, A_i$ be the projective limit of a sequence of commutative Banach algebras, and assume that the homomorphism $\pi_i : A \to A_i$ has dense image for every i. If I is a finitely generated ideal of A such that $\pi_i(I)$ generates the improper ideal of A_i for every i then $I = A$.*

The proof of Theorem 32.9 rests on the following theorem.

32.10. THEOREM. *Let* $X = \text{proj}\, X_i$ *be the projective limit of a sequence of complete metric spaces, and assume that the mapping* $\pi_{i,i+1} : X_{i+1} \to X_i$ *has dense image for every* i. *Then the mapping* $\pi_i : X \to X_i$ *has dense image for every* i.

PROOF. We show that π_1 has dense image. The proof that π_i has dense image is similar. Let $x_1 \in X_1$ and $\varepsilon > 0$ be given. Let d_i be the metric of X_i for each i. Since $\pi_{12} : X_2 \to X_1$ has dense image we can find $x_2 \in X_2$ such that

$$d_1(\pi_{12}(x_2), x_1) \leq \varepsilon/2.$$

Since $\pi_{23} : X_3 \to X_2$ has dense image and $\pi_{12} : X_2 \to X_1$ is continuous we can find $x_3 \in X_3$ such that

$$d_2(\pi_{23}(x_3), x_2) \leq \varepsilon/4 \quad \text{and} \quad d_1(\pi_{13}(x_3), \pi_{12}(x_2)) \leq \varepsilon/4.$$

Proceeding inductively we can find $x_j \in X_j$ for each $j = 2, 3, \ldots$ such that

$$d_i(\pi_{ij}(x_j), \pi_{i,j-1}(x_{j-1})) \leq \varepsilon/2^{j-1} \quad \text{for} \quad i = 1, \ldots, j-1.$$

We claim that $(\pi_{ij}(x_j))_{j=i}^{\infty}$ is a Cauchy sequence in X_i for every $i = 1, 2, 3, \ldots$. Indeed, for $k \geq j \geq i$ we have that

$$\text{(32.3)} \quad d_i(\pi_{ik}(x_k), \pi_{ij}(x_j)) \leq \sum_{r=j}^{\infty} d_i(\pi_{i,r+1}(x_{r+1}), \pi_{ir}(x_r))$$

$$\leq \sum_{r=j}^{\infty} \varepsilon \cdot 2^{-r} = \varepsilon \cdot 2^{-j+1}.$$

Thus the sequence $(\pi_{ij}(x_j))_{j=i}^{\infty}$ converges to an element $y_i \in X_i$ for each $i = 1, 2, \ldots$. Since $\pi_{ij} \circ \pi_{jk}(x_k) = \pi_{ik}(x_k)$ for $k \geq j \geq i$, letting $k \to \infty$ we get that $\pi_{ij}(y_j) = y_i$ for $j \geq i$. This shows that $y = (y_i)_{i=1}^{\infty}$ belongs to X. If follows from (32.3) that $d_1(\pi_{1k}(x_k), x_1) \leq \varepsilon$. Letting $k \to \infty$ we get that

$d_1(y_1, x_1) \leq \varepsilon$, and the proof is complete.

32.11. LEMMA. *Let A and B be two commutative Banach algebras, let $\pi : A \to B$ be a homomorphism with dense image, and suppose that A is generated by n elements $a_1, \ldots, a_n$. Then given $\varepsilon > 0$ and n elements $y_1, \ldots, y_n \in B$ such that*

$$\pi(a_1)y_1 + \ldots + \pi(a_n)y_n = e, \tag{32.4}$$

we can find $x_1, \ldots, x_n \in A$ such that

$$a_1x_1 + \ldots + a_nx_n = e \tag{32.5}$$

and

$$\| \pi(x_j) - y_j \| \leq \varepsilon \quad \text{for} \quad j = 1, \ldots, n. \tag{32.6}$$

PROOF. Since $a_1, \ldots, a_n$ generate A we can find $s_1, \ldots, s_n \in A$ such that

$$a_1s_1 + \ldots + a_ns_n = e \tag{32.7}$$

let $t_1, \ldots, t_n$ be arbitrary elements of A and define

$$x_j = t_j + s_j(e - \sum_{k=1}^{n} a_k t_k) \tag{32.8}$$

for $j = 1, \ldots, n$. It follows from (32.7) that

$$\sum_{j=1}^{n} a_j x_j = \sum_{j=1}^{n} a_j t_j + (e - \sum_{k=1}^{n} a_k t_k) = e,$$

so that $x_1, \ldots, x_m$ satisfy (32.5). On the other hand, it follows from (32.4) that

$$e - \sum_{k=1}^{n} \pi(a_k)\pi(t_k) = \sum_{k=1}^{n} \pi(a_k)(y_k - \pi(t_k)).$$

Hence

$$\|\pi(x_j) - y_j\| \le \|\pi(t_j) - y_j\| + \|\pi(s_j)\| \sum_{k=1}^{n} \|\pi(a_k)\| \, \|y_k - \pi(t_k)\|$$

for $j = 1,\dots,n$. Thus we can achieve (32.6) by choosing $\pi(t_j)$ sufficiently close to y_j for each $j = 1,\dots,n$. Since π has dense image, this is always possible.

PROOF OF THEOREM 32.9. Let $a_1,\dots,a_n$ be a set of generators for the ideal I. It follows from the hypothesis that $\pi_i(a_1),\dots,\pi_i(a_n)$ generate A_i for each $i \in \mathbb{N}$. Let X be the set of all $(x_1,\dots,x_n) \in A^n$ such that $\sum_{j=1}^{n} a_j x_j = e$. Likewise, for each $i \in \mathbb{N}$ let X_i be the set of all $(x_1,\dots,x_n) \in A_i^n$ such that $\sum_{j=1}^{n} \pi_i(a_j) x_j = e$.

Then each X_i is nonvoid by hypothesis and X_i is a complete matric space, as a closed subset of A_i^n. Since A is the projective limit of the Banach algebras A_i, the set X is in a natural way the projective limit of the metric spaces X_i. It follows from Lemma 32.11 that the natural mapping $X_{i+1} \to X_i$ has dense image for each $i \in \mathbb{N}$. And then it follows from Theorem 32.10 that the natural mapping $X \to X_i$ has dense image for every $i \in \mathbb{N}$, and in particular X is nonvoid. Hence $I = A$, as asserted.

32.12. DEFINITION. A complete metrizable locally m-convex algebra is called a *Fréchet algebra*.

With the aid of Theorem 32.6 and 32.9 we can extend Proposition 31.2 to commutative Fréchet algebras.

32.13. PROPOSITION. *Let $x_1,\dots,x_n$ be elements of a commutative Fréchet algebra A. Then*

$$\sigma(x_1,\dots,x_n) = \{(h(x_1),\dots,h(x_n)) : h \in S(A)\}.$$

PROOF. By Theorem 32.6 A is the projective limit of a sequence of commutative Banach algebras, say $A = \text{proj}\, A_i$. If $\lambda \in \sigma(x_1,\dots,x_n)$ then the ideal I generated by $x_1 - \lambda_1 e,\dots,x_n - \lambda_n e$

is a proper ideal. By Theorem 32.9 there exists $i \in \mathbb{N}$ such that the set $\pi_i(I)$ is contained in a proper ideal of A_i. Hence there exists $h_i \in S(A_i)$ such that $\pi_i(I) \subset Ker\, h_i$. If we set $h = h_i \circ \pi_i$ then $h \in S(A)$ and $I \subset Ker\, h$. Whence it follows that $(\lambda_1, \ldots, \lambda_n) = (h(x_1), \ldots, h(x_n))$, as asserted. To prove the converse it suffices to repeat the proof of Proposition 31.2.

As an application we can prove the following result.

32.14. PROPOSITION. *Let U be a polynomially convex open set in $\mathbb{C}^n$. Let $f_1, \ldots, f_m \in \mathcal{H}(U)$ be m functions without common zeros. Then there are functions $g_1, \ldots, g_m \in \mathcal{H}(U)$ such that $f_1 g_1 + \ldots + f_m g_m$ is identically one on U.*

PROOF. By Exercise 28.H, for each continuous complex homomorphism h of $(\mathcal{H}(U), \tau_c)$ there exists a unique point $a \in U$ such that $h(f) = f(a)$ for every $f \in \mathcal{H}(U)$. Hence the set $\{f_1, \ldots, f_m\}$ is not contained in the kernel of any $h \in S(\mathcal{H}(U), \tau_c)$. Then it follows from Proposition 32.13 that the point $(0, \ldots, 0)$ does not belong to $\sigma(f_1, \ldots, f_n)$. Hence the functions $f_1, \ldots, f_n$ generate the improper ideal.

EXERCISES

32.A. Let A be a topological algebra whose topology is given by a family P of seminorms satisfying (32.1). Let $\tilde{P} = \{\tilde{p} : p \in P\}$, where each $\tilde{p}$ is defined by

$$\tilde{p}(x) = \sup\{ p(xy) : y \in A,\ p(y) \leq 1\}.$$

(a) Show that $p(x)/p(e) \leq \tilde{p}(x) \leq p(x)$ for $x \in A$.

(b) Show that each $\tilde{p} \in \tilde{P}$ satisfies (32.1) and (32.2).

Conclude that a topological algebra is locally m-convex if and only if its topology is given by a family of seminorms satisfying (32.1).

32.B. A subset U of an algebra is said to be *idempotent* if $U^2 \subset U$. Show that a topological algebra is locally m-convex if and only if it has a base of convex, idempotent neighborhoods of zero.

32.C. Let $(A_n)_{n=1}^{\infty}$ be an increasing sequence of subalgebras of and algebra A satisfying $A = \bigcup_{n=1}^{\infty} A_n$. Suppose that each A_n is a Banach algebra and that each inclusion mapping $A_n \hookrightarrow A_{n+1}$ is continuous, with norm not greater than one. Endow A with the finest locally convex topology such that each inclusion mapping $A_n \hookrightarrow A$ is continuous. Show that A is a locally m-convex algebra.

32.D. Let A be a commutative, complete, Hausdorff locally m-convex algebra. Show that given $h \in S_a(A)$ and $x \in A$ one can find $\tilde{h} \in S(A)$ such that $\tilde{h}(x) = h(x)$.

32.E. Let A be a commutative Fréchet algebra. Show that given $h \in S_a(A)$ and $x_1,\ldots,x_n \in A$ one can find $\tilde{h} \in S(A)$ such that $\tilde{h}(x_j) = h(x_j)$ for $j = 1,\ldots,n$.

32.F. Show that if A is a commutative Fréchet algebra then the spectrum $S(A)$ of A is a hemicompact space for the Gelfand topology.

33. THE MICHAEL PROBLEM

This section is devoted to the study of an unsolved problem of E. Michael converning the continuity of the complex homomorphisms of a commutative Fréchet algebra. We show that to solve the Michael problem for all commutative Fréchet algebras, it is sufficient to solve the problem for a suitable Fréchet algebra of holomorphic functions on a Banach space. All Banach spaces in this section are assumed to be complex.

A linear mapping $T : X \to Y$ between two topological vector spaces is said to be *bounded* if it maps bounded sets into bounded sets. Clearly each continuous linear mapping $T : X \to Y$ is bounded. If X is metrizable then one can readily prove that each bounded linear mapping $T : X \to Y$ is continuous.

The following two problems, posed by E. Michael [1] in 1952, remain open.

33.1. PROBLEM. Is every complex homomorphism of a commutative Fréchet algebra continuous ?

33.2. PROBLEM. Is every complex homomorphism of a commutative, complete, Hausdorff locally m-convex algebra bounded ?

Clearly a positive solution to Problem 33.2 implies a positive solution to Problem 33.1. We shall soon prove that the reverse implication is also true.

To begin with we introduce some notation. Set $I = \bigcup_{n=1}^{\infty} I_n$, where

$$I_n = \{\alpha = (\alpha_j)_{j=1}^{\infty} \in \mathbb{N}_o^{\mathbb{N}} : \alpha_j = 0 \quad \text{for} \quad j > n\}$$

for every $n \in \mathbb{N}$. If A is any commutative ring then for each $a = (a_j)_{j=1}^{\infty} \in A^{\mathbb{N}}$ and $\alpha = (\alpha_j)_{j=1}^{\infty} \in I$ we define

$$a^{\alpha} = \prod_{j=1}^{\infty} a_j^{\alpha_j}.$$

Let E be a Banach space with a Schauder basis $(e_n)_{n=1}^{\infty}$. Let $z = (z_n)_{n=1}^{\infty} \in \mathcal{H}(E)^{\mathbb{N}}$ denote the sequence of coordinate functionals, let E_n denote the subspace generated by $e_1,\dots,e_n$, and let $T_n : E \to E_n$ denote the canonical projection. Let $f \in \mathcal{H}(E)$. Since the sequence (T_n) converges to the identity uniformly on compact sets, it follows that the sequence $(f \circ T_n)$ converges to f uniformly on compact sets. If for each $\alpha \in I_n$ we set

$$(33.1)\quad c_\alpha f = (2\pi i)^{-n} \int\limits_{\substack{|\zeta_1|=R_1 \\ \cdots\cdots \\ |\zeta_n|=R_n}} \frac{f(\zeta_1 e_1 + \ldots + \zeta_n e_n)}{\zeta_1^{\alpha_1+1} \ldots \zeta_n^{\alpha_n+1}} d\zeta_1 \ldots d\zeta_n ,$$

where $R_1 > 0, \ldots, R_n > 0$, then it follows from Corollary 7.8 that each $c_\alpha f$ is independent from the choice of $R_1, \ldots, R_n$, and the multiple series $\sum_{\alpha \in I_n} c_\alpha f\, z^\alpha$ converges absolutely and uniformly on compact subsets of E to the function $f \circ T_n$. Thus

$$f = \lim_{n\to\infty} f \circ T_n = \lim_{n\to\infty} \sum_{\alpha\in I_n} c_\alpha f z^\alpha$$

with uniform convergence on compact sets. After these preliminaries we can prove the following theorem.

33.3. THEOREM. *Let E be a Banach space with a normalized Schauder basis $(e_n)_{n=1}^\infty$, and let $(z_n)_{n=1}^\infty$ be the sequence of coordinate functionals. Let A be a commutative, complete, Hausdorff locally m-convex algebra. Let $(a_n)_{n=1}^\infty$ be a sequence in A such that $\sum_{n=1}^\infty \sqrt{p(a_n)} < \infty$ for every continuous seminorm p on A. Then there exists a continuous algebra homomorphism $T : (\mathcal{H}(E), \tau_c) \to A$ such that $T(1) = e$ and $T(z_n) = a_n$ for every $n \in \mathbb{N}$.*

PROOF. The topology of A is given by a family P of seminorms satisfying (32.1) and (32.2). Given $p \in P$ choose $\varepsilon > 0$ such that $\varepsilon \sum_{n=1}^\infty \sqrt{p(a_n)} < 1$. Set $r_n = \varepsilon \sqrt{p(a_n)}$, $R_n = \varepsilon^{-1} \sqrt{p(a_n)}$, $r = (r_n)_{n=1}^\infty$, $R = (R_n)_{n=1}^\infty$ and $p(a) = (p(a_n))_{n=1}^\infty$. Observe that $r_n R_n = p(a_n)$ for every n. Since $\sum_{n=1}^\infty R_n < \infty$ the set

$$L = \{ \sum_{n=1}^\infty \zeta_n e_n : \zeta_n \in \mathbb{C}, |\zeta_n| \le R_n \text{ for every } n\}$$

is a compact subset of E, since it is the continuous image of the compact set

$$K = \{(\zeta_n)_{n=1}^{\infty} \in \mathbb{C}^{\mathbb{N}} : |\zeta_n| \leq R_n \quad \text{for every} \quad n\}.$$

If $f \in \mathcal{H}(E)$ then it follows from (33.1) that

$$|c_\alpha f| \leq (R^\alpha)^{-1} \sup_L |f|$$

for every $\alpha \in I$. Then for each $n \in \mathbb{N}$ we have that

$$\sum_{\alpha \in I_n} p(c_\alpha f a^\alpha) \leq \sum_{\alpha \in I_n} |c_\alpha f|\, p(a)^\alpha$$

$$= \sum_{\alpha \in I_n} |c_\alpha f|\, R^\alpha r^\alpha \leq \sup_L |f| \sum_{\alpha \in I_n} r^\alpha$$

$$= \sup_L |f| (1 - r_1)^{-1} \ldots (1 - r_n)^{-1}.$$

Since $\sum_{n=1}^{\infty} r_n = \theta < 1$ we see that

$$\sum_{n=1}^{\infty} \frac{r_n}{1 - r_n} \leq \sum_{n=1}^{\infty} \frac{r_n}{1 - \theta} = \frac{\theta}{1 - \theta}$$

and hence the infinite product

$$\prod_{n=1}^{\infty} (1 - r_n)^{-1} = \prod_{n=1}^{\infty} \left(1 + \frac{r_n}{1 - r_n}\right)$$

converges. Hence there exists a constant $c > 0$ such that

$$(33.2) \qquad \sum_{\alpha \in I_n} p(c_\alpha f a^\alpha) \leq c \sup_L |f|$$

for every $f \in \mathcal{H}(E)$ and $n \in \mathbb{N}$. Since A is complete, it follows that the multiple series $\sum_{\alpha \in I} c_\alpha f\, a^\alpha$ converges absolutely in A for each $f \in \mathcal{H}(E)$. Define $T : (\mathcal{H}(E), \tau_c) \to A$ by $Tf = \sum_{\alpha \in I} c_\alpha f a^\alpha$ for each $f \in \mathcal{H}(E)$. Clearly T is linear. And T is actually an algebra homomorphism, since for all $f, g \in \mathcal{H}(E)$ we have that

$$T(fg) = \sum_{\gamma \in I} c_\gamma (fg) a^\gamma = \sum_{\gamma \in I} \Big(\sum_{\alpha+\beta=\gamma} c_\alpha f \cdot c_\beta g \Big) a^\gamma$$

$$= \Big(\sum_{\alpha \in I} c_\alpha f\, a^\alpha \Big)\Big(\sum_{\beta \in I} c_\beta g\, a^\beta \Big) = Tf \cdot Tg.$$

If follows from (33.2) that for each $p \in P$ the exist a compact set L in E and a constant $c > 0$ such that $p(Tf) \leq c \sup_L |f|$ for every $f \in \mathcal{H}(E)$. This shows that T is continuous. Since it is clear that $T(1) = e$ and $T(z_n) = a_n$ for every $n \in \mathbb{N}$, the proof of the theorem is complete.

33.4. DEFINITION. Let $\mathcal{H}_b(E)$ be the Fréchet algebra of all $f \in \mathcal{H}(E)$ which are bounded on bounded sets, with the topology given by the seminorms

$$p_n(f) = \sup \{ |f(x)| : \|x\| \leq n \} \qquad (n \in \mathbb{N}).$$

33.5. THEOREM. *Let E be a Banach space with a Schauder basis, and let A be a commutative, complete, Hausdorff locally m-convex algebra.*

(a) *If every complex homomorphism of $(\mathcal{H}(E), \tau_c)$ is bounded, then every complex homomorphism of A is bounded as well.*

(b) *If every complex homomorphism of $\mathcal{H}_b(E)$ is continuous, then every complex homomorphism of A is bounded.*

PROOF. Without loss of generality we may assume that E has a normalized Schauder basis $(e_n)_{n=1}^\infty$, in which case the sequence $(z_n)_{n=1}^\infty$ of coordinate functionals is equicontinuous. Suppose h is an unbounded complex homomorphism of A. Then there exists a bounded sequence (b_n) in A such that $|h(b_n)| > 8^n$ for every n. Set $a_n = 4^{-n} b_n$ for every n. Then for each continuous seminorm p on A there exists $c > 0$ such that $p(a_n) \leq 4^{-n} c$ for every n. Thus $\sum_{n=1}^\infty \sqrt{p(a_n)} < \infty$ for each continuous seminorm p on A. Then by Theorem 33.3 there is a continuous algebra homomorphism $T : (\mathcal{H}(E), \tau_c) \to A$ such that $T(1) = e$ and $T(z_n) = a_n$ for every n. Then $h \circ T$ is a complex homomorphism of $(\mathcal{H}(E), \tau_c)$ which is unbounded, since $|h \circ T(z_n)| = |h(a_n)| > 2^n$ for every n. For the same reason, the restriction of $h \circ T$ to $\mathcal{H}_b(E)$ is an unbounded complex homomorphism of $\mathcal{H}_b(E)$.

Theorem 33.5 shows in particular that Problems 33.1 and 33.2 are equivalent. It also shows that to solve Problem 33.1 for all commutative Fréchet algebras, it is sufficient to solve the problem for the algebra $\mathcal{H}_b(E)$, where E is some Banach space with a Schauder basis.

To conclude this section we shall prove that every complex homomorphism of the Fréchet algebra $(\mathcal{H}(\mathbb{C}^n), \tau_c)$ is continuous. This will be shown with the aid of the following proposition.

33.6. PROPOSITION. *Let A be an algebra of continuous complex-valued functions on a topological space X with the following properties:*

(a) *If $f_1, \ldots, f_m \in A$ are functions without common zeros then there are functions $g_1, \ldots, g_m \in A$ such that $f_1 g_1 + \ldots + f_m g_m = 1$.*

(b) *There are functions $\varphi_1, \ldots, \varphi_n \in A$ such that the set $\{x \in X : \varphi_j(x) = \lambda_j$ for $j = 1, \ldots, n\}$ is compact for each $(\lambda_1, \ldots, \lambda_n) \in \mathbb{C}^n$.*

Then for each complex homomorphism h of A there is a point $x \in X$ such that $h(f) = f(x)$ for every $f \in A$.

PROOF. Let h be a complex homomorphism of A and set $Z(f) = \{x \in X : f(x) = 0\}$ for every $f \in A$. Then it follows from condition (b) that the set $K = \bigcap_{j=1}^{n} Z(\varphi_j - h(\varphi_j))$ is compact. On the other hand it follows from condition (a) that $\bigcap_{j=1}^{m} Z(f_j)$ is nonvoid for all $f_1, \ldots, f_m \in \operatorname{Ker} h$. Whence it follows that the family $\mathcal{F} = \{K \cap Z(f) : f \in \operatorname{Ker} h\}$ is a collection of closed subsets of K with the finite intersection property. Hence $\cap \mathcal{F}$

is nonvoid, and this implies that $\cap\{Z(f) : f \in Ker\, h\}$ is nonvoid. If x belongs to this intersection then $f(x) = h(f)$ for every $f \in A$.

33.7. PROPOSITION. *Let U be a polynomially convex open set in $\mathbb{C}^n$. Then for each complex homomorphism h of $\mathcal{H}(U)$ there is a point $a \in U$ such that $h(f) = f(a)$ for every $f \in \mathcal{H}(U)$. In particular each complex homomorphism of $\mathcal{H}(U)$ is continuous.*

PROOF. It suffices to apply Proposition 33.6, for condition (a) is true by Proposition 32.14, whereas condition (b) is clearly satisfied by the coordinate functionals $z_1,\dots,z_n$.

EXERCISES

33.A. Let U be an open subset of E. Show that a complex homomorphism h of $(\mathcal{H}(U), \tau_c)$ is bounded if and only if for each increasing sequence (A_n) of open sets with $U = \bigcup_{n=1}^{\infty} A_n$ there exists $n \in \mathbb{N}$ such that $|h(f)| \le \sup_{A_n} |f|$ for every $f \in \mathcal{H}(U)$.

33.B. Let U be an open subset of a separable Banach space E, and let h be a bounded complex homomorphism of $(\mathcal{H}(U), \tau_c)$.

(a) Using Exercise 33.A show that the restriction of h to each equicontinuous subset of $\mathcal{H}(U)$ is continuous for the topology of pointwise convergence.

(b) Using the Banach-Dieudonné Theorem conclude that the restriction of h to $E'_c = (E', \tau_c)$ is continuous.

(c) Using the Mackey-Arens Theorem show the existence of a point $a \in E$ such that $h(P) = P(a)$ for every $P \in P_f(E)$.

33.C. By adapting the proof of the Banach - Dieudonné Theorem show that the topology of compact convergence is the finest topology on $P(^mE)$ which coincides with the topology of pointwise convergence on each equicontinuous subset of $P(^mE)$.

33.D. Let E be a separable Banach space with the approximation property, and let h be a bounded complex homomorphism of $(\mathcal{H}(E), \tau_c)$. Using Exercises 33.B and 33.C show the existence of a point $a \in E$ such that $h(f) = f(a)$ for every $f \in \mathcal{H}(E)$. In particular each bounded complex homomorphism of $(\mathcal{H}(E), \tau_c)$ is continuous.

33.E. Let A be a commutative Fréchet algebra, let $S(A)$ be endowed with the Gelfand topology, and let $\hat{A} \subset C(S(A))$ be the algebra of Gelfand transforms of elements of A.

(a) Using Proposition 32.13 show that the algebra $\hat{A}$ always satisfies condition (a) in Proposition 33.6.

(b) Suppose there are elements $x_1, \ldots, x_n \in A$ such that the set $\{h \in S(A) : h(x_j) = \lambda_j \text{ for } j = 1, \ldots, n\}$ is compact for each $(\lambda_1, \ldots, \lambda_n) \in \mathbb{C}^n$. Using Proposition 33.6 and Exercise 32.D show that each complex homomorphism of A is continuous.

NOTES AND COMMENTS

Section 29, 30 and 31 constitute a brief introduction to the theory of commutative Banach algebras. The interested reader is referred to the book of I. Gelfand, D. Raikov and G. Shilov [1] or to the book of W. Zelazko [1] for further reading. Theorem 31.7 was proved by G. Shilov [1] in the case of finitely generated algebras, and by R. Arens and A. Calderón [1] in the general case, using an integral formula of A. Weil [1]. The idea to use Oka's Extension Theorem 24.11 is due to L. Waelbroeck [1]. Theorem 31.8 is due to G. Shilov [1].

Locally m-convex algebras were extensively studied by E. Michael [1], who proved Corollary 32.4, Theorem 32.6 and Propositions 32.7, 32.8 and 33.6. Proposition 32.3 has been taken from the book of A. Guichardet [1]. Theorems 32.9 and 32.10, and Proposition 32.13 are due to R. Arens [1], though N. Bourbaki [1] refers to Theorem 32.10 as the Mittag - Leffler Theorem. I. Craw [1], D. Clayton [1] and M. Schottenloher

[6] have shown that to solve the Michael problem for arbitrary commutative Fréchet algebras it is sufficient to solve the problem for certain algebras of holomorphic functions. By adapting their ideas we have obtained Theorems 33.3 and 33.5. We point out that Theorems 33.3 and 33.5 are still true if E is any infinite dimensional Banach space. Indeed, R. Ovsepian and A. Pelczynski [1] have shown that if E is any infinite dimensional Banach space then there are a sequence (e_n) in E and a sequence (z_n) in E' such that $z_m(e_n) = \delta_{mn}$, $\|e_m\| = 1$ and $\|z_m\| \leq 20$ for all m and n in $\mathbb{N}$, and the proofs of Theorems 33.3 and 33.5 work equally well in this case. Finally we mention that the equivalence of Problems 33.1 and 33.2 was established first by P. Dixon and D. Fremlin [1].

CHAPTER VIII

PLURISUBHARMONIC FUNCTIONS

34. PLURISUBHARMONIC FUNCTIONS

In this section we introduce plurisubharmonic functions in Banach spaces and establish their basic properties. Throughout this chapter the letters E and F will represent complex Banach spaces.

We recall that if X is a topological space then a function $f : X \to [-\infty, \infty)$ is said to be *upper semicontinuous* if the set $\{x \in X : f(x) < c\}$ is open in X for each $c \in \mathbb{R}$. Likewise, a function $f : X \to (-\infty, \infty]$ is said to be *lower semicontinuous* if the set $\{x \in X : f(x) > c\}$ is open in X for each $c \in \mathbb{R}$. In the next proposition we give some properties of upper semicontinuous functions. The proofs are straightforward and are left to the reader as exercises.

34.1. PROPOSITION. *Let X be a topological space.*

(a) *The infimum of a family of upper semicontinuous functions on X is upper semicontinuous as well.*

(b) *If f is an upper semicontinuous function on X then for each compact subset K of X there exists $a \in K$ such that $f(x) \leq f(a)$ for every $x \in K$. The set $\{x \in K : f(x) = f(a)\}$ is compact.*

(c) *Let I be an open interval in $[-\infty, \infty)$. If $f : X \to I$ and $g : I \to [-\infty, \infty)$ are upper semicontinuous, and g is increasing, then $g \circ f$ is upper semicontinuous.*

34.2. PROPOSITION. *If X is a metric space then each upper semicontinuous* *function* $f : X \to [-\infty, \infty)$ *is the pointwise limit of a decreasing sequence of continuous functions* $f_n : X \to \mathbb{R}$.

PROOF. If $f \equiv -\infty$ then it suffices to take $f_n \equiv -n$ for each $n \in \mathbb{N}$. Thus we may assume $f \not\equiv -\infty$. Then consider the function $g = e^{-f} : X \to (0, \infty]$ and define

$$(34.1) \qquad g_n(x) = \inf_{z \in X} [g(z) + nd(x,z)]$$

for each $n \in \mathbb{N}$ and $x \in X$. It is clear that $g_n(x) \leq g_{n+1}(x) \leq g(x)$ for each $n \in \mathbb{N}$ and $x \in X$. Since $f \not\equiv -\infty$ we see that $g_n(x) < \infty$ for every $n \in \mathbb{N}$ and $x \in X$. We claim that $g_n(x) > 0$ for every $n \in \mathbb{N}$ and $x \in X$. Indeed, let $x \in X$ and $0 < c < g(x)$. Since g is lower semicontinuous there exists $\delta > 0$ such that $g(z) > c$ for every $z \in B(x;\delta)$. Then it follows from (34.1) that

$$(34.2) \qquad g_n(x) \geq \min \{c, n\delta\} > 0$$

for every $n \in \mathbb{N}$, as asserted. But it also follows from (34.2) that $g_n(x) \geq c$ for all sufficiently large n, and this shows that the sequence (g_n) converges pointwise to g. Next we claim that

$$(34.3) \qquad |g_n(x) - g_n(y)| \leq nd(x,y)$$

for all $n \in \mathbb{N}$ and $x, y \in X$. Indeed, for every $z \in X$ we have that

$$g_n(x) \leq g(z) + nd(x,z) \leq [g(z) + nd(y,z)] + nd(x,y).$$

Whence $g_n(x) \leq g_n(y) + nd(x,y)$, and after interchanging the roles of x and y, (34.3) follows. Then the sequence (f_n) defined by $f_n = -\log g_n$, has the required properties.

We recall that if U is an open set in $\mathbb{C}$ then a function $f : U \to [-\infty, \infty)$ is said to be *subharmonic* if f is upper semicontinuous and

$$f(a) \leq \frac{1}{2\pi} \int_0^{2\pi} f(a + re^{i\theta})d\theta$$

for each $a \in U$ and $r > 0$ such that $\overline{\Delta}(a;r) \subset U$. More generally, we have the following definition.

34.3. DEFINITION. Let U be an open subset of E. A function $f : U \to [-\infty, \infty)$ is said to be *plurisubharmonic* if f is upper semicontinuous and

$$f(a) \leq \frac{1}{2\pi} \int_0^{2\pi} f(a + e^{i\theta}b)d\theta$$

for each $a \in U$ and $b \in E$ such that $a + \overline{\Delta}b \subset U$. Observe that f is Borel measurable and, by Proposition 34.1, f is bounded above on each disc $a + \overline{\Delta}b \subset U$. Thus the integral $\int_0^{2\pi} f(a + e^{i\theta}b)d\theta$ is well defined, though it can be $-\infty$. We shall denote by $\mathcal{P}s(U)$ the set of all plurisubharmonic functions on U. We shall denote by $\mathcal{P}sc(U)$ the subset of all functions $f : U \to \mathbb{R}$ which are plurisubharmonic and continuous.

34.4. EXAMPLE. Let U be an open subset of E, and let $f \in \mathcal{H}(U)$. Then $Re f$, $Im f$ and $|f|$ belong to $\mathcal{P}sc(U)$.

PROOF. If $a + \overline{\Delta}b \subset U$ then

$$f(a) = \frac{1}{2\pi} \int_0^{2\pi} f(a + e^{i\theta}b)d\theta$$

by the Cauchy Integral Formula 7.1. Everything follows from this.

34.5. PROPOSITION. *Let U be an open subset of E.*

(a) *If $f, g \in \mathcal{P}s(U)$ and α, β are nonnegative constants then $\alpha f + \beta g \in \mathcal{P}s(U)$.*

(b) *If $f_i \in \mathcal{P}s(U)$ for every $i \in I$, $\sup_{i \in I} f_i < \infty$ on U and*

$\sup_{i\in I} f_i$ *is upper semicontinuous on* U, *then* $\sup_{i\in I} f_i \in \mathcal{P}s(U)$.

(c) *Let* $f : U \to [-\infty, \infty)$ *be upper semicontinuous. Then* f *is plurisubharmonic if and only if the restriction of* f *to* $U \cap M$ *is plurisubharmonic for each finite dimensional subspace* M *of* E.

(d) *If* f *is the pointwise limit of a decreasing sequence* $(f_n) \subset \mathcal{P}s(U)$ *the* $f \in \mathcal{P}s(U)$.

PROOF. Assertions (a), (b) and (c) are clear. To show (d) we first observe that f is upper semicontinuous by Proposition 34.1. If $a + \bar{\Delta}b \subset U$ then

$$(34.4) \qquad f(a) \leq f_n(a) \leq \frac{1}{2\pi}\int_0^{2\pi} f_n(a + e^{i\theta}b)d\theta$$

for every n. Since f_1 is bounded above on $a + \bar{\Delta}b$, an application of the Monotone Convergence Theorem shows that

$$\int_0^{2\pi} f(a + e^{i\theta}b)d\theta = \lim_{n\to\infty}\int_0^{2\pi} f_n(a + e^{i\theta}b)d\theta,$$

and the desired conclusion follows from (34.4).

34.6. COROLLARY. *Let* U *be an open subset of* E, *and let* $f \in \mathcal{H}(E;F)$. *Then* $\|f\|$ *belongs to* $\mathcal{P}sc(U)$.

PROOF. By the Hahn-Banach Theorem

$$\|f(x)\| = \sup\{|\psi \circ f(x)| : \psi \in F', \|\psi\| = 1\}$$

for each $x \in U$. Since $\psi \circ f \in \mathcal{H}(U)$ for each $\psi \in F'$ the conclusion follows from Example 34.4 and Proposition 34.5.

34.7. PROPOSITION. *Let* U *be a connected open set in* E, *and let* $f \in \mathcal{P}s(U)$.

(a) *If there exists* $a \in U$ *such that* $f(x) \leq f(a)$ *for every* $x \in U$ *then* $f(x) = f(a)$ *for every* $x \in U$.

(b) *If* $f \equiv -\infty$ *on a nonvoid open subset of* U *then* $f \equiv -\infty$ *on* U.

PROOF. (a) Since f is upper semicontinuous, the nonvoid set

$$A = \{x \in U : f(x) = f(a)\} = U \setminus \{x \in U : f(x) < f(a)\}$$

is closed in U. To show that $A = U$ it suffices to prove that A is open. Let $b \in A$ and let $r > 0$ such that $B(b;r) \subset U$. We claim that $B(b;r) \subset A$. Otherwise we can find $x \in B(b;r)$ such that $f(x) < f(a)$. Since f is upper semicontinuous we can find $\varepsilon > 0$ such that $B(x;\varepsilon) \subset B(a;r)$ and $f(y) < f(a)$ for every $y \in B(x;\varepsilon)$. This clearly implies that

$$\frac{1}{2\pi}\int_0^{2\pi} f[b + e^{i\theta}(x-b)]\,d\theta < \frac{1}{2\pi}\int_0^{2\pi} f(a)\,d\theta = f(a) = f(b),$$

a contradiction.

(b) Let A be the set of all $a \in U$ such that $f \equiv -\infty$ on a neighborhood of a. Then A is open and nonvoid. To complete the proof we show that A is closed in U. Let (a_n) be a sequence in A which converges to a point $a \in U$. First choose $r > 0$ such that $B(a;r) \subset U$, next choose $n \in \mathbb{N}$ such that $a_n \in B(a;r)$, and then choose $\varepsilon > 0$ such that $B(a_n;\varepsilon) \subset B(a;r)$ and $f \equiv -\infty$ on $B(a_n;\varepsilon)$. We claim that $f \equiv -\infty$ on $B(a;\varepsilon)$. Indeed, if $x \in B(a;\varepsilon)$ then $x + (a_n - a) \in B(a_n;\varepsilon)$ and it follows that

$$f(x) \le \frac{1}{2\pi}\int_0^{2\pi} f[x + e^{i\theta}(a_n - a)]\,d\theta = -\infty.$$

Thus $a \in A$ and the proof is complete.

Proposition 34.7(a) is often called the *maximum principle for plurisubharmonic functions*.

34.8. THEOREM. *Let* U *be an open subset of* E, *and let* $f : U \to [-\infty,\infty)$

be an upper semicontinuous function. Then the following conditions are equivalent:

(a) *For each* $a \in U$ *and* $b \in E$ *such that* $a + \bar{\Delta} b \subset U$ *we have that*

$$f(a) \leq \frac{1}{2\pi} \int_0^{2\pi} f(a + e^{i\theta} b) d\theta.$$

(b) *For each* $a \in U$, $b \in E$ *and* $\delta > 0$ *there exists* r *such that* $0 < r < \delta$, $a + r\bar{\Delta} b \subset U$ *and*

$$f(a) \leq \frac{1}{2\pi} \int_0^{2\pi} f(a + re^{i\theta} b) d\theta.$$

(c) *For each* $P \in \mathcal{P}(\mathbb{C})$ *and each* $a \in U$ *and* $b \in E$ *such that* $a + \bar{\Delta} b \subset U$ *the inequality* $f(a + \lambda b) \leq \operatorname{Re} P(\lambda)$ *for* $|\lambda| = 1$ *implies the same inequality for* $|\lambda| \leq 1$.

PROOF. The implication (a) $\Rightarrow$ (b) is obvious.

(b) $\Rightarrow$ (c): Let $P \in \mathcal{P}(\mathbb{C})$ and let $a \in U$ and $b \in E$ such that $a + \bar{\Delta} b \subset U$. Set $v(\lambda) = f(a + \lambda b) - \operatorname{Re} P(\lambda)$ for $|\lambda| \leq 1$. Suppose that $v(\lambda) \leq 0$ for every $\lambda \in \partial\Delta$ but $v(\lambda) > 0$ for some $\lambda \in \Delta$. Then $M = \sup_{\bar{\Delta}} v > 0$ and the set $A = \{\lambda \in \bar{\Delta} : v(\lambda) = M\}$ is a nonvoid compact subset of Δ, by Proposition 34.1. Thus $d(A, \partial\Delta) = \delta > 0$ and we can find $\lambda_o \in A$ such that $d(\lambda_o, \partial\Delta) = \delta$. By (b) we can find r such that $0 < r < \delta$, $a + \lambda_o b + r\bar{\Delta} b \subset U$ and

$$(34.5) \qquad f(a + \lambda_o b) \leq \frac{1}{2\pi} \int_0^{2\pi} f(a + \lambda_o b + re^{i\theta} b) d\theta.$$

By our choice of λ_o it is clear that

$$(34.6) \qquad \frac{1}{2\pi} \int_0^{2\pi} v(\lambda_o + re^{i\theta}) d\theta < \frac{1}{2\pi} \int_0^{2\pi} M \, d\theta = M.$$

From (34.5) and (34.6) it follows that

$$v(\lambda_o) \leq \frac{1}{2\pi} \int_0^{2\pi} v(\lambda_o + re^{i\theta}) d\theta < M,$$

a contradiction, since $\lambda_o \in A$.

(c) ⇒ (a): Let $a \in U$ and $b \in E$ such that $a + \bar{\Delta} b \subset U$. Let φ be a continuous, real-valued function on $[0, 2\pi]$ such that $\varphi(0) = \varphi(2\pi)$ and

$$(34.7) \qquad f(a + e^{i\theta} b) \leq \varphi(\theta) \quad \text{for} \quad 0 \leq \theta \leq 2\pi.$$

We shall prove that

$$(34.8) \qquad f(a) \leq \frac{1}{2\pi} \int_0^{2\pi} \varphi(\theta) d\theta.$$

Let $\varepsilon > 0$ be given. By the Stone-Weierstrass Theorem we can find a trigonometric polynomial $\psi(\theta) = \sum_{k=-n}^{n} c_k e^{ik\theta}$ such that $\varphi(\theta) \leq \psi(\theta) \leq \varphi(\theta) + \varepsilon$ for $0 \leq \theta \leq 2\pi$. Define $P \in P(\mathbb{C})$ by $P(\lambda) = c_o + 2 \sum_{k=1}^{n} c_k \lambda^k$. Since $c_k = \frac{1}{2\pi} \int_0^{2\pi} \psi(\theta) e^{ik\theta} d\theta$, we see that $\bar{c}_k = c_{-k}$ for $0 \leq k \leq n$, and it follows that $\psi(\theta) = \operatorname{Re} P(e^{i\theta})$ for $0 \leq \theta \leq 2\pi$. Hence $f(a + e^{i\theta} b) \leq \varphi(\theta) \leq \psi(\theta) = \operatorname{Re} P(e^{i\theta})$ for $0 \leq \theta \leq 2\pi$, and (c) implies that $f(a + \lambda b) \leq \operatorname{Re} P(\lambda)$ for $|\lambda| \leq 1$. Thus

$$f(a) \leq \operatorname{Re} P(0) = c_o = \frac{1}{2\pi} \int_0^{2\pi} \psi(\theta) d\theta \leq \frac{1}{2\pi} \int_0^{2\pi} \varphi(\theta) d\theta + \varepsilon,$$

and since $\varepsilon > 0$ was arbitrary, (34.8) follows. Now, by Proposition 34.2 we can find a decreasing sequence of continuous functions $\varphi_n : [0, 2\pi] \to \mathbb{R}$ such that $\varphi_n(0) = \varphi_n(2\pi)$ and $\lim_{n \to \infty} \varphi_n(\theta) = f(a + e^{i\theta} b)$ for every $\theta \in [0, 2\pi]$. Thus each φ_n satisfies (34.7) and therefore satisfies (34.8) as well. Then an application of the Monotone Convergence Theorem shows that

$$f(a) \leq \frac{1}{2\pi} \int_0^{2\pi} f(a + e^{i\theta} b) d\theta$$

and the proof of the theorem is complete.

34.9. COROLLARY. *Let U be an open subset of E. A function $f : U \to [-\infty, \infty)$ is plurisubharmonic if and only if for each $x \in U$ there exists an open neighborhood V of x in U such that $f|V$ is plurisubharmonic.*

34.10. COROLLARY. *Let U be an open subset of E.*

(a) *If $f \in \mathcal{H}(U)$ then $\log|f| \in \mathcal{P}s(U)$.*

(b) *If $f \in \mathcal{H}(U;F)$ then $\log\|f\| \in \mathcal{P}s(U)$.*

PROOF. (a) We verify condition (c) in Theorem 38.8. Let $a \in U$ and $b \in E$ such that $a + \bar{\Delta}b \subset U$ and let $P \in \mathcal{P}(\mathbb{C})$ such that $\log|f(a + \lambda b)| \leq Re\, P(\lambda)$ for $|\lambda| = 1$. Then $f(a + \lambda b) \leq e^{Re\, P(\lambda)} = |e^{P(\lambda)}|$ for $|\lambda| = 1$. Thus $|e^{-P(\lambda)} f(a + \lambda b)| \leq 1$ for $|\lambda| = 1$, and therefore for $|\lambda| \leq 1$, by the Maximum Principle. It follows that $\log|f(a + \lambda b)| \leq Re\, P(\lambda)$ for $|\lambda| \leq 1$, and (a) has been proved.

(b) Since

$$\log\|f(x)\| = \sup\{\log|\psi \circ f(x)| : \psi \in F', \|\psi\| = 1\},$$

it suffices to apply (a) and Proposition 34.5.

We recall that if I is an interval in $\mathbb{R}$ then a function $\varphi : I \to \mathbb{R}$ is said to be *convex* if

$$\varphi[(1 - \lambda)a + \lambda b] \leq (1 - \lambda)\varphi(a) + \lambda\varphi(b)$$

for all $a, b \in I$ and $0 \leq \lambda \leq 1$.

34.11. LEMMA. *Let I be an open interval in $\mathbb{R}$, and let $\varphi : I \to \mathbb{R}$ be a convex function. Then:*

(a) *φ is upper semicontinuous.*

(b) *For each $a \in I$ there exists $k \in \mathbb{R}$ such that $\varphi(x) - \varphi(a) \geq k(x - a)$ for every $x \in I$. In particular φ is lower* *semincontinuous and therefore continuous.*

PROOF. (a) Let $a \in I$ and suppose $\varphi(a) < c$. Choose $t > 0$ such that $[a - t, a + t] \subset I$. Then for $0 \leq \lambda \leq 1$ we have that

$$(34.9)\quad \varphi(a + \lambda t) = \varphi[(1 - \lambda)a + \lambda(a + t)] \leq (1 - \lambda)\varphi(a) + \lambda\varphi(a + t)$$

and

$$(34.10)\quad \varphi(a - \lambda t) = \varphi[(1 - \lambda)a + \lambda(a - t)] \leq (1 - \lambda)\varphi(a) + \lambda\varphi(a - t).$$

Using (34.9) we can find δ_1 with $0 < \delta_1 < 1$ such that $\varphi(a + \lambda t) < c$ for $0 < \lambda < \delta_1$. And using (34.10) we can find δ_2 with $0 < \delta_2 < 1$ such that $\varphi(a - \lambda t) < c$ for $0 < \lambda < \delta_2$. If we set $\delta = min\{\delta_1, \delta_2\}$ then $\varphi(a + \lambda t) < c$ for $-\delta < \lambda < \delta$.

(b) Let $x, y \in I$ with $x < a < y$. Then we can write

$$a = \frac{y - a}{y - x}\, x + \frac{a - x}{y - x}\, y.$$

Since φ is convex we have that

$$\varphi(a) \leq \frac{y - a}{y - x}\, \varphi(x) + \frac{a - x}{y - x}\, \varphi(y),$$

and it follows that

$$\frac{\varphi(x) - \varphi(a)}{x - a} \leq \frac{\varphi(y) - \varphi(a)}{y - a}.$$

Hence we can find $k \in \mathbb{R}$ such that

$$\sup_{\substack{x \in I \\ x < a}} \frac{\varphi(x) - \varphi(a)}{x - a} \leq k \leq \inf_{\substack{x \in I \\ x > a}} \frac{\varphi(x) - \varphi(a)}{x - a}$$

and (b) follows.

34.12. PROPOSITION. *Let U be an open subset of E, let $f \in Ps(U)$ and let $\varphi : \mathbb{R} \to \mathbb{R}$ be a convex increasing function. If* *we define* $\varphi(-\infty) = \lim_{t \to -\infty} \varphi(t)$ *then* $\varphi \circ f \in \mathcal{P}s(U)$.

PROOF. By Lemma 34.11 and Proposition 34.1 the function $\varphi \circ f$ is upper semicontinuous. Let $a \in U$ and $b \in E$ such that $a + \bar{\Delta} b \subset U$. By Lemma 34.11 for each $t_o \in \mathbb{R}$ there exists $k \in \mathbb{R}$ such that $\varphi(t) \geq \varphi(t_o) + k(t - t_o)$ for every $t \in \mathbb{R}$. In particular

$$\varphi[f(a + e^{i\theta}b)] \geq \varphi(t_o) + k[f(a + e^{i\theta}b) - t_o]$$

for every $\theta \in [0, 2\pi]$, and therefore

$$\frac{1}{2\pi}\int_0^{2\pi} \varphi[f(a + e^{i\theta}b)]d\theta \geq \varphi(t_o) + k[\frac{1}{2\pi}\int_0^{2\pi} f(a + e^{i\theta}b)dt - t_o].$$

If we choose $t_o = \frac{1}{2\pi}\int_0^{2\pi} f(a + e^{i\theta}b)dt \geq f(a)$ then we obtain

$$\frac{1}{2\pi}\int_0^{2\pi} \varphi[f(a + e^{i\theta}b)]d\theta \geq \varphi(t_o) \geq \varphi(f(a)),$$

and the proof is complete.

EXERCISES

34.A. Let X and Y be topological spaces. Show that if $f : X \to Y$ is continuous and $g : Y \to [-\infty, \infty)$ is upper semicontinuous then $g \circ f$ is upper semicontinuous.

34.B. Let U be an open subset of E, and let $f : U \to [-\infty, \infty)$ be an upper semicontinuous function. Show that given a compact set $K \subset U$ and $\varepsilon > 0$ we can find $\delta > 0$ such that $K + B(0;\delta) \subset U$ and $f(y) < f(x) + \varepsilon$ for each $x \in K$ and $y \in B(x;\delta)$.

34.C. Let U be an open subset of E, and let $f, g : U \to [0,\infty)$ be two functions such that $\log f$ and $\log g$ belong to $\mathcal{P}s(U)$. Show that $\log(f + g) \in \mathcal{P}s(U)$.

34.D. Let U be an open subset of E and let $f \in \mathcal{H}(U;F)$.

(a) Show that each of the functions $u_m(x) = \sup_{k \geq m} \| P^k(x) \|$ is real-valued and continuous on U.

(b) Show that the function $-\log r_c f(x)$ is plurisubharmonic on U.

35. REGULARIZATION OF PLURISUBHARMONIC FUNCTIONS

This section is devoted to the study of plurisubharmonic functions of class C^2. We show that each plurisubharmonic function on an open subset of $\mathbb{C}^n$ is the pointwise limit of a decreasing sequence of plurisubharmonic functions of class C^∞.

35.1. LEMMA. *Let U be an open set in $\mathbb{C}$. Then for each function $f \in C^2(U;\mathbb{R})$ the following conditions are equivalent:*

(a) *f is subharmonic on U.*

(b) $$4 \frac{\partial^2 f}{\partial z \partial \bar{z}} = \frac{\partial^2 f}{\partial x^2} + \frac{\partial^2 f}{\partial y^2} \geq 0 \quad \text{on } U.$$

(c) *For each $a \in U$ the integral*

$$M(r) = \frac{1}{2\pi} \int_0^{2\pi} f(a + re^{i\theta}) d\theta \qquad (r \geq 0)$$

is an increasing function of r.

PROOF. (a) ⇒ (b): Let $\bar{\Delta}(a;r) \subset U$. Since f is subharmonic,

$$\int_0^{2\pi} [f(a + re^{i\theta}) - f(a)] d\theta \geq 0.$$

Now, by Taylor's formula,

$$f(a + re^{i\theta}) - f(a) = r \cos\theta \frac{\partial f}{\partial x}(a) + r \sin\theta \frac{\partial f}{\partial y}(a)$$

$$+ \frac{r^2}{2} \cos^2\theta \frac{\partial^2 f}{\partial x^2}(c_{r\theta}) + \frac{r^2}{2} \sin^2\theta \frac{\partial^2 f}{\partial y^2}(c_{r\theta}) + r^2 \cos\theta \sin\theta \frac{\partial^2 f}{\partial x \partial y}(c_{r\theta}),$$

where $c_{r\theta} \in [a, a + re^{i\theta}]$. Thus

$$\frac{r^2}{2}\int_0^{2\pi} [\cos^2\theta \frac{\partial^2 f}{\partial x^2}(c_{r\theta}) + \sin^2\theta \frac{\partial^2 f}{\partial y^2}(c_{r\theta}) + 2\cos\theta\sin\theta \frac{\partial^2 f}{\partial x \partial y}(c_{r\theta})]\, d\theta \geq 0,$$

and after dividing by $r^2/2$, an application of the Dominated Convergence Theorem shows that $\pi[\frac{\partial^2 f}{\partial x^2}(a) + \frac{\partial^2 f}{\partial y^2}(a)] \geq 0$, as we wanted.

(b) ⇒ (c): One can readily verify that

$$\frac{\partial^2 f}{\partial x^2} + \frac{\partial^2 f}{\partial y^2} = \frac{\partial^2 f}{\partial r^2} + \frac{1}{r}\frac{\partial f}{\partial r} + \frac{1}{r^2}\frac{\partial^2 f}{\partial \theta^2}.$$

Then condition (b) implies that

$$\int_0^{2\pi} (\frac{\partial^2}{\partial r^2} + \frac{1}{r}\frac{\partial}{\partial r}) f(a + re^{i\theta})d\theta \geq 0,$$

that is, $M''(r) + r^{-1}M'(r) \geq 0$. Thus $[rM'(r)]' = rM''(r) + M'(r) \geq 0$ and $rM'(r)$ is an increasing function of r. Since it is clear that $rM'(r) \to 0$ when $r \to 0$, we conlude that $rM'(r) \geq 0$ for every $r > 0$. Hence $M'(r) \geq 0$ for every $r > 0$ and $M(r)$ is an increasing function of r.

Since clearly (c) ⇒ (a), the proof of the lemma is complete.

If U is an open subset of E and $f \in C^2(U;\mathbb{R})$ then it follows from Exercise 35.A that $D'D''f(a)$ is an Hermitian form for each $a \in U$. In other words, $D'D''f(a)(s,t)$ is $\mathbb{C}$-linear in s, $\mathbb{C}$-antilinear in t, and $D'D''f(a)(s,t) = \overline{D'D''f(a)(t,s)}$. If $E = \mathbb{C}^n$ then one can readily prove that

$$D'D''f(a)(s,t) = \sum_{j,k=1}^{n} \frac{\partial^2 f}{\partial z_j \partial \bar{z}_k}(a)s_j\bar{t}_k$$

for each $a \in U$ and $s, t \in \mathbb{C}^n$. The Hermitian form $D'D''f(a)$ serves to characterize plurisubharmonic functions as follows.

35.2. PROPOSITION. *Let U be an open subset of E, and let $f \in C^2(U;\mathbb{R})$. Then f is plurisubharmonic on U if and only if the Hermitian form $D'D''f(a)$ is positive for each $a \in U$, that is, $D'D''f(a)(t,t) \geq 0$ for each $a \in U$ and $t \in E$.*

PROOF. Given $a \in U$ and $t \in E$ consider the functions $u(\zeta) = f(a + \zeta t)$, which is defined on a suitable disc $\Delta(0;r)$. Then it follows from Exercises 35.B and 35.D that

$$\frac{\partial^2 u}{\partial\zeta\partial\bar{\zeta}}(\zeta) = D'D''f(a + \zeta t)(t,t).$$

Hence the desired conclusion follows from Lemma 35.1.

35.3. DEFINITION. Let U be an open subset of E. A function $f \in C^2(U;\mathbb{R})$ is said to be *strictly plurisubharmonic* on U if the Hermitian form $D'D''f(a)$ is strictly positive for each $a \in U$, that is, $D'D''f(a)(t,t) > 0$ for each a in U and each $t \neq 0$ in E.

The following lemma will be useful later on.

35.4. LEMMA. *Let U be an open subset of E, and let f be a strictly plurisubharmonic function on U. Then there exists a strictly positive function $\varphi \in C^\infty(U;\mathbb{R})$ such that*

$$\sum_{j,k=1}^{n} \frac{\partial^2 f}{\partial z_j \partial \bar{z}_k}(a) t_j \bar{t}_k \geq \varphi(a) \sum_{j=1}^{n} |t_j|^2$$

for each $a \in U$ and $t \in \mathbb{C}^n$.

PROOF. Let $S = \{t \in \mathbb{C}^n : \sum_{j=1}^{n} |t_j|^2 = 1\}$. Let $(U_i)_{i=1}^{\infty}$ be an increasing sequence of relatively compact open subsets of U such that $U = \bigcup_{i=1}^{\infty} U_i$. Since f is strictly plurisubharmonic on U it follows that

$$\inf \{ \sum_{j,k=1}^{n} \frac{\partial^2 f}{\partial z_j \partial \bar{z}_k}(a) t_j \bar{t}_k : a \in \bar{U}_i,\ t \in S\} = c_i > 0$$

for every $i \in \mathbb{N}$. If we set $U_o = \phi$ then by Exercise 15.C there exists a function $\psi \in C^\infty(U;\mathbb{R})$ such that $\psi \geq 1/c_i$ on $U_i \setminus U_{i-1}$ for every $i \in \mathbb{N}$. If we define $\varphi = 1/\psi$ then $\varphi \in C^\infty(U;\mathbb{R})$, φ is strictly positive and

$$\sum_{j,k=1}^{n} \frac{\partial^2 f}{\partial z_j \partial \bar{z}_k}(a) t_j \bar{t}_k \geq c_i \geq \varphi(a)$$

for every $a \in U_i \setminus U_{i-1}$ and $t \in S$. The desired conclusion follows.

35.5. PROPOSITION. *Let U be a connected open subset of $\mathbb{C}^n$ and let $f \in \mathcal{P}s(U)$. If $f \not\equiv -\infty$ then $f \in L^1(U, loc)$.*

PROOF. We first show that

$$(35.1) \qquad f(a) \leq \frac{1}{(\pi r^2)^n} \int_{\bar{\Delta}^n(a;r)} f(\zeta) d\lambda(\zeta)$$

for each compact plydisc $\bar{\Delta}^n(a;r) \subset U$. Indeed, if $(e_1, \ldots, e_n)$ denotes the canonical basis of $\mathbb{C}^n$ then for $0 < \rho \leq r$ we have that

$$f(a) \leq \frac{1}{2\pi} \int_0^{2\pi} f(a + \rho e^{i\theta} e_1) d\theta,$$

and it follows that

$$f(a) \leq \frac{1}{\pi r^2} \int_0^r \rho d\rho \int_0^{2\pi} f(a + \rho e^{i\theta} e_1) d\theta.$$

Since f is bounded above on each compact subset of U, Fubini's Theorem allows us to replace the iterated integral by a double integral. Thus

$$f(a) \leq \frac{1}{\pi r^2} \int_{\bar{\Delta}(0;r)} f(a + \zeta_1 e_1) d\lambda(\zeta_1).$$

By iteration we get that

$$f(a) \leq \frac{1}{(\pi r^2)^n} \int_{\overline{\Delta}(0;r)} d\lambda(\zeta_1) \ldots \int_{\overline{\Delta}(0;r)} f(a + \zeta_1 e_1 + \ldots + \zeta_n e_n) d\lambda(\zeta_n),$$

and another application of Fubini's Theorem yields (35.1).

Now we show that f is Lebesgue integrable on a neighborhood of every point. Indeed, let $a \in U$ and choose $r > 0$ so that $\overline{\Delta}^n(a;2r) \subset U$. Then we can find a point $b \in \Delta^n(a;r)$ such that $f(b) > -\infty$, for otherwise f would be identically $-\infty$ on U, by Proposition 34.7. Since $\overline{\Delta}^n(b;r) \subset \overline{\Delta}^n(a;2r) \subset U$, the function f is Lebesgue integrable on $\overline{\Delta}^n(b;r)$, by (35.1). Since $a \in \Delta^n(b;r)$, the proof is complete.

35.6. THEOREM. *Let U be a connected open subset of $\mathbb{C}^n$, and let $f \in \mathcal{Ps}(U)$, $f \not\equiv -\infty$. Then:*

(a) $f * \rho_\delta \in C^\infty(U_\delta) \cap \mathcal{Ps}(U_\delta)$ *for each* $\delta > 0$.

(b) $f * \rho_\delta(z)$ *decreases to* $f(z)$ *when* δ *decreases to zero, for each* $z \in U$.

PROOF. Since $f \in L^1(U, loc)$, by Proposition 35.5, the convolution $f * \rho_\delta$ is well defined and belongs to $C^\infty(U_\delta)$, By Proposition 16.4. To show that $f * \rho_\delta \in \mathcal{Ps}(U_\delta)$ let $a \in U_\delta$ and $b \in \mathbb{C}^n$ such that $a + \overline{\Delta} b \subset U_\delta$. Then an application of Fubini's Theorem shows that

$$(f * \rho_\delta)(a) = \int_{\mathbb{C}^n} f(a - \zeta)\rho_\delta(\zeta) d\lambda(\zeta)$$

$$\leq \frac{1}{2\pi} \int_{\mathbb{C}^n} \rho_\delta(\zeta) d\lambda(\zeta) \int_0^{2\pi} f(a - \zeta + e^{i\theta} b) d\theta$$

$$= \frac{1}{2\pi} \int_0^{2\pi} (f * \rho_\delta)(a + e^{i\theta} b) d\theta,$$

proving that $f * \rho_\delta$ is plurisubharmonic. Next we show that $f * \rho_\delta$ converges pointwise to f when $\delta \to 0$. Indeed, given $a \in U$ and $c > f(a)$ we choose $r > 0$ such that $\overline{\Delta}^n(a;r) \subset U$ and

$f \leq c$ on $\overline{\Delta}^n(a;r)$. We claim that

(35.2) $$f(a) \leq (f * \rho_\delta)(a) \leq c$$

for $0 < \delta \leq r$. Indeed, since the function $\rho_\delta(\zeta)$ depends only of $|\zeta_1|,\ldots,|\zeta_n|$, the proof of (35.2) is almost a repetition of the proof of (35.1). We leave the details to the reader.

We still have to show that $(f * \rho_\delta)(a)$ is an increasing function of δ for each $a \in U$. We first show that for each $\varepsilon > 0$ the integral

(35.3) $$\int_{\mathbb{C}^n} (f * \rho_\varepsilon)(a - \delta\zeta)\rho(\zeta)d\lambda(\zeta)$$

is an increasing function of δ. Indeed, this follows from repeated applications of the condition (c) in Lemma 35.1 to the function $f * \rho_\varepsilon$, which is plurisubharmonic and of class C^∞ on U_ε. Next we claim that for each $\delta > 0$ the integral in (35.3) converges to the integral

(35.4) $$\int_{\mathbb{C}^n} f(a - \delta\zeta)\rho(\zeta)d\lambda(\zeta) = (f * \rho_\delta)(a)$$

when $\varepsilon \to 0$. Indeed, on one hand $f * \rho_\varepsilon$ converges pointwise to f when $\varepsilon \to 0$. On the other hand, the estimates in (35.2) allow us to apply the Dominated Convergence Theorem to get the desired conclusion. Since the integral in (35.3) is an increasing function of δ for each $\varepsilon > 0$, the integral in (35.4) is also an increasing function of δ. This completes the proof of the theorem.

35.7. THEOREM. *Let $U \subset E$ and $V \subset F$ be open sets, let $f \in \mathcal{H}(U;F)$ and $g \in \mathcal{Ps}(V)$, with $f(U) \subset V$. Then $g \circ f \in \mathcal{Ps}(U)$.*

PROOF. (a) First assume that g is of class C^2. Then by Exercise 35.D we have that

$$D'D''(g \circ f)(x)(t,t) = D'D''g(f(x))(D'f(x)(t),D'f(x)(t)),$$

and the desired conclusion follows from Proposition 35.2.

(b) Next assume that F is finite dimensional. By Theorem 35.6 we can find an increasing sequence of open sets V_j with $V = \bigcup_{j=1}^{\infty} V_j$, and a decreasing sequence of functions $g_j \in \mathcal{P}s(V_j) \cap C^{\infty}(V_j)$ which converges pointwise to g in V. If we set $U_j = f^{-1}(V_j)$ then $g_j \circ f \in \mathcal{P}s(U_j)$ for every j by part (a). Since the sequence $(g_j \circ f)$ is decreasing and converges pointwise to $g \circ f$ in U, we conclude that $g \circ f \in \mathcal{P}s(U)$ by Proposition 34.5.

(c) We finally deal with the general case. By Exercise 34.A the function $g \circ f$ is upper semicontinuous. We still have to show that

$$(35.5) \qquad g \circ f(a) \leq \frac{1}{2\pi} \int_0^{2\pi} g \circ f(a + e^{i\theta} b)\, d\theta$$

for every $a \in U$ and $b \in E$ such that $a + \overline{\Delta} b \subset U$. Since $f(a + \lambda b)$ is a holomorphic function of λ on a neighborhood of the disc $\overline{\Delta}$, we have a series expansion $f(a + \lambda b) = \sum_{m=0}^{\infty} c_m \lambda^m$, where the coefficients c_m lie in F and the series converges uniformly on $\overline{\Delta}$. Consider the partial sums $S_m(\lambda) = \sum_{j=0}^{m} c_j \lambda^j$. Since $f(a + \overline{\Delta} b) \subset V$ we can find $m_0 \in \mathbb{N}$ such that $S_m(\overline{\Delta}) \subset V$ for every $m \geq m_0$. Let F_m be the vector subspace of F generated by $c_0, c_1, \ldots, c_m$. By part (b), the function $g \circ S_m$ is subharmonic on an open neighborhood of $\overline{\Delta}$, for each $m \geq m_0$. Hence

$$(35.6) \qquad g \circ f(a) = g \circ S_m(0) \leq \frac{1}{2\pi} \int_0^{2\pi} g \circ S_m(e^{i\theta})\, d\theta$$

for every $m \geq m_0$. Let $\varepsilon > 0$ be given. By Exercise 34.B we can find $\delta > 0$ such that $f(a + \overline{\Delta} b) + B(0;\delta) \subset V$ and $g(y) < g(x) + \varepsilon$ for every $x \in f(a + \overline{\Delta} b)$ and $y \in B(x;\delta)$. Choose $m_1 \geq m_0$ such that $\| S_m(\lambda) - f(a + \lambda b) \| < \delta$ for all $\lambda \in \overline{\Delta}$ and $m \geq m_1$. Then

(35.7) $$g \circ S_m(\lambda) < g \circ f(a + \lambda b)$$

for all $\lambda \in \overline{\Delta}$ and $m \geq m_1$. From (35.6) and (35.7) it follows that

$$g \circ f(a) \leq \frac{1}{2\pi} \int_0^{2\pi} g \circ f(a + e^{i\theta} b) d\theta + \varepsilon,$$

and since $\varepsilon > 0$ was arbitrary, (35.5) follows. The proof of the theorem is now complete.

35.8. **PROPOSITION.** *Let U be an open subset of $\mathbb{C}^n$. Let (f_j) be a sequence of plurisubharmonic functions on U such that:*

(a) *The sequence (f_j) is bounded above on each compact subset of U.*

(b) *There exists $c \in \mathbb{R}$ such that $\limsup_{j \to \infty} f_j(z) \leq c$ for every $z \in U$.*

Then for each compact set $K \subset U$ and $\varepsilon > 0$ there exists j_o such that $f_j(z) \leq c + \varepsilon$ for every $j \geq j_o$.

PROOF. Let K and ε be given. Let $a \in K$ and choose $r > 0$ such that $\overline{\Delta}^n(a;3r) \subset U$. By (a) we may assume without loss of generality that $f_j \leq 0$ on $\overline{\Delta}^n(a;3r)$ for every j. Then an application of Fatou's Lemma shows that

$$\limsup_{j \to \infty} \int_{\overline{\Delta}^n(a;r)} f_j d\lambda \leq \int_{\overline{\Delta}^n(a;r)} (\limsup_{j \to \infty} f_j) d\lambda \leq c(\pi r^2)^n.$$

Hence we can find j_o such that

$$\int_{\overline{\Delta}^n(a;r)} f_j d\lambda \leq (c + \varepsilon)(\pi r^2)^n$$

for every $j \geq j_o$. Let $0 < \delta < r$ and let $z \in \overline{\Delta}^n(a;\delta)$. Then

$$\overline{\Delta}^n(a;r) \subset \overline{\Delta}^n(z;r + \delta) \subset \overline{\Delta}^n(a;r + 2\delta) \subset U.$$

Now, by (35.1) we have that

$$f_j(z)\pi^n(r+\delta)^{2n} \leq \int_{\overline{\Delta}^n(z;r+\delta)} f_j \, d\lambda$$

for every j. Since $f_j \leq 0$ on $\overline{\Delta}^n(a;3r)$ it follows that

$$f_j(z)\pi^n(r+\delta)^{2n} \leq \int_{\overline{\Delta}^n(z;r+\delta)} f_j \, d\lambda \leq \int_{\overline{\Delta}^n(a;r)} f_j \, d\lambda \leq (c+\varepsilon)\pi^n r^{2n}$$

for every $j \leq j_o$. Thus

$$f_j(z) \leq (c+\varepsilon)\left(\frac{r}{r+\delta}\right)^{2n} \leq c + 2\varepsilon$$

for every $z \in \overline{\Delta}^n(a;\delta)$ and $j \geq j_o$, provided $\delta > 0$ is sufficiently small. Since K can be covered by finitely many polydiscs, the proof is complete.

EXERCISES

35.A. Let U be an open subset of E, and let $f \in C^2(U;F)$. Show that

$$4D'D''f(a)(s,t) = D^2f(a)(s,t) + D^2f(a)(is,it)$$
$$+ D^2f(a)(s,it) - iD^2f(a)(is,t)$$

for all $a \in U$ and $s, t \in E$.

35.B. Let U be an open subset of $\mathbb{C}^n$, and let $f \in C^2(U;F)$. Show that

$$D'D''f(a)(s,t) = \sum_{j,k=1}^{n} \frac{\partial^2 f}{\partial z_j \partial \overline{z}_k}(a) s_j \overline{t}_k$$

for all $a \in U$ and $s, t \in \mathbb{C}^n$.

35.C. Let E, F, G be complex Banach spaces, let $U \subset E$ and $V \subset F$ be open sets, and let $f \in C^1(U;F)$ and $g \in C^1(V;F)$ with $f(U) \subset V$. Show that

$$D'(g \circ f)(x) = D'g(f(x)) \circ D'f(x) + D''g(f(x)) \circ D''f(x)$$

and

$$D''(g \circ f)(x) = D'g(f(x)) \circ D''f(x) + D''g(f(x)) \circ D'f(x)$$

for every $x \in U$.

35.D. Let E, F, G be complex Banach spaces, let $U \subset E$ and $V \subset F$ be open sets, and let $f \in \mathcal{H}(U;F)$ and $g \in C^2(V;F)$ with $f(U) \subset V$. Show that

$$D'D''(g \circ f)(x)(s,t) = D'D''g(f(x))(D'f(x)(s), D'f(x)(t))$$

for all $x \in U$ and $s, t \in E$.

35.E. Let $U \subset E$ and $V \subset \mathbb{R}$ be open sets, and let $f \in C^2(U;\mathbb{R})$ and $g \in C^2(V;\mathbb{R})$ with $f(U) \subset V$.

(a) Show that

$$D'D''(g \circ f)(x)(s,t) = g''(f(x))D'f(x)(s)\,\overline{D'f(x)(t)}$$
$$+ g'(f(x)D'D''f(x)(s,t)$$

for all $x \in U$ and $s, t \in E$.

(b) Show that if f is plurisubharmonic and g is convex and increasing then $g \circ f$ is plurisubharmonic.

(c) Show that if f is strictly plurisubharmonic and g is convex and strictly increasing then $g \circ f$ is strictly plurisubharmonic.

35.F. Let U be a connected open subset of $\mathbb{C}^n$.

(a) Show that if $f \in \mathcal{P}\!s(U)$ is not identically $-\infty$ then the set $\{z \in U : f(x) = -\infty\}$ has Lebesgue measure zero.

(b) Show that if $f \in \mathcal{H}(U;F)$ is not identically zero then the set $\{z \in U : f(x) = 0\}$ has Lebesgue measure zero.

36. SEPARATELY HOLOMORPHIC MAPPINGS

In this section we prove that each separately holomorphic mapping is holomorphic, a result which had been promised in Section 8.

36.1. THEOREM. *Let U be an open subset of $\mathbb{C}^n$. Then a mapping $f : U \to F$ is holomorphic if and only if f is separately holomorphic.*

To prove this theorem we need two auxiliary lemmas.

36.2. LEMMA. *Let E, F, G be Banach spaces, let $U \subset E$ and $V \subset F$ be open sets, and let $f : U \times V \to G$ be a separately continuous mapping. Then for each compact set $L \subset V$ there are nonvoid open sets A and B, with $A \subset U$ and $L \subset B \subset V$, such that f is bounded on $A \times B$.*

PROOF. Consider the sets

$$A_n = \{x \in U : |f(x,y)| \leq n \quad \text{for every} \quad y \in L + B_F(0;1/n)\}.$$

Since $f(x,y)$ is a continuous function of x when y is held fixed, each A_n is closed in U. And since $f(x,y)$ is a continuous function of y when x is held fixed, $U = \bigcup_{n=1}^{\infty} A_n$. By the Baire Category Theorem some A_n contains an open ball $B_E(a;r)$. Thus $\|f\| \leq n$ on $B_E(a;r) \times [L + B_F(0;1/n)]$, as asserted.

36.3. LEMMA. *Let $n \geq 2$ and let $f : \Delta^n(a;R) \to F$ be a mapping which is holomorphic in $z' = (z_1,\dots,z_{n-1})$ for z_n fixed,* *and holomorphic in* z_n *for* z' *fixed. If* f *is bounded on a polydisc* $\Delta^{n-1}(a';r) \times \Delta(a_n;R)$, *with* $0 < r < R$, *then* f *is holomorphic on* $\Delta^n(a;R)$.

PROOF. Since f is holomorphic in z' for z_n fixed, for each $z \in \Delta^n(a;R)$ we have a series expansion

$$(36.1) \qquad f(z) = f(z',z_n) = \sum_\alpha c_\alpha(z_n)(z'-a')^\alpha,$$

where the summation is taken over all multi-indices $\alpha = (\alpha_1,\ldots,\alpha_{n-1}) \in \mathbb{N}_0^{n-1}$, and

$$c_\alpha(z_n) = \frac{1}{(2\pi i)^{n-1}} \int_{\partial_0 \Delta^{n-1}(a';r)} \frac{f(\zeta,z_n)d\zeta_1 \ldots d\zeta_{n-1}}{(\zeta_1 - a_1)^{\alpha_1+1} \ldots (\zeta_{n-1} - a_{n-1})^{\alpha_{n-1}+1}}.$$

Since f is holomorphic in z_n for z' fixed, we see that each $c_\alpha(z_n)$ is a holomorphic function of z_n. Thus, to show that f is holomorphic on $\Delta^n(a;R)$ it is sufficient to show that the series in (36.1) converges uniformly for $z \in \Delta^n(a,R_1)$ whenever $0 < R_1 < R$.

Let $0 < R_1 < R_2 < R_3 < R$. By hypothesis there exists $c \geq 1$ such that $\|f\| \leq c$ on $\Delta^{n-1}(a';r) \times \Delta(a_n;R)$. Hence

$$(36.2) \qquad \|c_\alpha(z_n)\| r^{|\alpha|} \leq c \quad \text{for} \quad |z_n| < R.$$

Since for each z_n fixed the series in (36.1) converges uniformly on $\Delta^{n-1}(a';R_3)$ it follows that

$$(36.3) \qquad \lim_{|\alpha|\to\infty} \|c_\alpha(z_n)\| R_3^{|\alpha|} = 0 \quad \text{for} \quad |z_n| < R.$$

Consider the subharmonic functions $u_\alpha(z_n) = \frac{1}{|\alpha|} \log \|c_\alpha(z_n)\|$. From (36.2) and (36.3) it follows that

$$u_\alpha(z_n) \leq \log \frac{c}{r} \quad \text{for} \quad |z_n| < R$$

and

$$\limsup_{|\alpha|\to\infty} u_\alpha(z_n) \leq -\log R_3 \quad \text{for} \quad |z_n| < R.$$

Then an application of Proposition 35.8 shows that $u_\alpha(z_n) \leq -\log R_2$ for $|z_n| \leq R_3$ and $|\alpha|$ sufficiently large. This means that $\|c_\alpha(z_n)\| R_2^{|\alpha|} \leq 1$ for $|z_n| \leq R_3$ and $|\alpha|$ sufficiently large. Whence it follows that the series in (36.1) converges uniformly for $z \in \Delta^n(a;R_1)$. This completes the proof.

PROOF OF THEOREM 36.1. We proceed by induction on n, the theorem being obviously true for $n = 1$. Let $n \geq 2$ and assume the theorem is true for $n - 1$. Thus we have a mapping $f: U \to F$ which is holomorphic in $z' = (z_1, \ldots, z_{n-1})$ for z_n fixed, and holomorphic in z_n for z' fixed. Let $a \in U$ and choose $R > 0$ such that $\Delta^n(a;2R) \subset U$. By Lemma 36.2 there is an open polydisc $\Delta^{n-1}(b';r) \subset \Delta^{n-1}(a';R)$ such that f is bounded on $\Delta^{n-1}(b';r) \times \overline{\Delta}(a_n;R)$. By Lemma 36.3 f is holomorphic on $\Delta^n(b;R)$, where $b = (b',a_n)$. Since clearly $a \in \Delta^n(b;R)$, the proof of the theorem is complete.

Now we can prove a result promised in Section 8.

36.4. THEOREM. *Let U be an open subset of E, and let $f \in \mathcal{H}_G(U;F)$. Then for each $a \in U$ there is a unique power series $\sum_{m=0}^{\infty} P^m f(a)(x - a)$ from E into F which converges pointwise to $f(x)$ for every $x \in U_a$, where U_a denotes the largest a-balanced open set which is contained in U. The polynomials $P^m f(a)$ are given by the formulas*

$$P^m f(a)(t) = \frac{1}{2\pi i} \int_{|\zeta|=r} \frac{f(a + \zeta t)}{\zeta^{m+1}} d\zeta \tag{36.4}$$

where for each $t \in E$, $r > 0$ is chosen so that $a + \zeta t \in U$ for every $\zeta \in \overline{\Delta}(0;r)$.

PROOF. In view of Proposition 8.4 it is sufficient to prove that the mapping $P^m f(a) : E \to F$ defined by (36.4) belongs to

$P_a({}^mE;F)$. Fix $s, t \in E$ and choose $r > 0$ such that $a + \lambda s + \mu t \in U$ for $|\lambda| < r$ and $|\mu| < r$. Then the mapping $\varphi(\lambda,\mu) = f(a + \lambda s + \mu t)$ is separately holomorphic, and therefore holomorphic, for $|\lambda| < r$ and $|\mu| < r$, by Theorem 36.1. Thus we have a series expansion

$$f(a + \lambda s + \mu t) = \sum_{j,k} c_{jk} \lambda^j \mu^k,$$

with absolute and uniform convergence on each polydisc $|\lambda| \leq \rho$, $|\mu| \leq \rho$, with $0 < \rho < r$. Given $\eta \in \mathbb{C}$ chooose $\rho > 0$ such that $0 < \rho < r$ and $\rho|\eta| < r$. Then a term by term integration shows that

$$P^m f(a)(s + \eta t) = \frac{1}{2\pi i} \int_{|\zeta|=\rho} \frac{f(a + \zeta s + \zeta\eta t)}{\zeta^{m+1}} d\zeta$$

$$= \frac{1}{2\pi i} \sum_{j,k} c_{jk} \eta^k \int_{|\zeta|=\rho} \frac{\zeta^{j+k}}{\zeta^{m+1}} d\zeta = \sum_{j+k=m} c_{jk} \eta^k.$$

This shows that $P^m f(a)(s + \eta t)$ is a polynomial in η of degree at most m for each $s, t \in E$. By Theorem 3.6 we conclude that $P^m f(a)$ is a polynomial of degree at most m. Now, by Proposition 8.4 we have that $P^m f(a)(\mu t) = \mu^m P^m f(a)(t)$ for all $t \in E$ and $\mu \in \mathbb{C}$. Thus $P^m f(a) \in P_a({}^mE;F)$ and the proof is complete.

We observe that in view of Theorem 36.1 the conclusions of Theorem 7.7 and Corollaries 7.8 and 7.9 are valid for G-holomorphic mappings. If $f \in \mathcal{H}_G(U;F)$ and $a \in U$ then we shall denote by $A^m f(a)$ the unique member of $\mathcal{L}_a^s({}^mE;F)$ such that $[A^m f(a)]\hat{} = P^m f(a)$.

We next want to extend Theorem 36.1 to the case of Banach spaces. A key tool in the proof is the following theorem.

36.5. THEOREM. *Let U be a connected open subset of E, and let $f \in \mathcal{H}_G(U;F)$. If f is locally bounded at some point of U then f is locally bounded at every point of U, and is therefore holomorphic.*

Before proving this theorem we give two preparatory results.

36.6. **LEMMA.** *Let* $f : B(a;R) \to F$ *be a G-holomorphic mapping. Then for each* $b \in B(a;R)$, $k \in \mathbb{N}$ *and* $t \in E$ *we have that*

$$(36.5) \qquad P^k f(b)(t) = \sum_{m=k}^{\infty} \binom{m}{k} A^m f(a)(b-a)^{m-k} t^k .$$

PROOF. If $0 < r < R - \|b - a\|$ then $B(b;r) \subset B(a;R)$ and hence for each $t \in B(0;r)$ we have that

$$(36.6) \qquad f(b+t) = \sum_{m=0}^{\infty} P^m f(a)(b-a+t)$$

$$= \sum_{m=0}^{\infty} \sum_{k=0}^{m} \binom{m}{k} A^m f(a)(b-a)^{m-k} t^k .$$

Choose $\rho > 1$ such that $\rho(\|b - a\| + r) < R$. Then $\lambda(b - a) + \mu t \in B(0;R)$ for $|\lambda| \leq \rho$ and $|\mu| \leq \rho$. Since Corollary 7.9 applies to G-holomorphic mappings we have that

$$A^m f(a)(b-a)^{m-k} t^k = \frac{(m-k)!\, k!}{m!\,(2\pi i)^2} \int_{\substack{|\lambda|=\rho \\ |\mu|=\rho}} \frac{f[a + \lambda(b-a) + \mu t]}{\lambda^{m-k+1}\,\mu^{k+1}}\, d\lambda d\mu .$$

By Theorem 36.1 there is a constant $c > 0$ such that $\|f[a + \lambda(b - a) + \mu t]\| \leq c$ for $|\lambda| \leq \rho$ and $|\mu| \leq \rho$. Whence it follows that

$$\sum_{m=0}^{\infty} \sum_{k=0}^{m} \binom{m}{k} \|A^m f(a)(b-a)^{m-k} t^k\| \leq \sum_{k=0}^{\infty} \sum_{m=k}^{\infty} \frac{c}{\rho^{m-k}\,\rho^k} = \frac{c\rho^2}{(\rho-1)^2} .$$

This shows that we can interchange the order of summation in (36.6). Thus

$$(36.7) \qquad f(b+t) = \sum_{k=0}^{\infty} \sum_{m=k}^{\infty} \binom{m}{k} A^m f(a)(b-a)^{m-k} t^k .$$

Since for every $\eta \in \mathbb{C}$ we have that

$$\sum_{m=k}^{\infty} \binom{m}{k} \|A^m f(a)(b-a)^{m-k} (\eta t)^k\| \leq \sum_{m=k}^{\infty} \frac{c|\eta|^k}{\rho^{m-k}\,\rho^k} = \left(\frac{|\eta|}{\rho}\right)^k \frac{c\rho}{\rho - 1},$$

the series $\sum_{m=k}^{\infty} \binom{m}{k} A^m f(a)(b-a)^{m-k}$ converges pointwise to an element of $P_a({}^kE;F)$. Then comparison of (36.7) with the series expansion in Theorem 36.4 yields (36.5).

36.7. PROPOSITION. *Let* $f : B(a;R) \to F$ *be a G-holomorphic mapping. If* $P^m f(a)$ *is continuous for every* $m \in \mathbb{N}$ *then* f *is holomorphic.*

PROOF. Let $b \in B(a;R)$ and let $0 < r < R - \|b - a\|$. By Lemma 36.6, for each $k \in \mathbb{N}$ and $t \in E$ we have the series expansion

$$(36.8) \qquad P^k f(b)(t) = \sum_{m=k}^{\infty} \binom{m}{k} A^m f(a)(b-a)^{m-k} t^k .$$

Since the mappings $A^m f(a)$ are continuous, an application of Lemma 2.7 yields an open set V in E where the partial sums of the series in (36.8) are uniformly bounded. Hence the polynomial $P^k f(b)$ is bounded on V and is therefore continuous. Now, for every $t \in B(0;r)$ we have the series expansion

$$(36.9) \qquad f(b+t) = \sum_{k=0}^{\infty} P^k f(b)(t).$$

Then another application of Lemma 2.7 shows the existence of a ball $B(c;\rho) \subset B(0;r)$ where all the polynomials $P^k f(b)$ are uniformly bounded. By Corollary 7.6 the polynomials $P^k f(b)$ are uniformly bounded on the ball $B(0;\rho)$. Then it follows from the Cauchy-Hadamard Formula that the series in (36.9) has radius of convergence greater than or equal to ρ. This completes the proof.

PROOF OF THEOREM 36.5. To prove the theorem it is clearly sufficient to show that the set of points where f is locally bounded is closed in U. Let $a \in U$ be the limit of a sequence of points where f is locally bounded. Choose $r > 0$ such that $B(a;2r) \subset U$, and choose $b \in B(a;r)$ such that f is locally bounded at b. This implies that $P^m f(b)$ is continuous for every

$m \in \mathbb{N}$. Since $B(b;r) \subset B(a;2r) \subset U$, an application of Proposition 36.7 shows that f is holomorphic on the ball $B(b;r)$. Since $a \in B(b;r)$ we conclude that f is locally bounded at a, and the proof is complete.

Now we can generalize Theorem 36.1 as follows.

36.8. THEOREM. *Let* $E_1, \ldots, E_n, F$ *be Banach spaces, and let* U *be an open subset of* $E_1 \times \ldots \times E_n$. *Then a mapping* $f : U \to F$ *is holomorphic if and only if* f *is separately holomorphic.*

PROOF. To prove the nontrivial implication assume $f : U \to F$ is separately holomorphic. It clearly suffices to prove the theorem for $n = 2$. And without loss of generality we may assume that $U = U_1 \times U_2$, where U_j is an open ball in E_j for $j = 1,2$. We claim that f is G-holomorphic on U. Indeed, given $(a_1,a_2) \in U_1 \times U_2$ and $(b_1,b_2) \in E_1 \times E_2$, the mapping $(\lambda_1,\lambda_2) \to f(a_1 + \lambda_1 b_1, a_2 + \lambda_2 b_2)$ is separately holomorphic, and hence holomorphic by Theorem 36.1. Hence the mapping $\lambda \to f(a_1 + \lambda b_1, a_2 + \lambda b_2)$ is holomorphic, and f is therefore G-holomorphic, as asserted. Since f is separately continuous, an application of Lemma 36.2 shows that f is bounded on some nonvoid open subset of $U_1 \times U_2$. By Theorem 36.5 we conclude that f is holomorphic on $U_1 \times U_2$, completing the proof.

We conclude this section with a characterization of holomorphic mappings in Banach spaces with a Schauder basis.

36.9. THEOREM. *Let* E *be a Banach space with a Schauder basis* (e_n), *let* E_n *be the subspace generated by* $e_1, \ldots, e_n$, *and let* U *be an open subset of* E. *Then a mapping* $f : U \to F$ *is holomorphic if and only if* f *is continuous and* $f | U \cap E_n$ *is holomorphic for each* $n \in \mathbb{N}$.

PROOF. To prove the nontrivial implication, suppose that f is continuous and $f | U \cap E_n$ is holomorphic for every $n \in \mathbb{N}$. To prove that f is holomorphic it suffices to show that f is

G-holomorphic. Let $a \in U$ and $b \in E$ and choose $r > 0$ such that the compact set $K = a + 2r\bar{\Delta}b$ is contained in U. Let $T_n : E \to E_n$ be the canonical projection for each $n \in \mathbb{N}$. Since $T_n x$ converges to x uniformly on compact sets we can find $n_0 \in \mathbb{N}$ such that $T_n(K) \subset U$ for every $n \geq n_0$. By Corollary 7.2, for each $n \geq n_0$ and $|\lambda| \leq 2r$ we have a series expansion

$$(36.10) \qquad f \circ T_n(a + \lambda b) = \sum_{k=0}^{\infty} \lambda^k c_{nk},$$

where

$$c_{nk} = \frac{1}{2\pi i} \int_{|\zeta|=2r} \frac{f \circ T_n(a + \zeta b)}{\zeta^{k+1}} \, d\zeta$$

for every $n \in \mathbb{N}_0$. To complete the proof we shall show that for $|\lambda| \leq r$ we have a series expansion

$$(36.11) \qquad f(a + \lambda b) = \sum_{k=0}^{\infty} \lambda^k c_k,$$

where

$$c_k = \frac{1}{2\pi i} \int_{|\zeta|=2r} \frac{f(a + \zeta b)}{\zeta^{k+1}} \, d\zeta$$

for every $n \in \mathbb{N}_0$. In view of (36.10), to prove (36.11) it is sufficient to show that

$$(36.12) \qquad \lim_{n \to \infty} f \circ T_n(a + \lambda b) = f(a + \lambda b)$$

and

$$(36.13) \qquad \lim_{n \to \infty} \sum_{k=0}^{\infty} \lambda^k c_{nk} = \sum_{k=0}^{\infty} \lambda^k c_k$$

for $|\lambda| \leq r$. Let $\varepsilon > 0$ be given. Since f is continuous and K is compact we can find $\delta > 0$ such that $K + B(0;\delta) \subset U$ and $\| f(y) - f(x) \| \leq \varepsilon$ for $x \in K$ and $y \in B(x;\delta)$. If we choose $n_1 \geq n_0$ such that $\| T_n x - x \| < \delta$ for every $x \in K$ then $\| f \circ T_n(x) - f(x) \| < \varepsilon$ for every $x \in K$ and $n \geq n_1$. This already shows (36.12). Furthermore

$$\sum_{k=0}^{\infty} |\lambda|^k \| c_{nk} - c_k \| \leq \sum_{k=0}^{\infty} r^k \varepsilon (2r)^{-k} = 2\varepsilon$$

for every $n \geq n_1$, and (36.13) follows. This completes the proof of the theorem.

37. PSEUDOCONVEX DOMAINS

In this section we introduce the notion of pseudoconvex domain in a Banach space and give several necessary and sufficient conditions for a domain to be pseudoconvex.

We recall that if U is an open subset of a Banach space and $x \in U$ then $d_U(x)$ denotes the distance from x to the boundary of U. We now introduce another distance function.

37.1. DEFINITION. Let U be an open subset of E. Then the function $\delta_U : U \times E \to (0, \infty]$ is defined by

$$\delta_U(x,t) = \sup \{ r > 0 : x + r\overline{\Delta}t \subset U \}.$$

Thus, for each $t \in E$, $\delta_U(x,t)$ may be regarded as the distance from x to the boundary of U along the direction t. Observe that if $U = E$ then $d_U \equiv \infty$ and $\delta_U \equiv \infty$.

37.2. PROPOSITION. *Let U be a proper open subset of E. Then:*

(a) $|d_U(x) - d_U(y)| \leq \|x - y\|$ *for all* $x, y \in U$. *In particular, the function* d_U *is continuous on* U.

(b) *The function* δ_U *is lower semicontinuous on* $U \times E$.

(c) $d_U(x) = \inf \{ \delta_U(x,t) : t \in E, \|t\| = 1\}$ *for each* $x \in U$.

PROOF. (a) This is easy: it was left as Exercise 7.D.

(b) Let $(x_0, t_0) \in U \times E$ and let $0 < r < s < \delta_U(x_0, t_0)$.

Then $x_o + \lambda t_o \in U$ for $|\lambda| \leq s$ and a standard compactness argument yields a neighborhood V of x_o and a neighborhood W of t_o such that $x + \lambda t \in U$ for every $x \in V$, $t \in W$ and $|\lambda| \leq s$. Hence $\delta_U(x,t) \geq s > r$ for every $(x,t) \in V \times W$.

(c) Since $\overline{B}(x;r) = \cup \{x + r\overline{\Delta}t : t \in E, \|t\| = 1\}$, the desired conclusion follows at once.

37.3. DEFINITION. An open subset U of E is said to be *pseudoconvex* if the function $-\log \delta_U$ is plurisubharmonic on $U \times E$.

In order to give other characterizations of pseudoconvex domains we introduce the following definition.

37.4. DEFINITION. Let U be an open subset of E. For each set $A \subset U$ we define

$$\hat{A}_{\mathcal{P}s(U)} = \{x \in U : f(x) \leq \sup_A f \quad \text{for all} \quad f \in \mathcal{P}s(U)\}$$

and

$$\hat{A}_{\mathcal{P}sc(U)} = \{x \in U : f(x) \leq \sup_A f \quad \text{for all} \quad f \in \mathcal{P}sc(U)\}.$$

Note that the set $\hat{A}_{\mathcal{P}sc(U)}$ is always closed in U, but it is not clear whether the same is true for $\hat{A}_{\mathcal{P}s(U)}$. Note also that since $|f| \in \mathcal{P}sc(U)$ whenever $f \in \mathcal{H}(U)$, we always have the inclusions $A \subset \hat{A}_{\mathcal{P}s(U)} \subset \hat{A}_{\mathcal{P}sc(U)} \subset \hat{A}_{\mathcal{H}(U)}$.

37.5. THEOREM. *For an open subset U of E the following conditions are equivalent:*

(a) *U is pseudoconvex.*

(b) *The function $-\log \delta_U(\cdot, t)$ is plurisubharmonic on U for each $t \in E$.*

(c) *The function $-\log d_U$ is plurisubharmonic on U.*

(d) *$d_U(\hat{A}_{\mathcal{P}sc(U)}) = d_U(A)$ for each set $A \subset U$.*

(e) *The set* $\hat{K}_{Psc(U)}$ *is compact for each compact set* $K \subset U$.

(f) *The set* $\hat{K}_{Ps(U)}$ *is relatively compact in* U *for each compact set* $K \subset U$.

(g) *If* (H,D) *is a Hartogs figure in* $\mathbb{C}^2$ *then each* $f \in \mathcal{H}(H;E)$ *with* $f(H) \subset U$ *has an extension* $\tilde{f} \in \mathcal{H}(D;E)$ *with* $\tilde{f}(D) \subset U$.

PROOF. The implications (a) ⇒ (b), (c) ⇒ (d), (d) ⇒ (e) and (e) ⇒ (f) are clear. The implication (b) ⇒ (c) follows from Proposition 37.2(c). Thus it only remains to prove the implications (f) ⇒ (g) and (g) ⇒ (a).

(f) ⇒ (g): We have that

$$D = \{(\lambda,\zeta) \in \mathbb{C}^2 : |\lambda| < R \quad \text{and} \quad |\zeta| < R\}$$

and

$$H = \{(\lambda,\zeta) \in D : |\lambda| > r \quad \text{or} \quad |\zeta| < s\},$$

where $0 < r < R$ and $0 < s < R$. Let $f \in \mathcal{H}(H;E)$ with $f(H) \subset U$ be given. By Example 10.2 and Exercise 8.I there exists $\tilde{f} \in \mathcal{H}(D;E)$ such that $\tilde{f} = f$ on H. We still have to show that $\tilde{f}(D) \subset U$. Fix b with $r < b < R$. For each $t \in (0,R]$ consider the set

$$D_t = \{(\lambda,\zeta) \in \mathbb{C}^2 : |\lambda| < b \quad \text{and} \quad |\zeta| < t\},$$

and let

$$T = \{t \in (0, R] : \tilde{f}(D_t) \subset U\}.$$

We shall prove that $T = (0, R]$. This means that $\tilde{f}(D_R) \subset U$. And since $\tilde{f}(D \setminus D_R) \subset \tilde{f}(H) = f(H) \subset U$, (g) will follow. The set T is nonvoid, since $(0, s] \subset T$. Thus to show that $T = (0, R]$ if suffices to prove that T is open and closed in $(0, R]$. Note that $t \in T$ implies $(0, t] \subset T$. The set T is closed in

$(0, R]$, for $t_n \in T$ for each $n \in \mathbb{N}$ implies that $\sup t_n \in T$. To show that T is open in $(0, R]$, fix $t \in T$ with $t < r$, fix a with $r < a < b$, and consider the set

$$J = \{(\lambda,\zeta) \in \mathbb{C}^2 : a \leq |\lambda| \leq b \quad \text{and} \quad |\zeta| \leq t\}.$$

Then J is a compact subset of H, and $K = f(J)$ is a compact subset of U. If $u \in \mathcal{P}s(U)$ then $u \circ \tilde{f} \in \mathcal{P}s(D_t)$ by Theorem 35.7. Thus, if we fix ζ with $|\zeta| < t$, $u \circ \tilde{f}(\lambda,\zeta)$ is a subharmonic function of λ for $|\lambda| < b$. Then for each c with $a < c < b$ it follows from the maximum principle for subharmonic functions that

$$\sup_{|\lambda| \leq c} u \circ \tilde{f}(\lambda,\zeta) = \sup_{|\lambda|=c} u \circ \tilde{f}(\lambda,\zeta) = \sup_{|\lambda|=c} u \circ f(\lambda,\zeta) \leq \sup_K u.$$

Whence it follows that $\tilde{f}(\lambda,\zeta) \in \hat{K}_{\mathcal{P}s(U)}$ for every $(\lambda,\zeta) \in D_t$. Then by (f) there exists $\varepsilon > 0$ such that

$$(37.1) \qquad d_U(\tilde{f}(D_t)) > \varepsilon.$$

Since $\tilde{f}$ is uniformly continuous on each compact subset of D, we can find $\delta > 0$ with $t + \delta < R$ such that

$$(37.2) \qquad |\tilde{f}(\lambda,\zeta) - \tilde{f}(\lambda',\zeta')| < \varepsilon$$

for all $(\lambda,\zeta), (\lambda',\zeta') \in D_{t+\delta}$ such that $|\lambda - \lambda'| < \delta$ and $|\zeta - \zeta'| < \delta$. From (37.1) and (37.2) it follows that $\tilde{f}(D_{t+\delta}) \subset U$. Thus T is open in $(0, R]$ and (g) follows.

(g) ⇒ (a): Assuming (g) we have to show that the function $-\log \delta_U$ is plurisubharmonic on $U \times E$. Fix $(a,b) \in U \times E$ and $(s,t) \in E \times E$ such that

$$(37.3) \qquad a + \lambda s \in U \quad \text{for} \quad |\lambda| \leq 1.$$

Let $P \in \mathcal{P}(\mathbb{C})$ such that $-\log \delta_U(a + \lambda s, b + \lambda t) \leq \operatorname{Re} P(\lambda)$ for $|\lambda| = 1$. Since $e^{\operatorname{Re} P(\lambda)} = |e^{P(\lambda)}|$, this means that

$$(37.4) \quad \delta_U(a + \lambda s,\ e^{-P(\lambda)}(b + \lambda t)) \geq 1 \quad \text{for} \quad |\lambda| = 1.$$

We have to show that (37.4) holds for $|\lambda| \leq 1$. And it is clearly sufficient to show that

$$(37.5) \quad \delta_U(a + \lambda s,\ \rho e^{-P(\lambda)}(b + \lambda t)) \geq 1 \quad \text{for} \quad |\lambda| \leq 1 \quad \text{and} \quad 0 < \rho < 1.$$

For a fixed ρ with $0 < \rho < 1$ consider the function $f \in \mathcal{H}(\mathbb{C}^2;E)$ defined by

$$f(\lambda,\zeta) = a + \lambda s + \rho\zeta e^{-P(\lambda)}(b + \lambda t).$$

Then (37.3) guarantees that $f(\overline{\Delta} \times \{0\}) \subset U$, whereas (37.4) guarantees that $f(\partial\Delta \times \overline{\Delta}) \subset U$. By a compactness argument we can find $\varepsilon > 0$ such that if we set

$$H_1 = \{(\lambda,\zeta) \in \mathbb{C}^2 : |\lambda| < 1 + \varepsilon \quad \text{and} \quad |\zeta| < \varepsilon\},$$

$$H_2 = \{(\lambda,\zeta) \in \mathbb{C}^2 : 1 - \varepsilon < |\lambda| < 1 + \varepsilon \quad \text{and} \quad |\zeta| < 1 + \varepsilon\}$$

and $H = H_1 \cup H_2$, then $f(H) \subset U$. If we set

$$D = |(\lambda,\zeta) \in \mathbb{C}^2 : |\lambda| < 1 + \varepsilon \quad \text{and} \quad |\zeta| < 1 + \varepsilon\}$$

then (H,D) is a Hartogs figure in $\mathbb{C}^2$, and by (g) there is a function $\tilde{f} \in \mathcal{H}(D;E)$ such that $\tilde{f} = f$ on H and $\tilde{f}(D) \subset U$. Then $\tilde{f} = f$ on D by the Identity Principle, and therefore $f(D) \subset U$. Whence (37.5) follows and the proof of the theorem is complete.

37.6. COROLLARY. *An open subset U of E is pseudoconvex if and only if $U \cap M$ is a pseudoconvex open subset of M for each finite dimensional subspace M of E.*

PROOF. This follows from Proposition 34.5(c) and Theorem 37.5(b) or (c).

37.7. COROLLARY. *Every holomorphically convex open set in a*

Banach space is pseudoconvex. In particular every domain of holomorphy in a Banach space is pseudoconvex.

PROOF. Since $\hat{K}_{Psc(U)} \subset \hat{K}_{\mathcal{H}(U)}$ for every set $K \subset U$, the conclusion follows from Theorem 37.5(e).

37.8. PROPOSITION. *Let V be a pseudoconvex open set in F, let $T \in \mathcal{L}(E;F)$ and let $U = T^{-1}(V)$. Then U is a pseudoconvex open set in E.*

PROOF. Let K be a compact subset of U. Then one can readily verify that

$$(37.6) \qquad \hat{K}_{Ps(U)} \subset T^{-1}[(T(K))\hat{}_{Ps(V)}].$$

By Theorem 37.5(f) there is a 0-neighborhood B in F such that $(T(K))\hat{}_{Ps(V)} + B \subset V$. Then $A = T^{-1}(B)$ is a 0-neighborhood in E and it follows from (37.6) that $\hat{K}_{Ps(U)} + A \subset U$. By Theorem 37.5(f) again, U is pseudoconvex.

37.9. PROPOSITION. *Let $U \subset E$ and $V \subset F$ be open sets. Then $U \times V$ is pseudoconvex if and only if U and V are pseudoconvex.*

PROOF. If $(x,y) \in U \times V$ and $(s,t) \in E \times F$ then one can readily verify that

$$\delta_{U \times V}((x,y),(x,t)) = min\{\delta_U(x,s), \delta_V(y,t)\}.$$

Whence it follows that $-log\, \delta_{U \times V}$ is plurisubharmonic if and only if $-log\, \delta_U$ and $-log\, \delta_V$ are plurisubharmonic.

By Corollary 37.7 every domain of holomorphy in a Banach space is pseudoconvex. The converse is not true in general, for as we have mentioned already, B. Josefson [1] has given an example of a holomorphically convex open set in a nonseparable Banach space which is not a domain of holomorphy. But the following problem remains open.

37.10. PROBLEM. Let E be a separable Banach space. Is every pseudoconvex open set in E a domain of existence, or at least a domain of holomorphy ?

Problem 37.10 is called the *Levi problem*, for it was posed by E. Levi [1] in 1911 for $E = \mathbb{C}^n$. In Section 43 we shall solve the Levi problem in the case where $E = \mathbb{C}^n$, and in Section 45 we shall solve the Levi problem in the case of separable Banach spaces with the bounded approximation property.

EXERCISES

37.A. Show that an open subset U of E is pseudoconvex if and only if each connected component of U is pseudoconvex.

37.B. Show that if $(U_i)_{i \in I}$ is a family of pseudoconvex open subsets of E then the open set $U = int \bigcap_{i \in I} U_i$ is pseudoconvex as well.

37.C. Show that if $(U_j)_{j=1}^{\infty}$ is an increasing sequence of pseudoconvex open subsets of E then the open set $U = \bigcup_{j=1}^{\infty} U_j$ is pseudoconvex as well.

37.D. Let U be a pseudoconvex open subset of E, let $f \in \mathcal{Psc}(U)$ and let $V = \{x \in U : f(x) < 0\}$. Show that V is pseudoconvex.

38. PLURISUBHARMONIC FUNCTIONS ON PSEUDOCONVEX DOMAINS

In this section we obtain several results of existence of plurisubharmonic functions on pseudoconvex domains. We begin with the following lemma.

38.1. LEMMA. *Let U be a pseudoconvex open set in $\mathbb{C}^n$, let K be a compact subset of U, and let V be an open neighborhood of $\hat{K}_{\mathcal{Ps}(U)}$ in U. Then there exists a function $f \in \mathcal{Psc}(U)$ such* *that:*

(a) *The set* $\{z \in U : f(z) \leq c\}$ *is compact for each* $c \in \mathbb{R}$.

(b) $f(z) < 0$ *for every* $z \in K$.

(c) $f(z) > 0$ *for every* $z \in U \setminus V$.

PROOF. Let $u \in Psc(U)$ be defined by

$$u(z) = \max \{\|z\|, -\log d_U(z)\} - k,$$

where the constant k is chosen so that $\sup_K u < 0$. Clearly the set $K_c = \{z \in U : u(z) \leq c\}$ is compact for each $c \in \mathbb{R}$. If $K_o \subset V$ then the function $f = u$ has the required properties. Suppose then that $K_o \not\subset V$. Since K_2 is compact we can find $\delta > 0$ such that $K_2 \subset U_\delta$, where $U_\delta = \{z \in U : d_U(z) > \delta\}$. For each $a \in K_o \setminus V$ we can find $\varphi \in Ps(U)$ such that $\sup_K \varphi < 0 < \varphi(a)$. By Theorem 35.6 there is a decreasing sequence (φ_j) in $Psc(U_\delta)$ which converges pointwise to φ in U_δ. Hence

$$K \subset \bigcup_{j=1}^{\infty} \{z \in U_\delta : \varphi_j(z) < 0\}$$

and thus we can find a function $\psi \in Psc(U_\delta)$ such that $\sup_K \psi < 0 < \psi(a)$. Since $K_o \setminus V$ is compact we can thus find $\psi_1, \ldots, \psi_m \in Psc(U_\delta)$ such that $\sup_K \psi_j < 0$ for $j = 1, \ldots, m$ and

$$K_o \setminus V \subset \bigcup_{j=1}^{m} \{z \in U_\delta : \psi_j(z) > 0\}.$$

Set $v = \max \{\psi_1, \ldots, \psi_m\}$. Then $v \in Psc(U_\delta)$, $v(z) < 0$ for every $z \in K$ and $v(z) > 0$ for every $z \in K_o \setminus V$. Set $M = \sup_{K_2} v > 0$ and define $f : U \to \mathbb{R}$ by

$$f(z) = \begin{cases} \max \{M \cdot u(z), v(z)\} & \text{if} \quad u(z) < 2 \\ M \cdot u(z) & \text{if} \quad u(z) > 1. \end{cases}$$

If $1 < u(z) < 2$ then $v(z) \leq M < M \cdot u(z)$, so that f is well defined and belongs to $Psc(U)$. Clearly $f(z) < 0$ for every $z \in K$ and $f(z) > 0$ for every $z \in U \setminus V$. And since

$$\{z \in U : f(z) \leq c\} \subset \{z \in U : M \cdot u(z) \leq c\},$$

we conclude that the set $\{z \in U : f(z) \leq c\}$ is compact for every $c \in \mathbb{R}$.

38.2. **PROPOSITION.** *If U is a pseudoconvex open set in $\mathbb{C}^n$ then $\hat{K}_{Ps(U)} = \hat{K}_{Psc(U)}$ for each compact set $K \subset U$. In particular $\hat{K}_{Ps(U)}$ is compact for each compact set $K \subset U$.*

PROOF. The inclusion $\hat{K}_{Ps(U)} \subset \hat{K}_{Psc(U)}$ is always true. To show the opposite inclusion let $a \in U$ with $a \notin \hat{K}_{Ps(U)}$. By applying Lemma 38.1 with $V = U \setminus \{a\}$ we can find a function $f \in Psc(U)$ such that $\sup_K f < 0 < f(a)$. Thus $a \notin \hat{K}_{Psc(U)}$ and the proof is complete.

38.3. **PROPOSITION.** *Let U be a pseudoconvex open set in E, let $f \in Ps(U)$ and let $V = \{x \in U : f(x) < 0\}$. Then V is pseudoconvex as well.*

PROOF. If suffices to show that $V \cap M$ is pseudoconvex for each finite dimensional subspace M of E. Let K be a compact subset of $V \cap M$. Then there is a constant $c < 0$ such that $f \leq c$ on K. Hence $f \leq c < 0$ on $\hat{K}_{Ps(U \cap M)}$ and it follows that $\hat{K}_{Ps(V \cap M)} \subset \hat{K}_{Ps(U \cap M)} \subset V \cap M$. Since $\hat{K}_{Ps(U \cap M)}$ is compact, by Proposition 38.2, we conclude that $\hat{K}_{Ps(V \cap M)}$ is relatively compact in $V \cap M$. Thus $V \cap M$ is pseudoconvex and the proof is complete.

38.4. **THEOREM.** *Let U be a pseudoconvex open subset of E. Then each $f \in Ps(U)$ is the pointwise limit of a decreasing sequence $(f_n) \subset Psc(U)$.*

PROOF. If $f \equiv -\infty$ then we may take $f_n \equiv -n$ for every $n \in \mathbb{N}$.

Thus we may assume $f \not\equiv -\infty$. Then set $g = e^{-f}$ and define $g_n(x) = \inf_{z \in U} [g(z) + n\|x - z\|]$ and $f_n(x) = -\log g_n(x)$ for every $n \in \mathbb{N}$ and $x \in U$. Then the proof of Proposition 34.2 shows that each $f_n : U \to \mathbb{R}$ is continuous, and that the sequence (f_n) is decreasing and converges pointwise to f. Thus it only remains to prove that each f_n is plurisubharmonic on U. Then set $U \times \mathbb{C}$ is pseudoconvex, by Proposition 37.9, and the set

$$V = \{(x, \lambda) \in U \times \mathbb{C} : f(x) + \log|\lambda| < 0\}$$

is pseudoconvex as well, by Proposition 38.3. Observe that $(x, 0) \in V$ for every $x \in U$. To show that the function f_n is plurisubharmonic on U we shall prove that

$$g_n(x) = d_V^n((x, 0)) \quad \text{for every} \quad x \in U,$$

where d_V^n denotes the distance to the boundary of V with respect to the norm

$$\|(x, \lambda)\|_n = n\|x\| + |\lambda|.$$

Observe that all these norms are equivalent and induce the product topology on $E \times \mathbb{C}$. Since

$$V = \{(x, \lambda) \in U \times \mathbb{C} : |\lambda| < g(x)\}$$

it is clear that

$$d_V^n((x, 0)) = \inf\{n\|x - z\| + |\lambda| : (x, \lambda) \notin V\} = \min\{\alpha\ \beta\},$$

where

$$\alpha = \inf\{n\|x - z\| + |\lambda| : z \notin U\} = d_{U \times \mathbb{C}}^n((x, 0))$$

and

$$\beta = \inf\{n\|x - z\| + |\lambda| : z \in U \text{ and } |\lambda| \geq g(z)\}.$$

Since clearly $d^n_V((x,0)) \leq \alpha$ we conclude that

$$d^n_V((x,0)) = \beta = \inf\{n\|x - z\| + g(z) : z \in U\} = g_n(x).$$

This completes the proof.

Now we can extend Proposition 38.2 to the case of Banach spaces.

38.5. PROPOSITION. *If U is a pseudoconvex open set in E then $\hat{K}_{Ps(U)} = \hat{K}_{Psc(U)}$ for each compact set $K \subset U$. In particular $\hat{K}_{Ps(U)}$ is compact for each compact set $K \subset U$.*

PROOF. If $a \in U$ with $a \notin \hat{K}_{Ps(U)}$ then there is $f \in Ps(U)$ such that $\sup_K f < 0 < f(a)$. By Theorem 38.4 there is a decreasing sequence $(f_n) \subset Psc(U)$ which converges pointwise to f. Thus

$$K \subset \bigcup_{n=1}^{\infty} \{x \in U : f_n(x) < 0\}$$

and we can find $n \in \mathbb{N}$ such that $\sup_K f_n < 0 < f_n(a)$. Thus $a \notin \hat{K}_{Psc(U)}$ and the proof is complete.

To conclude this section we improve Lemma 38.1 as follows.

38.6. THEOREM. *Let U be a pseudoconvex open set in $\mathbb{C}^n$, let K be a compact subset of U, and let V be an open neighborhood of $\hat{K}_{Ps(U)}$ in U. Then there exists a strictly plurisubharmonic function $f \in C^\infty(U)$ such that:*

(a) *The set $\{z \in U : f(z) \leq c\}$ is compact for each $c \in \mathbb{R}$.*

(b) *$f(z) < 0$ for every $z \in K$.*

(c) *$f(z) > 0$ for every $z \in U \setminus V$.*

PROOF. By Lemma 38.1 there is a function $u \in Psc(U)$ satisfying the conditions (a), (b) and (c) of the theorem. Set $K_c = \{z \in U : u(z) \leq c\}$ for every $c \in \mathbb{R}$, and consider the functions

$u_j \in C^\infty(U)$ $(j = 0,1,2,\dots)$ defined by

$$u_j(z) = \int_U u(\zeta)\rho_{\delta_j}(z - \zeta)d\lambda(\zeta) + \delta_j \sum_{k=1}^{n} |z_k|^2 \qquad \text{for} \quad z \in U.$$

Observe that

$$u_j(z) = (u * \rho_{\delta_j})(z) + \delta_j \sum_{k=1}^{n} |z_k|^2 \qquad \text{for} \quad z \in U_{\delta_j}.$$

If $\delta_j > 0$ is sufficiently small then by Theorem 35.6 the function u_j is strictly plurisubharmonic and $u < u_j < u + 1$ on a neighborhood of K_j. In addition $\delta_o > 0$ and $\delta_1 > 0$ can be chosen so small that $u_o < 0$ and $u_1 < 0$ on a neighborhood of K. If we define $\chi(t) = 0$ for $t \leq 0$ and $\chi(t) = \int_0^t e^{-1/s}ds$ for $t > 0$ then it follows from Exercise 15.A that $\chi \in C^\infty(\mathbb{R};\mathbb{R})$ and it is clear that $\chi'(t) > 0$ and $\chi''(t) > 0$ for $t > 0$. Then it follows from Exercise 35.E that the function $\chi \circ (u_j - j + 1)$ is plurisubharmonic on a neighborhood of K_j and strictly plurisubharmonic on a neighborhood of $K_j \setminus \overset{\circ}{K}_{j-1}$. For each $j \in \mathbb{N}_o$ let $f_j \in C^\infty(U;\mathbb{R})$ be defined by

$$f_j = u_o + \sum_{r=1}^{j} c_r \chi \circ (u_r - r + 1)$$

where the coefficients $c_r > 0$ will be chosen later. Observe that $\chi \circ (u_r - r + 1) = 0$ on K_i for $r \geq i + 2$, and hence $f_j = f_j$ on K_i for $j, k \geq i + 1$. Since $U = \bigcup_{i=0}^{\infty} \overset{\circ}{K}_i$ we conclude that $f = \lim f_j \in C^\infty(U;\mathbb{R})$.

We shall inductively choose the coefficients c_r so that each f_j is $> u$ and is strictly plurisubharmonic on a neighborhood of K_j. This is clear for $f_o = u_o$. Let $j > 0$ and suppose that f_{j-1} is $> u$ and is strictly plurisubharmonic on a neighborhood of K_{j-1}. Set $S = \{t \in \mathbb{C}^n : \sum_{k=1}^{n} |t_k|^2 = 1\}$. Since $\chi \circ (u_j - j + 1)$ is > 0 and is strictly plurisubharmonic on a neighborhood of $K_j \setminus \overset{\circ}{K}_{j-1}$, we can find $c_j > 0$ such that

$$c_j \, inf \{\chi \circ (u_j - j + 1)(a) : a \in K_j \setminus \overset{\circ}{K}_{j-1}\}$$

$$> sup \{ |f_{j-1}(a)| : a \in K_j \setminus \overset{\circ}{K}_{j-1}\} + sup \{ |u(a)| : a \in K_j \setminus \overset{\circ}{K}_{j-1}\}$$

and

$$c_j \, inf \{ D'D''\chi \circ (u_j - j + 1)(a)(t,t) : a \in K_j \setminus \overset{\circ}{K}_{j-1}, \ t \in S\}$$

$$> sup \{ D'D''f_{j-1}(a)(t,t) : a \in K_j \setminus \overset{\circ}{K}_{j-1}, \ t \in S\}.$$

Since $f_j = f_{j-1} + \chi \circ (u_j - j + 1)$, it follows that f_j is $> u$ and f_j is strictly plurisubharmonic on a neighborhood of $K_j \setminus \overset{\circ}{K}_{j-1}$. Using the induction hypothesis we conclude that f_j is $> u$ and f_j is strictly plurisubharmonic on a neighborhood of K_j, as we wanted.

Since $f = f_{i+1}$ on K_i for every i we conclude that f is $> u$ and f is strictly plurisubharmonic on all of U. Hence $f > u > 0$ on $U \setminus V$. We also have that $f = u_0 < 0$ on K. And since $f > u$ it follows that

$$\{z \in U : f(z) \leq c\} \subset \{z \in U : u(z) \leq c\}$$

and hence the set $\{z \in U : f(z) \leq c\}$ is compact for each $c \in \mathbb{R}$. The proof is now complete.

NOTES AND COMMENTS

The results in Section 34 are straightforward generalizations of the corresponding results for subharmonic functions. Theorem 35.7 is due to P. Lelong, who obtained the result for finite dimensional spaces in P. Lelong [1] and for infinite dimensional spaces in P. Lelong [2]. Theorem 36.1 on separate analyticity is due to F. Hartogs [1], and its generalization to Banach spaces, Theorem 36.8, is due to M. Zorn [1]. Theorem

36.5 is also due to M. Zorn [2]. The proof of Theorem 36.5 given here is due to P. Noverraz [1]. Theorem 36.4 can be found in the book of E. Hille and R. Phillips [1]. Theorem 36.9, due to M. Matos [1], extends an old result of A. Taylor [1]. Our presentation in Section 37 follows essentially a paper of H. Bremermann [2], who was the first to consider plurisubharmonic functions and pseudoconvex domains in infinite dimensional spaces. Another reference for Section 37 is the book of P. Noverraz [3]. Theorem 38.4 is essentially due to J. P. Ferrier and N. Sibony [1], who only stated the result for $E = \mathbb{C}^n$. M. Estévez and C. Hervés [1] noticed that the result is valid for Banach spaces, and used it to establish Proposition 38.5. Theorem 38.6 is taken from the book of L. Hörmander [3] and will play a fundamental role in the solution of the $\bar{\partial}$ equation in pseudoconvex domains.

CHAPTER IX

THE $\overline{\partial}$ EQUATION IN PSEUDOCONVEX DOMAINS

39. DENSELY DEFINED OPERATORS IN HILBERT SPACES

This section is devoted to the study of densely defined operators in Hilbert spaces. This is an indispensable tool to solve the $\overline{\partial}$ equation in pseudoconvex domains.

39.1. DEFINITION. Let E and F be Hilbert spaces, and let $A : D_A \to F$ be a linear operator whose domain is a subspace D_A of E. We shall denote by N_A the kernel of A, by R_A the range of A, and by G_A the graph of A, that is

$$N_A = \{x \in D_A : Ax = 0\},\ R_A = \{Ax : x \in D_A\},\ G_A = \{(x,Ax) : x \in D_A\}.$$

The operator A is said to be *densely defined* if D_A is dense in E. The operator A is said to be *closed* if G_A is closed in $E \times F$.

39.2. DEFINITION. Let E and F be Hilbert spaces, and let A be a densely defined operator from E into F. Let D_{A^*} denote the subspaces of all $y \in F$ for which there exists $y^* \in E$ such that

$$(Ax \mid y) = (x \mid y^*) \quad \text{for every} \quad x \in D_A .$$

Since D_A is dense in E, the vector y^* is unique whenever it exists. The operator $A^* : D_{A^*} \to E$ defined by $A^*y = y^*$ is called the *adjoint* of A. Thus we have that

$$(Ax \mid y) = (x \mid A^*y) \quad \text{for all} \quad x \in D_A \quad \text{and} \quad y \in D_{A^*} .$$

39.3. PROPOSITION. *Let E and F be Hilbert spaces, and let A be a densely defined operator from E into F. Then the adjoint operator A^* is closed.*

PROOF. Let (y_n) be a sequence in D_{A^*} such that (y_n) converges to y_o in F and (A^*y_n) converges to x_o in E. Then $(Ax \mid y_n) = (x \mid A^*y_n)$ for every $x \in D_A$ and $n \in \mathbb{N}$. Letting $n \to \infty$ we get that $(Ax \mid y_o) = (x \mid x_o)$ for every $x \in D_A$. Hence $y_o \in D_{A^*}$ and $A^*y_o = x_o$.

39.4. PROPOSITION. *Let E and F be Hilbert spaces, and let A be a densely defined operator from E into F. Then $N_{A^*} = (R_A)^\perp$ and $(N_{A^*})^\perp = \overline{R}_A$.*

PROOF. It suffices to prove the first identity. The second one is an immediate consequence. If $y \in N_{A^*}$ then $(Ax \mid y) = (x \mid A^*y) = (x \mid 0) = 0$ for every $x \in D_A$, and therefore $y \in (R_A)^\perp$. Conversely, if $y \in (R_A)^\perp$ then $(Ax \mid y) = 0 = (x,0)$ for every $x \in D_A$. Hence $y \in D_{A^*}$ and $A^*y = 0$, that is, $y \in N_{A^*}$.

39.5. THEOREM. *Let E and F be Hilbert spaces, and let A be a closed, densely defined operator from E into F. Then the adjoint A^* is a closed, densely defined operator from F into E, and $A^{**} = A$.*

PROOF. We already know that A^* is closed. To show that A^* is densely defined and that $A^{**} = A$ we proceed as follows. Observe that $E \times F$ and $F \times E$ are isometric Hilbert spaces under the canonical inner products, that is,

$$((x_1,y_1) \mid (x_2,y_2)) = (x_1 \mid x_2) + (y_1 \mid y_2)$$

and

$$((y_1,x_1) \mid (y_2,x_2)) = (y_1 \mid y_2) + (x_1 \mid x_2)$$

for $x_1, x_2 \in E$ and $y_1, y_2 \in F$. Consider the isometric operators $S : E \times F \to F \times E$ and $T : F \times E \to E \times F$ defined by

$$S(x,y) = (-y,x) \quad \text{and} \quad T(y,x) = (-x,y).$$

Observe that $T \circ S = -id_{E \times F}$ and $S \circ T = -id_{F \times E}$. Since the equation $(Ax \mid y) = (x \mid y^*)$ is equivalent to the equation $((-Ax,x) \mid (y,y^*)) = 0$ we see that

$$(39.1) \qquad G_{A^*} = (SG_A)^{\perp}.$$

Since G_A is closed and S is an isometry we get that $(G_{A^*})^{\perp} = \overline{SG_A} = SG_A$. Since T is an isometry it follows that

$$(39.2) \qquad (TG_{A^*})^{\perp} = TSG_A = G_A.$$

Let $y_o \in (D_{A^*})^{\perp}$. Then $(0,y_o)$ is orthogonal to $(-A^*y,y)$ for every $y \in D_{A^*}$. Thus $(0,y_o) \in (TG_{A^*})^{\perp} = G_A$, by (39.2), and therefore $y_o = A(0) = 0$. Thus D_{A^*} is dense in F and consequently A^{**} exists. Applying (39.1) with A^* in place of A and T in place of S we obtain, with the help of (39.2), that $G_{A^{**}} = (TG_{A^*})^{\perp} = G_A$. Hence $A^{**} = A$ and the proof is complete.

From Theorem 39.5 and Proposition 39.4 we obtain:

39.6. COROLLARY. *Let E and F be Hilbert spaces, and let A be a closed, densely defined operator from E into F. Then $N_A = (R_{A^*})^{\perp}$ and $(N_A)^{\perp} = \overline{R}_{A^*}$.*

39.7. THEOREM. *Let E and F be Hilbert spaces, and let A be a closed, densely defined operator from E into F. Then A is surjective if and only if its adjoint A^* has a continuous inverse. In this case R_{A^*} is closed in E.*

PROOF. Assume first that A is surjective. Then $N_{A^*} = (R_A)^{\perp} = \{0\}$ and A^* is injective. To show that $(A^*)^{-1}$ is continuous let (y_n) be a sequence in D_{A^*} such that the sequence (A^*y_n) is bounded in E. Then there is a constant $c > 0$ such that $|(Ax \mid y_n)| = |(x \mid A^*y_n)| \leq \|x\| \, \|A^*y_n\| \leq c\|x\|$ for every $x \in D_A$

and $n \in \mathbb{N}$. Since $R_A = F$ we conclude that the sequence (y_n) is weakly bounded, and therefore norm bounded, by the Principle of Uniform Boundedness. This shows that $(A^*)^{-1}$ is continuous.

Conversely, assume that A^* has a continuous inverse. Then there is a constant $c > 0$ such that

$$(39.3) \qquad \|y\| \leq c\, \|A^*y\| \quad \text{for every} \quad y \in D_{A^*}.$$

Let $y_o \in F$. Then $|(y \mid y_o)| \leq c\, \|A^*y\|\, \|y_o\|$ for every $y \in D_{A^*}$ and hence the mapping $A^*y \to (y \mid y_o)$ is a well defined continuous linear functional on R_{A^*}. By the Riesz Representation Theorem there is a unique vector $x_o \in \overline{R}_{A^*}$ such that $\|x_o\| \leq c\,\|y_o\|$ and $(A^*y \mid x_o) = (y \mid y_o)$ for every $y \in D_{A^*}$. Hence $x_o \in D_{A^{**}} = D_A$ and $y_o = A^{**}x_o = Ax_o$, proving that $F = R_A$.

We still have to show that R_{A^*} is closed in E. Let (y_n) be a sequence in D_{A^*} such that (A^*y_n) converges to a point x in E. It follows from (39.3) that (y_n) is a Cauchy sequence in F and hence converges to a point y. Since A^* is closed it follows that $y \in D_{A^*}$ and $A^*y = x$. Hence $x \in R_{A^*}$ and the proof is complete.

This theorem will be frequently used in the following form.

39.8. COROLLARY. *Let E and F be Hilbert spaces, let A be a closed, densely defined operator from E into F, and let F_o be a closed subspace of F such that $R_A \subset F_o$. Then $R_A = F_o$ if and only if there is a constant $c > 0$ such that*

$$\|y\| \leq c\, \|A^*y\| \quad \textit{for every} \quad y \in D_{A^*} \cap F_o .$$

In this case for each $y \in F_o$ there exists $x \in D_A$ such that $\|x\| \leq c\, \|y\|$ and $Ax = y$. Furthermore, for each $x \in (N_A)^{\perp}$ there exists $y \in D_{A^*} \cap F_o$ such that $\|y\| \leq c\, \|x\|$ and $A^*y = x$.

PROOF. The first assertion follows from Theorem 39.7 applied to the Hilbert spaces E and F_o. The second assertion follows from the proof of Theorem 39.7. Finally, by Corollary 39.6 and Theorem 39.7 we have that $(N_A)^{\perp} = R_{A^*|D_{A^*} \cap F_o}$, and the last assertion in the corollary follows.

EXERCISES

39.A. Let U be an open subset of $\mathbb{R}^n$. Show that for each multi-index α the operator $\partial^{\alpha}/\partial x^{\alpha}$ in the sense of distributions defines a closed, densely defined operator in $L^2(U)$. Determine its adjoint.

40. THE $\bar{\partial}$ OPERATOR FOR L^2 DIFFERENTIAL FORMS

In this section we shall extend the $\bar{\partial}$ operator from the spaces of C^{∞} differential forms to certain spaces of L^2 differential forms. Our aim is to apply the Hilbert space techniques studied in Section 39 to the solution of the $\bar{\partial}$ equation in pseudoconvex domains.

To begin with we introduce some notation. Let U be an open subset of $\mathbb{C}^n$. Given differential forms $f = \sum_{J,K} f_{JK} dz_J \wedge d\bar{z}_K$ and $g = \sum_{J,K} g_{JK} dz_J \wedge d\bar{z}_K$ in $\Omega_{pq}(U)$ we shall set

$$f \cdot g = \sum_{J,K} f_{JK} \overline{g_{JK}}$$

and

$$|f|^2 = f \cdot f = \sum_{J,K} |f_{JK}|^2 .$$

Next we introduce several spaces of differential forms. If U is an open subset of $\mathbb{C}^n$ then we shall denote by $L^2_{pq}(U)$ (resp. $L^2_{pq}(U, loc)$) the vector space of all $f = \sum f_{JK} dz_J \wedge d\bar{z}_K$ in $\Omega_{pq}(U)$ such that each f_{JK} belongs to $L^2(U)$ (resp. $L^2(U, loc)$). If $\varphi \in C^{\infty}(U; \mathbb{R})$ then the space $L^2_{pq}(U,\varphi)$ is defined in a similar manner. Observe that $L^2_{pq}(U,\varphi)$ is a Hilbert space for the inner product

$$(f \mid g)_\varphi = \int_U f \cdot \bar{g}\, e^{-\varphi}\, d\lambda = \sum_{J,K} \int_U f_{JK}\, \overline{g_{JK}}\, e^{-\varphi}\, d\lambda.$$

The function φ is called a *weight function*. We shall set

$$\| f \|_\varphi^2 = (f \mid f)_\varphi = \int_U |f|^2 e^{-\varphi}\, d\lambda = \sum_{J,K} \int_U |f_{JK}|^2 e^{-\varphi}\, d\lambda.$$

By Exercise 16.C, $L^2(U, loc)$ is the union of the spaces $L^2(U,\varphi)$, with $\varphi \in C^\infty(U;\mathbb{R})$, and whence if follows that $L^2_{pq}(U, loc)$ is the union of the spaces $L^2_{pq}(U,\varphi)$, with $\varphi \in C^\infty(U,\mathbb{R})$.

40.1. DEFINITION. Let U be an open subset of $\mathbb{C}^n$.

(a) Let D_0 be the subspace of all f in $L^2(U, loc)$ such that the distribution $\dfrac{\partial f}{\partial \bar{z}_j}$ belongs to $L^2(U, loc)$ for $j = 1,\dots,n$. Let $\bar{\partial} : D_0 \to L^2_{01}(U, loc)$ be defined by

$$\bar{\partial} f = \sum_j \frac{\partial f}{\partial \bar{z}_j}\, d\bar{z}_j .$$

(b) Let D_1 be the subspace of all $g = \sum_j g_j\, d\bar{z}_j$ in $L^2_{01}(U, loc)$ such that the distribution $\dfrac{\partial g_j}{\partial \bar{z}_k} - \dfrac{\partial g_k}{\partial \bar{z}_j}$ belongs to $L^2(U, loc)$ for $j, k = 1,\dots,n$. Let $\bar{\partial} : D_1 \to L^2_{02}(U, loc)$ be defined by

$$\bar{\partial} g = \sum_{k<j} \left(\frac{\partial g_j}{\partial \bar{z}_k} - \frac{\partial g_j}{\partial \bar{z}_j} \right) d\bar{z}_k \wedge d\bar{z}_j .$$

It is clear that $C^\infty(U) \subset D_0$ and when restricted to $C^\infty(U)$ the new definition of $\bar{\partial}$ coincides with the usual definition. Likewise, $C^\infty_{01}(U) \subset D_1$ and when restricted to $C^\infty_{01}(U)$, the new and the old definition of $\bar{\partial}$ coincide.

40.2. PROPOSITION. *Let U be an open subset of $\mathbb{C}^n$, and let $f \in D_0$. Then $\bar{\partial} f \in D_1$ and $\bar{\partial}^2 f = 0$.*

PROOF. If $f \in D_0$ then $\overline{\partial} f = \sum_j g_j \, d\bar{z}_j$, where $g_j = \frac{\partial f}{\partial \bar{z}_j} \in L^2(U, loc)$ for every j. Hence

$$\frac{\partial g_j}{\partial \bar{z}_k} - \frac{\partial g_k}{\partial \bar{z}_j} = \frac{\partial^2 f}{\partial \bar{z}_k \, \partial \bar{z}_j} - \frac{\partial^2 f}{\partial \bar{z}_j \, \partial \bar{z}_k} = 0$$

for all j and k. Hence $\overline{\partial} f \in D_1$ and $\overline{\partial}^2 f = 0$.

In the next section we shall prove that whenever U is a pseudoconvex open set in $\mathbb{C}^n$ then the equation $\overline{\partial} f = g$ has a solution $f \in D_0$ for each $g \in D_1$ such that $\overline{\partial} g = 0$. In order to apply the Hilbert space technique studied in Section 39 we shall reduce this problem to a similar problem for the Hilbert spaces $L^2_{pq}(U,\varphi)$.

40.3. DEFINITION. Let U be an open subset of $\mathbb{C}^n$, and let $\alpha, \beta, \gamma \in C^\infty(U;\mathbb{R})$ be three weight functions.

(a) Let D_A be the subspace of all f in $L^2(U,\alpha)$ such that the distribution $\frac{\partial f}{\partial \bar{z}_j}$ belongs to $L^2(U,\beta)$ for $j=1,\ldots,n$. Let $A : D_A \to L^2_{01}(U,\beta)$ be defined by

$$Af = \overline{\partial} f = \sum_j \frac{\partial f}{\partial \bar{z}_j} \, d\bar{z}_j .$$

(b) Let D_B be the subspace of all $g = \sum_j g_j \, d\bar{z}_j$ in $L^2_{01}(U,\beta)$ such that the distribution $\frac{\partial g_j}{\partial \bar{z}_k} - \frac{\partial g_k}{\partial \bar{z}_j}$ belongs to $L^2_{02}(U,\gamma)$ for $j, k = 1,\ldots,n$. Let $B : D_B \to L^2_{02}(U,\gamma)$ be defined by

$$Bg = \overline{\partial} g = \sum_{k<j} \left(\frac{\partial g_j}{\partial \bar{z}_k} - \frac{\partial g_k}{\partial \bar{z}_j} \right) d\bar{z}_k \wedge d\bar{z}_j .$$

If U is an open subset of $\mathbb{C}^n$ then we shall denote by $\mathcal{D}_{pq}(U)$ the vector space of all $f = \sum f_{JK} dz_J \wedge d\bar{z}_K$ in $\Omega_{pq}(U)$ such that each f_{JK} belongs to $\mathcal{D}(U)$. It follows from Exercise

16.A that $\mathcal{D}(U)$ is dense in $L^2(U,\varphi)$ for each weight function φ, and whence it follows that $\mathcal{D}_{pq}(U)$ is dense in $L^2_{pq}(U,\varphi)$ for each weight function φ.

40.4. PROPOSITION. *Let U be an open subset of $\mathbb{C}^n$, let $\alpha, \beta, \gamma \in C^\infty(U;\mathbb{R})$ be three weight functions, and let A and B be the operators introduced in Definition 40.3. Then:*

(a) *A is a closed, densely defined operator from $L^2(U,\alpha)$ into $L^2_{01}(U,\beta)$.*

(b) *B is a closed, densely defined operator from $L^2_{01}(U,\beta)$ into $L^2_{02}(U,\gamma)$.*

(c) $R_A \subset N_B$.

PROOF. (a) It is clear that $\mathcal{D}(U) \subset D_A$ and hence D_A is dense in $L^2(U,\alpha)$. To show that A is closed let (f^m) be a sequence in D_A such that (f^m) converges to f in $L^2(U,\alpha)$ and (Af^m) converges to g in $L^2_{01}(U,\beta)$. It follows from the Cauchy-Schwarz inequality that (f^m) converges to f in $\mathcal{D}'(U)$, and then it follows from Proposition 17.6 that $(\partial f^m/\partial \bar{z}_j)$ converges to $\partial f/\partial \bar{z}_j$ in $\mathcal{D}'(U)$ for $j = 1,\ldots,n$. On the other hand, since $Af^m = \sum_j \frac{\partial f^m}{\partial \bar{z}_j} d\bar{z}_j$ and $g = \sum_j g_j d\bar{z}_j$, we have that $(\partial f^m/\partial \bar{z}_j)$ converges to g_j in $L^2(U,\beta)$, and therefore in $\mathcal{D}'(U)$, for $j = 1,\ldots,n$. Hence $\partial f/\partial \bar{z}_j = g_j \in L^2(U,\beta)$ for $j = 1,\ldots,n$. Thus $f \in D_A$ and $Af = g$.

The proof of (b) is similar to (a) and is left to the reader as an exercise. And to prove (c) it suffices to repeat the proof of Proposition 40.2.

The preceding proposition tells us that $R_A \subset N_B$ and we would like to prove that $R_A = N_B$ for suitable weight functions α, β, γ. To achieve the desired conclusion we intend to apply Corollary 39.8. Thus we want to find a constant $c > 0$ such that

(40.1) $\|g\|_\beta \leq c\|A^*g\|_\alpha$ for every $g \in D_{A^*} \cap N_B$.

Clearly if will be sufficient to find a constant $c > 0$ such that

(40.2) $\|g\|_\beta^2 \leq c^2(\|A^*g\|_\alpha^2 + \|Bg\|_\gamma^2)$ for every $g \in D_{A^*} \cap D_B$.

This will be achieved in the next section. In this section we shall prove that for suitable weight functions α, β, γ, the space $\mathcal{D}_{01}(U)$ is a dense subspace of $D_{A^*} \cap D_B$ with respect to the norm

$$|||\, g\, ||| = \|g\|_\beta + \|A^*g\|_\alpha + \|Bg\|_\gamma,$$

and then it will be sufficient to prove (40.2) for every $g \in \mathcal{D}_{01}(U)$. Before proving this density result we give three preparatory lemmas furnishing information about the operators A, B and A^*.

40.5. LEMMA. *Let U be an open subset of $\mathbb{C}^n$, let $\alpha, \beta \in C^\infty(U;\mathbb{R})$ be two weight functions, and let A be the operator introduced in Definition 40.3. Then $\mathcal{D}_{01}(U) \subset D_{A^*}$. If $g = \sum_j g_j d\bar{z}_j \in D_{A^*}$ then A^*g is given by the formula*

$$A^*g = -e^\alpha \sum_j \frac{\partial}{\partial z_j}(g_j e^{-\beta}) = -e^{\alpha-\beta}\sum_j\left(\frac{\partial g_j}{\partial z_j} - g_j\frac{\partial \beta}{\partial z_j}\right).$$

PROOF. To show that $\mathcal{D}_{01}(U) \subset D_{A^*}$ let $f \in D_A$ and let $g = \sum_j g_j d\bar{z}_j \in \mathcal{D}_{01}(U)$. Then

$$(Af\,|\,g)_\beta = \sum_j \int_U \frac{\partial f}{\partial \bar{z}_j}\bar{g}_j e^{-\beta} d\lambda = -\sum_j \int_U f \frac{\partial}{\partial \bar{z}_j}(\bar{g}_j e^{-\beta}) d\lambda$$

$$= -\sum_j \int_U f\, \overline{\frac{\partial}{\partial z_j}(g_j e^{-\beta})} d\lambda = -\left(f, e^\alpha \sum_j \frac{\partial}{\partial z_j}(g_j e^{-\beta})\right)_\alpha.$$

This shows that $g \in D_{A^*}$ and that $A^*g = -e^\alpha \sum_j \frac{\partial}{\partial z_j}(g_j e^{-\beta})$.

To prove that this formula is valid for every $g = \sum_j g_j d\bar{z}_j$ in D_{A^*}, we proceed similarly. Indeed, for every $f \in \mathcal{D}(U) \subset D_A$ we have that

$$(f \mid A^*g)_\alpha = (Af \mid g)_\beta = \sum_j \int_U \frac{\partial f}{\partial \bar{z}_j} \bar{g}_j e^{-\beta} d\lambda$$

$$= -\sum_j \frac{\partial}{\partial \bar{z}_j}(\bar{g}_j e^{-\beta})(f) = -\sum_j \overline{\frac{\partial}{\partial z_j}(g_j e^{-\beta})(f)}$$

and this implies that $A^*g \cdot e^{-\alpha} = -\sum_j \frac{\partial}{\partial z_j}(g_j e^{-\beta})$, as we wanted.

40.6. LEMMA. *Let U be an open subset of $\mathbb{C}^n$, let α, β, γ be three weight functions, and let A and B be the operators introduced in Definition 40.3.*

(a) *If $\varphi \in \mathcal{D}(U)$ and $f \in D_A$ then $\varphi f \in D_A$ and*

$$A(\varphi f) = \varphi A f + f \bar{\partial}\varphi.$$

(b) *If $\varphi \in \mathcal{D}(U)$ and $g \in D_B$ then $\varphi g \in D_B$ and*

$$B(\varphi g) = \varphi B g + \bar{\partial}\varphi \wedge g.$$

(c) *If $\varphi \in \mathcal{D}(U)$ and $g \in D_{A^*}$ then $\varphi g \in D_{A^*}$ and*

$$A^*(\varphi g) = \varphi A^* g - e^{\alpha-\beta} g \cdot \overline{\bar{\partial}\varphi}.$$

PROOF. (a) By Proposition 17.8

$$\frac{\partial}{\partial \bar{z}_j}(\varphi f) = \varphi \frac{\partial f}{\partial \bar{z}_j} + f \frac{\partial \varphi}{\partial \bar{z}_j}$$

for $j = 1,\ldots,n$, and the desired conclusion follows.

The proof of (b) is similar to (a) and is left as an exercise. To prove (c) let $f \in D_A$ and let $g = \sum g_j d\bar{z}_j \in D_{A^*}$. Then, using (a) we get that

$$\begin{aligned}(Af \mid \varphi g)_\beta &= (\bar{\varphi} Af \mid g)_\beta = (A(\bar{\varphi} f) - f\,\bar{\partial}\bar{\varphi} \mid g)_\beta \\ &= (\bar{\varphi} f \mid A^* g)_\alpha - (f\,\bar{\partial}\bar{\varphi} \mid g)_\beta \\ &= \int_U \bar{\varphi} f\,\overline{A^* g}\, e^{-\alpha} d\lambda - \sum_j \int_U f \frac{\partial \bar{\varphi}}{\partial \bar{z}_j} \bar{g}_j\, e^{-\beta} d\lambda \\ &= (f \mid \varphi A^* g - e^{\alpha-\beta} g \cdot \bar{\partial}\bar{u})_\alpha\end{aligned}$$

and the desired conclusion follows.

If $f = \sum_{J,K} f_{JK}\, dz_J \wedge d\bar{z}_K \in L^2_{pq}(U, loc)$ and $\varphi \in \mathcal{D}(\bar{B}(0;\delta))$ then we shall denote by $f * \varphi$ the differential form $f * \varphi = \sum_{J,K} (f_{JK} * \varphi) dz_J \wedge d\bar{z}_K$, which belongs to $C^\infty_{pq}(U_\delta)$, by Proposition 16.4. If the support of f is a compact subset of U and $\delta > 0$ is sufficiently small then $f * \varphi \in \mathcal{D}^\infty_{pq}(U)$. As a direct consequence of Proposition 17.14 we obtain the following result.

40.7. LEMMA. *Let U be an open subset of $\mathbb{C}^n$, let $\alpha, \beta, \gamma \in C^\infty(U;\mathbb{R})$ be three weight functions, and let A and B be the operators introduced in Definition 40.3.*

(a) *If $f \in D_A$ has compact support, and $\varphi \in \mathcal{D}(\bar{B}(0;\delta))$, with $\delta > 0$ sufficiently small, then $f * \varphi \in \mathcal{D}(U)$ and $A(f * \varphi) = (Af) * \varphi$.*

(b) *If $g \in D_B$ has compact support, and $\varphi \in \mathcal{D}(\bar{B}(0;\delta))$, with $\delta > 0$ sufficiently small, then $g * \varphi \in \mathcal{D}_{01}(U)$ and $B(g * \varphi) = (Bg) * \varphi$.*

40.8. PROPOSITION. *Let U be an open subset of $\mathbb{C}^n$. Let $(\eta_i)_{i=1}^\infty$ be a sequence in $\mathcal{D}(U)$ such that $0 \le \eta_i \le 1$ on U for every i, and $\eta_i \equiv 1$ on each compact subset of U for all sufficiently large i. Let $\alpha, \beta, \gamma \in C^\infty(U;\mathbb{R})$ be three weight functions such that $|\bar{\partial}\eta_i|^2 \le e^{\beta-\alpha}$ and $|\bar{\partial}\eta_i|^2 \le e^{\gamma-\beta}$ for every i. Then $\mathcal{D}_{01}(U)$ is dense in $D_{A^*} \cap D_B$ with respect to the norm*

(40.3) $$|||\, g \,||| = \|g\|_\beta + \|A^*g\|_\alpha + \|Bg\|_\gamma$$

PROOF. We shall proceed in two stages. We shall first prove that each $g \in D_{A^*} \cap D_B$ can be approximated by members of $D_{A^*} \cap D_B$ with compact support. Next we shall prove that each $g \in D_{A^*} \cap D_B$ with compact support can be approximated by members of $\mathcal{D}_{01}(U)$.

(a) Let $g = \sum_j g_j\, d\bar{z}_j \in D_{A^*} \cap D_B$. Then $\eta_i g \in D_{A^*} \cap D_B$ by Lemma 40.6, and we shall prove that

(40.4) $$\lim_{i \to \infty} |||\, \eta_i g - g \,||| = 0.$$

It follows from the Dominated Convergence Theorem that $\lim_{i\to\infty} \|\eta_i g - g\|_\beta = 0$, $\lim_{i \to \infty} \|\eta_i A^*g - A^*g\|_\alpha = 0$ and $\lim_{i \to \infty} \|\eta_i Bg - Bg\|_\gamma = 0$. Thus to prove (40.4) it is clearly sufficient to prove that

(40.5) $$\lim_{i \to \infty} \|A^*(\eta_i g) - \eta_i A^*g\|_\alpha = 0$$

and

(40.6) $$\lim_{i \to \infty} \|B(\eta_i g) - \eta_i Bg\|_\gamma = 0.$$

By Lemma 40.6,

$$A^*(\eta_i g) - \eta_i A^*g = - e^{\alpha-\beta} g \cdot \bar{\partial}\eta_i .$$

Hence

$$|A^*(\eta_i g) - \eta_i A^*g|^2 e^{-\alpha} \le e^{2\alpha-2\beta} |g|^2 \, |\bar{\partial}\eta_i|^2 \, e^{-\alpha} \le |g|^2 e^{-\beta}$$

and (40.5) follows from the Dominated Covergence Theorem. On the other hand, again by Lemma 40.6,

$$B(\eta_i g) - \eta_i Bg = \bar{\partial}\eta_i \wedge g = \sum_{k,j} \frac{\partial \eta_i}{\partial \bar{z}_k} g_j d\bar{z}_k \wedge d\bar{z}_j$$

$$= \sum_{k<j} \left(\frac{\partial \eta_i}{\partial \bar{z}_k} g_j - \frac{\partial \eta_i}{\partial \bar{z}_j} g_k \right) d\bar{z}_k \wedge d\bar{z}_j .$$

Hence

$$|B(\eta_i g) - \eta_i Bg|^2 = \sum_{k<j} \left| \frac{\partial \eta_i}{\partial \bar{z}_k} g_j - \frac{\partial \eta_i}{\partial \bar{z}_j} g_k \right|^2$$

$$= \sum_{k<j} \left(\frac{\partial \eta_i}{\partial \bar{z}_k} g_j - \frac{\partial \eta_i}{\partial \bar{z}_j} g_k \right) \left(\overline{\frac{\partial \eta_i}{\partial \bar{z}_k}} \bar{g}_j - \overline{\frac{\partial \eta_i}{\partial \bar{z}_j}} \bar{g}_k \right)$$

$$= \sum_{k<j} \left(\left| \frac{\partial \eta_i}{\partial \bar{z}_j} \right|^2 |g_j|^2 + \left| \frac{\partial \eta_i}{\partial \bar{z}_j} \right|^2 |g_k|^2 - \frac{\partial \eta_i}{\partial \bar{z}_k} g_j \overline{\frac{\partial \eta_i}{\partial \bar{z}_j}} \bar{g}_k - \frac{\partial \eta_i}{\partial \bar{z}_j} g_k \overline{\frac{\partial \eta_i}{\partial \bar{z}_j}} \bar{g}_k \right)$$

$$= \sum_{k,j} \left| \frac{\partial \eta_i}{\partial \bar{z}_k} \right|^2 |g_j|^2 - \sum_{k,j} \frac{\partial \eta_i}{\partial \bar{z}_k} \bar{g}_k \overline{\frac{\partial \eta_i}{\partial \bar{z}_j}} g_j$$

$$= |\bar{\partial} \eta_i|^2 |g|^2 - \left| \sum_k \frac{\partial \eta_i}{\partial \bar{z}_k} \bar{g}_k \right|^2 .$$

Thus

$$|B(\eta_i g) - \eta_i Bg|^2 e^{-\gamma} \le |\bar{\partial} \eta_i|^2 |g|^2 e^{-\gamma} \le |g|^2 e^{-\beta}$$

and another application of the Dominated Convergence Theorem shows (40.6). Thus (40.4) has been proved.

(b) Now let $g = \sum_j g_j d\bar{z}_j$ be a member of $D_{A^*} \cap D_B$ with compact support. By Lemma 40.7, $g * \rho_\delta \in \mathcal{D}_{01}(U)$ if $\delta > 0$ is sufficiently small, and we shall prove that

$$\text{(40.7)} \qquad \lim_{\delta \to 0} ||| g * \rho_\delta - g ||| = 0.$$

If follows from Exercise 16.A and Lemma 40.7 that $\lim_{\delta \to 0} \| g * \rho_\delta - g \|_\beta = 0$ and $\lim_{\delta \to 0} \| B(g * \rho_\delta) - Bg \|_\gamma = 0$. Thus to show (40.7) it suffices to prove that

$$(40.8) \qquad \lim_{\delta \to 0} \| A^*(g * \rho_\delta) - A^* g \|_\alpha = 0$$

Now, by Lemma 40.5, A^* can be decomposed as a sum $A^* = S + T$, where

$$Sg = - e^{\alpha-\beta} \sum_j \frac{\partial g_j}{\partial z_j} \quad \text{and} \quad Tg = e^{\alpha-\beta} \sum_j g_j \frac{\partial \beta}{\partial z_j} .$$

Thus to show (40.8) it suffices to prove that

$$(40.9) \qquad \lim_{\delta \to 0} \| S(g * \rho_\delta) - Sg \|_\alpha = 0$$

and

$$(40.10) \qquad \lim_{\delta \to 0} \| T(g * \rho_\delta) - Tg \|_\alpha = 0$$

Since clearly $S(g * \rho_\delta) = (Sg) * \rho_\delta$, (40.9) follows from Exercise 16.A. On the other hand,

$$T(g * \rho_\delta) - Tg = e^{\alpha-\beta} \sum_j (g_j * \rho_\delta - g_j) \frac{\partial \beta}{\partial z_j} .$$

Hence

$$|T(g * \rho_\delta) - Tg|^2 \le e^{2\alpha-2\beta} |g * \rho_\delta - g|^2 |\bar{\partial}\beta|^2$$

and (40.10) follows from Exercise 16.A. This shows (40.7) and the proof of the proposition is complete.

41. L^2 SOLUTIONS OF THE $\bar{\partial}$ EQUATION

In this section we solve the $\bar{\partial}$ equation in the sense of distributions for L^2 differential forms. The key result is the following.

41.1. PROPOSITION. *Let U be an open subset of $\mathbb{C}^n$. Let $(\eta_i)_{i=1}^\infty$ be a sequence in $\mathcal{D}(U)$ such that $0 \le \eta_i \le 1$ on U for every i, and $\eta_i \equiv 1$ on each compact subset of U for all sufficiently large i. Suppose there are functions $\varphi, \psi \in C^\infty(U; \mathbb{R})$ such that*

$|\bar\partial\eta_i|^2 \leq e^{\psi}$ *on* U *for every* i *and*

$$(41.1) \qquad \sum_{j,k} \frac{\partial^2\varphi}{\partial z_j \partial \bar{z}_j}(a) b_j \bar{b}_k \geq 2(|\bar\partial\psi(a)|^2 + e^{\psi(a)}) \sum_k |b_k|^2$$

for all $a \in U$ *and* $b \in \mathbb{C}^n$. *Let* α, β, γ *be the weight functions defined by* $\alpha = \varphi - 2\psi$, $\beta = \varphi - \psi$, $\gamma = \varphi$, *and let* A *and* B *be the operators introduced in Definition 40.3. Then*

$$(41.2) \quad \|g\|_\beta^2 \leq \|A^*g\|_\alpha^2 + \|Bg\|_\gamma^2 \quad \textit{for every} \quad g \in D_{A^*} \cap D_B.$$

For each $g \in N_B$ *there exists* $f \in D_A$ *such that* $\|f\|_\alpha \leq \|g\|_\beta$ *and* $Af = g$. *For each* $f \in (N_A)^\perp$ *there exists* $g \in D_{A^*} \cap D_B$ *such that* $\|g\|_\beta \leq \|f\|_\alpha$ *and* $A^*g = f$.

To prove this proposition we shall need the following lemma.

41.2. LEMMA. *Let* U *be an open subset of* $\mathbb{C}^n$, *and let* $\varphi \in C^\infty(U;\mathbb{R})$. *For each* $j = 1,\ldots,n$, *let* $\delta_j : \mathcal{D}(U) \to \mathcal{D}(U)$ *be defined by*

$$\delta_j f = e^{\varphi} \frac{\partial}{\partial z_j}(f e^{-\varphi}) = \frac{\partial f}{\partial z_j} - f \frac{\partial \varphi}{\partial z_j}.$$

Then for every $f, g \in \mathcal{D}(U)$ *we have that*

$$\int_U \frac{\partial f}{\partial \bar{z}_j} \bar{g} e^{-\varphi} d\lambda = - \int_U f \overline{\delta_j g} e^{-\varphi} d\lambda$$

and

$$\delta_j \frac{\partial f}{\partial \bar{z}_k} - \frac{\partial}{\partial \bar{z}_k} \delta_j u = \frac{\partial^2 \varphi}{\partial z_j \partial \bar{z}_k} f.$$

An integration by parts yields the first identity. And the second one follows readily from the definition of δ_j. The details are left to the reader as an exercise.

PROOF OF PROPOSITION 41.1. It is sufficient to prove (41.2) for every $g \in \mathcal{D}_{01}(U)$, for Proposition 40.8 will then imply the validity of (41.2) for every $g \in D_{A^*} \cap D_B$. Let $g = \sum_j g_j \, d\bar{z}_j \in \mathcal{D}_{01}(U)$. Then

$$Bg = \sum_{k<j} \left(\frac{\partial g_j}{\partial \bar{z}_k} - \frac{\partial g_k}{\partial \bar{z}_j} \right) d\bar{z}_k \wedge d\bar{z}_j ,$$

hence

$$|Bg|^2 = \sum_{k<j} \left| \frac{\partial g_j}{\partial \bar{z}_k} - \frac{\partial g_k}{\partial \bar{z}_j} \right|^2$$

$$= \sum_{k<j} \left(\frac{\partial g_j}{\partial \bar{z}_k} - \frac{\partial g_k}{\partial \bar{z}_j} \right) \left(\overline{\frac{\partial g_j}{\partial \bar{z}_k}} - \overline{\frac{\partial g_k}{\partial \bar{z}_j}} \right)$$

$$= \sum_{k<j} \left(\left| \frac{\partial g_j}{\partial \bar{z}_k} \right|^2 + \left| \frac{\partial g_k}{\partial \bar{z}_j} \right|^2 - \frac{\partial g_j}{\partial \bar{z}_k} \overline{\frac{\partial g_k}{\partial \bar{z}_j}} - \frac{\partial g_k}{\partial \bar{z}_j} \overline{\frac{\partial g_j}{\partial \bar{z}_k}} \right)$$

$$= \sum_{j,k} \left| \frac{\partial g_j}{\partial \bar{z}_k} \right|^2 - \sum_{j,k} \frac{\partial g_j}{\partial \bar{z}_k} \overline{\frac{\partial g_k}{\partial \bar{z}_j}} ,$$

and therefore

$$(41.3) \qquad - \sum_{j,k} \frac{\partial g_j}{\partial \bar{z}_k} \overline{\frac{\partial g_k}{\partial \bar{z}_j}} \leq |Bg|^2 .$$

On the other hand, by Lemmas 40.5 and 41.2 we have that

$$A^*g = - e^{\alpha-\beta} \sum_j \left(\frac{\partial g_j}{\partial z_j} - g_j \frac{\partial \beta}{\partial z_j} \right) = - e^{-\psi} \sum_j \left(\delta_j g_j + g_j \frac{\partial \psi}{\partial z_j} \right),$$

and hence

$$- \sum_j \delta_j g_j = e^{\psi} A^*g + \sum_j g_j \frac{\partial \psi}{\partial z_j} = e^{\psi} A^*g + g \cdot \bar{\partial} \psi .$$

It follows from the inequality $\| x + y \|^2 \leq 2 \| x \|^2 + 2 \| y \|^2$ that

$$\int_U \left| \sum_j \delta_j g_j \right|^2 e^{-\varphi} d\lambda \leq 2 \int_U e^{2\psi} |A^* g|^2 e^{-\varphi} d\lambda + 2 \int_U |g \cdot \bar{\partial}\psi|^2 e^{-\varphi} d\lambda ,$$

and therefore

$$(41.4)\quad \int_U \sum_{j,k} \delta_j g_j \overline{\delta_k g_k}\, e^{-\varphi} d\lambda \le 2\,\|A^* g\|_\alpha^2 + 2\int_U |g|^2 |\bar{\partial}\psi|^2\, e^{-\varphi} d\lambda.$$

It follows from (41.3) and (41.4) that

$$(41.5)\quad \int_U \sum_{j,k} \left(\delta_j g_j \overline{\delta_k g_k} - \frac{\partial g_j}{\partial \bar{z}_k}\overline{\frac{\partial g_k}{\partial \bar{z}_j}}\right) e^{-\varphi} d\lambda$$

$$\le \|Bg\|_\gamma^2 + 2\,\|A^* g\|_\alpha^2 + 2\int_U |g|^2\, |\bar{\partial}\psi|^2 e^{-\varphi} d\lambda.$$

Now, using Lemma 41.2 we have that

$$\int_U \delta_j g_j \overline{\delta_k g_k}\, e^{-\varphi} d\lambda - \int_U \frac{\partial g_j}{\partial \bar{z}_k}\overline{\frac{\partial g_k}{\partial \bar{z}_j}}\, e^{-\varphi} d\lambda$$

$$= -\int_U \frac{\partial}{\partial \bar{z}_k}\, \delta_j g_j \bar{g}_k\, e^{-\varphi} d\lambda + \int_U \delta_j \frac{\partial g_j}{\partial \bar{z}_k}\, \bar{g}_k\, e^{-\varphi} d\lambda$$

$$= \int_U \frac{\partial^2 \varphi}{\partial z_j\, \partial \bar{z}_k}\, g_j \bar{g}_k\, e^{-\varphi} d\lambda.$$

Hence substituting into (41.5) yields the inequality

$$\int_U \left(\sum_{j,k} \frac{\partial^2 \varphi}{\partial z_j\, \partial \bar{z}_k}\, g_j \bar{g}_k - 2|g|^2 |\bar{\partial}\psi|^2\right) e^{-\varphi} d\lambda \le \|Bg\|_\gamma^2 + 2\,\|A^* g\|_\alpha^2$$

and using (41.1) we get that

$$2\,\|g\|_\beta^2 = 2\int_U |g|^2 e^{\psi-\varphi} d\lambda \le \|Bg\|_\gamma^2 + 2\,\|A^* g\|_\alpha^2.$$

This shows (41.2) for every $g \in \mathcal{D}_{01}(U)$ and therefore for every $g \in D_{A^*} \cap D_B$. The remaining assertions in the proposition follow from Corollary 39.8.

41.3. THEOREM. *Let U be a pseudoconvex open set in $\mathbb{C}^n$. Then for each $g \in L^2_{01}(U, loc)$ such that $\bar{\partial} g = 0$ there exists $f \in L^2(U, loc)$*

such that $\bar{\partial} f = g$.

Before proving this theorem we give three auxiliary lemmas.

41.4. **LEMMA.** *Let* U *be an open subset of* $\mathbb{C}^n$. *Let* $(\eta_i)_{i=1}^{\infty}$ *be a sequence in* $\mathcal{D}(U)$ *such that* $\eta_i \equiv 1$ *on each compact subset of* U *for all sufficiently large* i. *Then there is a function* $\psi \in C^{\infty}(U;\mathbb{R})$ *such that* $|\partial \eta_i|^2 \leq e^{\psi}$ *on* U *for every* i.

PROOF. Let $(U_j)_{j=1}^{\infty}$ be an increasing sequence of relatively compact open subsets of U which cover U. Let $c_j = \sup_{i\in\mathbb{N}} \sup_{x\in U_j} |\bar{\partial}\eta_i(x)|^2$ for each $j \in \mathbb{N}$. Since $\bar{\partial}\eta_i \equiv 0$ on U_j for all sufficiently large i we see that $c_j < \infty$ for every j. If we set $U_o = \phi$ then by Exercise 15.C there is a function $\tau \in C^{\infty}(U;\mathbb{R})$ such that $\tau > c_j$ on $U_j \setminus U_{j-1}$ for every $j \in \mathbb{N}$. Then the func- $\psi = \log \tau$ has the required properties.

41.5. **LEMMA.** *Let* $f : \mathbb{R} \to \mathbb{R}$ *be a function which is bounded above on each interval of the form* $(-\infty, b]$. *Then there is an increasing function* $g \in C^{\infty}(\mathbb{R};\mathbb{R})$ *such that* $g(t)$ *is constant for* $t \leq 0$ *and* $g(t) \geq f(t)$ *for every* $t \in \mathbb{R}$.

PROOF. Let $c_j = \sup\{f(t) : t \leq j\}$ for each $j \in \mathbb{N}$. Observe that (c_j) is an increasing sequence. By Lemma 15.1 for each $j \in \mathbb{N}$ there is an increasing function $\varphi_j \in C^{\infty}(\mathbb{R};\mathbb{R})$ such that $\varphi_j(t) = 0$ for $t \leq j - 1$ and $\varphi_j(t) = 1$ for $t \geq j$. Let $g : \mathbb{R} \to \mathbb{R}$ be defined by

$$g(t) = c_1 + \sum_{j=1}^{\infty} (c_{j+1} - c_j)\varphi_j(t).$$

Since $\varphi_j(t) = 0$ for $t \leq n$ and $j \geq n + 1$ we see that g is well defined, increasing and of class C^{∞}. If $t \leq 0$ then $g(t) = c_1$. And if $n - 1 \leq t \leq n$, with $n \in \mathbb{N}$, then

$$g(t) = c_1 + \sum_{j=1}^{n-1} (c_{j+1} - c_j) + (c_{n+1} - c_n)\varphi_n(t) \geq c_n \geq f(t).$$

Thus g has the required properties.

41.6. LEMMA. *Let $f : \mathbb{R} \to \mathbb{R}$ be a function which is bounded above on each interval of the form $(-\infty, b]$. Then there is a convex, increasing function $g \in C^\infty(\mathbb{R};\mathbb{R})$ such that $g(t) \geq f(t)$ and $g'(t) \geq f(t)$ for every $t \in \mathbb{R}$.*

PROOF. By Lemma 41.5 there is an increasing function $\varphi \in C^\infty(\mathbb{R};\mathbb{R})$ such that $\varphi(t)$ is constant for $t \leq 0$ and $\varphi(t) \geq f(t)$ for every $t \in \mathbb{R}$. Since φ is constant on the interval $(-\infty, 0]$, the derivative φ' is bounded on each interval of the form $(-\infty, b]$. Then another application of Lemma 41.5 yields an increasing function $\psi \in C^\infty(\mathbb{R};\mathbb{R})$ such that

$$\psi(t) \geq max\,\{\varphi'(t),\ f(t),\ 0\}$$

for every $t \in \mathbb{R}$. Let $g \in C^\infty(\mathbb{R};\mathbb{R})$ be defined by

$$g(t) = \varphi(0) + \int_0^t \psi(s)\,ds.$$

Then for each $t \in \mathbb{R}$ we have that

$$g(t) \geq \varphi(0) + \int_0^t \varphi'(s)\,ds = \varphi(t) \geq f(t)$$

and $g'(t) = \psi(t) \geq f(t)$. Moreover, $g'(t) = \psi(t) \geq 0$ and $g''(t) = \psi'(t) \geq 0$, proving that g is convex and increasing, as we wanted.

PROOF OF THEOREM 41.3. Let $g_0 \in L^2_{01}(U, loc)$ with $\bar{\partial} g_0 = 0$ be given. With the help of Proposition 41.1 we shall construct weight functions $\alpha, \beta, \gamma \in C^\infty(U;\mathbb{R})$ such that $g_0 \in L^2_{01}(U,\beta)$ and

$$(41.6)\qquad \|g\|^2_\beta \leq \|A^*g\|^2_\alpha + \|Bg\|^2_\gamma \quad \text{for every} \quad g \in D_{A^*} \cap D_B.$$

Let $(\eta_i)_{i=1}^\infty$ be a sequence in $\mathcal{D}(U)$ such that $0 \leq \eta_i \leq 1$ on U for every i, and $\eta_i \equiv 1$ on each compact subset of U for all sufficiently large i. By Lemma 41.4 there is a function $\psi \in C^\infty(U;\mathbb{R})$ such that $|\bar\partial\eta_i|^2 \leq e^\psi$ on U for every i. Since $ge^\psi \in L^2_{01}(U,loc)$ there is a weight function $\sigma \in C^\infty(U;\mathbb{R})$ such that $ge^\psi \in L^2_{01}(U,\sigma)$, that is, $g \in L^2_{01}(U,\sigma - \psi)$. In order to apply Proposition 41.1 we should find a function $\varphi \in C^\infty(U;\mathbb{R})$ such that

$$\varphi \geq \sigma \quad \text{on} \quad U \tag{41.7}$$

and

$$\sum_{j,k} \frac{\partial^2\varphi}{\partial z_j \partial \bar z_k}(a) b_j \bar b_k \geq 2(|\partial\psi(a)|^2 + e^{\psi(a)}) \sum_k |b_k|^2 \tag{41.8}$$

for all $a \in U$ and $b \in \mathbb{C}^n$. To find such a function φ we proceed as follows. By Theorem 38.6 there is a strictly plurisubharmonic function $u \in C^\infty(U;\mathbb{R})$ such that the set $K_t = \{z \in U : u(z) \leq t\}$ is compact for each $t \in \mathbb{R}$. By Lemma 35.4 there is a strictly positive function $\tau \in C^\infty(U;\mathbb{R})$ such that

$$\sum_{j,k} \frac{\partial^2 u}{\partial z_j \partial \bar z_k}(a) b_j \bar b_k \geq \tau(a) \sum_k |b_k|^2$$

for every $a \in U$ and $b \in \mathbb{C}^n$. If $\chi \in C^\infty(\mathbb{R};\mathbb{R})$ is a convex, increasing function then it follows from Exercises 35.B and 35.E that

$$\sum_{j,k} \frac{\partial^2(\chi \circ u)}{\partial z_j \partial \bar z_k}(a) b_j \bar b_k \geq \chi'(u(a)) \sum_{j,k} \frac{\partial^2 u}{\partial z_j \partial \bar z_k}(a) b_j \bar b_k$$

$$\geq \chi'(u(a))\tau(a) \sum_k |b_k|^2$$

for every $a \in U$ and $b \in \mathbb{C}^n$. Thus it is sufficient to find a convex, increasing function $\chi \in C^\infty(\mathbb{R};\mathbb{R})$ such that

$$\chi \circ u \geq \sigma \tag{41.9}$$

and

$$(\chi' \circ u)\tau \geq 2(|\bar\partial\psi|^2 + e^\psi) \tag{41.10}$$

on U, for then the function $\varphi = \chi \circ u$ will satisfy (41.7) and (41.8). To find such a function χ consider the functions v, $w : \mathbb{R} \to \mathbb{R}$ defined by

$$v(t) = \sup_{K_t} \sigma \quad \text{and} \quad w(t) = \sup_{K_t} \frac{2(|\bar{\partial}\psi|^2 + e^{\psi})}{\tau}$$

for $t > \inf_U u$, and $v(t) = w(t) = 0$ for $t \leq \inf_U u$. Since the functions v and w are increasing for $t > \inf_U u$, an application of Lemma 41.6 yields a convex, increasing function $\chi \in C^{\infty}(\mathbb{R};\mathbb{R})$ such that $\chi(t) \geq v(t)$ and $\chi'(t) \geq w(t)$ for every $t \in \mathbb{R}$. Then for each $z \in U$ we have that

$$\chi(u(z)) \geq v(u(z)) \geq \sigma(z)$$

and

$$\chi'(u(z)) \geq w(u(z)) \geq \frac{2(|\bar{\partial}\psi(z)|^2 + e^{\psi(z)})}{\tau(z)} .$$

Thus χ satisfies (41.9) and (41.10) and $\varphi = \chi \circ u$ satisfies (41.7) and (41.8). If we set $\alpha = \varphi - 2\psi$, $\beta = \varphi - \psi$ and $\gamma = \varphi$, then $g_o \in L^2_{01}(U,\beta)$, and (41.6) follows from Proposition 41.1. Hence there exists $f_o \in L^2(U,\alpha) \subset L^2(U,loc)$ such that $\bar{\partial} f_o = g_o$ and the proof of the theorem is complete.

42. C^{∞} SOLUTIONS OF THE $\bar{\partial}$ EQUATION

In this section we solve the $\bar{\partial}$ equation for C^{∞} differential forms.

42.1. DEFINITION. Let U be an open subset of $\mathbb{C}^n$. For each $k \in \mathbb{N}_o$ we shall denote by $W^k(U)$ the vector space of all $f \in L^2(U)$ whose derivatives (in the sense of distributions) of order $\leq k$ belong to $L^2(U)$. Likewise, we shall denote by $W^k(U,loc)$ the vector space of all $f \in L^2(U,loc)$ whose derivatives of order $\leq k$ belong to $L^2(U,loc)$. The spaces of differential forms $W^k_{pq}(U)$ and $W^k_{pq}(U,loc)$ are defined in the obvious way.

42.2. PROPOSITION. *Let* U *be a pseudoconvex open set in* $\mathbb{C}^n$, *and let* $g \in W^k_{01}(U, loc)$ *with* $\bar{\partial} g = 0$. *Then each solution* $f \in L^2(U, loc)$ *of the equation* $\bar{\partial} f = g$ *belongs to* $W^{k+1}(U, loc)$.

To prove Proposition 42.2 we need the following lemma.

42.3. LEMMA. *Let* $f \in L^2(\mathbb{C}^n)$ *be a function with compact support. If* $\partial f/\partial \bar{z}_j \in L^2(\mathbb{C}^n)$ *for* $j = 1,\ldots,n$ *then* $f \in W^1(\mathbb{C}^n)$.

PROOF. Two integrations by parts immediately show that

$$\int_{\mathbb{C}^n} \left|\frac{\partial \varphi}{\partial z_j}\right|^2 d\lambda = \int_{\mathbb{C}^n} \left|\frac{\partial \varphi}{\partial \bar{z}_j}\right|^2 d\lambda \tag{42.1}$$

for every $\varphi \in \mathcal{D}(\mathbb{C}^n)$. Applying (42.1) with $\varphi = f * \rho_\delta - f * \rho_\varepsilon$ gives that

$$\int_{\mathbb{C}^n} \left|\frac{\partial f}{\partial z_j} * \rho_\delta - \frac{\partial f}{\partial z_j} * \rho_\varepsilon\right|^2 d\lambda = \int_{\mathbb{C}^n} \left|\frac{\partial}{\partial z_j} (f * \rho_\delta - f * \rho_\varepsilon)\right|^2 d\lambda$$

$$= \int_{\mathbb{C}^n} \left|\frac{\partial}{\partial \bar{z}_j}(f * \rho_\delta - f * \rho_\varepsilon)\right|^2 d\lambda = \int_{\mathbb{C}^n} \left|\frac{\partial f}{\partial \bar{z}_j} * \rho_\delta - \frac{\partial f}{\partial \bar{z}_j} * \rho_\varepsilon\right|^2 d\lambda.$$

Since $\partial f/\partial \bar{z}_j \in L^2(\mathbb{C}^n)$, the last written integral tends to zero when $\delta, \varepsilon \to 0$ by Exercise 16.A. Hence for each $j = 1, \ldots, n$ there exists $g_j \in L^2(\mathbb{C}^n)$ such that $(\frac{\partial f}{\partial z_j} * \rho_\delta)$ converges to g_j in $L^2(\mathbb{C}^n)$. It follows from the Cauchy-Schwarz inequality that $(\frac{\partial f}{\partial z_j} * \rho_\delta)$ converges to g_j in $\mathcal{D}'(\mathbb{C}^n)$. On the other hand $(\frac{\partial f}{\partial z_j} * \rho_\delta)$ converges to $\partial f/\partial z_j$ in $\mathcal{D}'(\mathbb{C}^n)$, by Proposition 17.16. Thus $\partial f/\partial z_j = g_j \in L^2(\mathbb{C}^n)$ for $j = 1,\ldots,n$ and the lemma has been proved.

PROOF OF PROPOSITION 42.2. Let $g = \sum_j g_j d\bar{z}_j \in W^k_{01}(U, loc)$ with $\bar{\partial} g = 0$, and let $f \in L^2(U, loc) = W^o(U, loc)$ be a solution of the equation $\bar{\partial} f = g$. To show that $f \in W^{k+1}(U, loc)$ we shall prove that if $f \in W^r(U, loc)$ and $0 \leq r \leq k$ then $f \in W^{r+1}(U, loc)$.

Let $\eta \in \mathcal{D}(U)$. Then

$$\frac{\partial}{\partial \bar{z}_j}(\eta f) = \frac{\partial \eta}{\partial \bar{z}_j} f + \eta \frac{\partial f}{\partial \bar{z}_j} = \frac{\partial \eta}{\partial \bar{z}_j} f + \eta g_j$$

and therefore $\frac{\partial}{\partial \bar{z}_j}(\eta f) \in W^r(\mathbb{C}^n)$. Let u be a derivative of order $\leq r$ of ηf. Then $\partial u/\partial \bar{z}_j$ is a derivative of order $\leq r$ of $\frac{\partial}{\partial \bar{z}_j}(\eta f)$. Since $\frac{\partial}{\partial \bar{z}_j}(\eta f) \in W^r(\mathbb{C}^n)$, it follows that $\partial u/\partial \bar{z}_j \in L^2(\mathbb{C}^n)$, and by Lemma 42.3 we may conclude that $u \in W^1(\mathbb{C}^n)$. This shows that $\eta f \in W^{r+1}(\mathbb{C}^n)$, and since $\eta \in \mathcal{D}(U)$ was arbitrary we conclude that $f \in W^{r+1}(U)$, completing the proof.

42.4. PROPOSITION. *Let U be an open subset of $\mathbb{C}^n$. Then $W^{2n+k}(U, loc) \subset \mathcal{C}^k(U)$ for every $k \in \mathbb{N}_o$.*

PROOF. If $\varphi \in \mathcal{D}(\mathbb{C}^n)$ then it follows from the Fundamental Theorem of Calculus that

$$\sup_{\mathbb{C}^n} |\varphi| \leq \int_{\mathbb{C}^n} \left| \frac{\partial^{2n} \varphi}{\partial x_1 \partial y_1 \cdots \partial x_n \partial y_n} \right| d\lambda. \tag{42.2}$$

(a) Let $f \in W^{2n}(\mathbb{C}^n)$ be a function with compact support. Applying (42.2) with $\varphi = f * \rho_\delta - f * \rho_\varepsilon$ gives that

$$\sup_{\mathbb{C}^n} |f * \rho_\delta - f * \rho_\varepsilon| \leq \int_{\mathbb{C}^n} \left| \frac{\partial^{2n}(f * \rho_\delta - f * \rho_\varepsilon)}{\partial x_1 \partial y_1 \cdots \partial x_n \partial y_n} \right| d\lambda$$

$$\leq c \int_{\mathbb{C}^n} \left| \frac{\partial^{2n}(f * \rho_\delta - f * \rho_\varepsilon)}{\partial x_1 \partial y_1 \cdots \partial x_n \partial y_n} \right|^2 d\lambda.$$

Since $f \in W^{2n}(\mathbb{C}^n)$, the last written integral converges to zero when $\delta, \varepsilon \to 0$, by Exercise 16.A. Hence $(f * \rho_\delta)$ converges uniformly on $\mathbb{C}^n$ to a function $g \in C(\mathbb{C}^n)$ with compact support.

Hence it follows that $(f * \rho_\delta)$ converges to g in $L^2(\mathbb{C}^n)$. On the other hand, we know that $(f * \rho_\delta)$ converges to f in $L^2(\mathbb{C}^n)$, and thus we conclude that $f = g$ almost everywhere, that is, $f \in C(\mathbb{C}^n)$.

(b) Let $f \in W^{2n}(U, loc)$. If $\eta \in \mathcal{D}(U)$ then it follows from (a) that $\eta f \in C(\mathbb{C}^n)$. Since $\eta \in \mathcal{D}(U)$ was arbitrary we may conclude that $f \in C(U)$. Thus we have shown that $W^{2n}(U, loc) \subset C(U)$.

(c) Finally let $f \in W^{2n+k}(U, loc)$. If u is a derivative of order $\leq r$ of f then $u \in W^{2n}(U, loc) \subset C(U)$ by (b). Hence $f \in C^k(U)$ and the proof of the proposition is complete.

From Propositions 42.2 and 42.4 we obtain at once the following theorem.

42.5. THEOREM. *Let U be a pseudoconvex open set in $\mathbb{C}^n$, and let $g \in C^\infty_{01}(U)$ with $\bar{\partial} g = 0$. Then each solution $f \in L^2(U, loc)$ of the equation $\bar{\partial} f = g$ belongs to $C^\infty(U)$.*

NOTES AND COMMENTS

The idea to use the weighted spaces $L^2_{pq}(U, \varphi)$ to solve the $\bar{\partial}$ equation in pseudoconvex domains is due to L. Hörmander [2]. Our presentation in Section 39 follows the book of F. Riesz and B. Sz. Nagy [1]. Our presentation in Sections 40, 41 and 42 follows the book of L. Hörmander [3].

CHAPTER X

THE LEVI PROBLEM

43. THE LEVI PROBLEM IN $\mathbb{C}^n$

In this section we solve Problem 37.10 in the case where $E = \mathbb{C}^n$.

43.1. THEOREM. *Each pseudoconvex open set in* $\mathbb{C}^n$ *is a domain of holomorphy.*

The proof of this theorem rests on the following lemma.

43.2. LEMMA. *Let* U *be a pseudoconvex open set in* $\mathbb{C}^n$, *where* $n \geq 2$, *and let* $U_1 = U \cap \mathbb{C}^{n-1}$. *Then the restriction mapping* $\mathcal{H}(U) \to \mathcal{H}(U_1)$ *is surjective.*

PROOF. Let $\pi : \mathbb{C}^n \to \mathbb{C}^{n-1}$ denote the canonical projection. Since the set U_1 is closed in U, but is open in $\mathbb{C}^{n-1}$, we see that U_1 and $U \setminus \pi^{-1}(U_1)$ are two disjoint closed subsets of U. By Corollary 15.5 there is a function $\varphi \in C^\infty(U)$ such that $\varphi \equiv 1$ on a neighborhood of U_1 and $\varphi \equiv 0$ on a neighborhood of $U \setminus \pi^{-1}(U_1)$. Given $f_1 \in \mathcal{H}(U_1)$ we define $f \in C^\infty(U)$ by

$$f = \varphi(f_1 \circ \pi) - z_n u,$$

where $u \in C^\infty(U)$ will be chosen so that $\bar{\partial} f = 0$. The equation $\bar{\partial} f = 0$ is equivalent to the equation $\bar{\partial} u = v$, where

$$v = \frac{f_1 \circ \pi}{z_n} \bar{\partial} \varphi .$$

Since v is a well defined member of $C^\infty_{01}(U)$ satisfying $\bar{\partial} v = 0$,

Theorem 42.5 guarantees the existence of $u \in C^{\infty}(U)$ such that $\bar{\partial}u = v$. Thus $f \in \mathcal{H}(U)$ and it is clear that $f = f_1$ on U_1.

PROOF OF THEOREM 43.1. By induction on n. By Corollary 10.6 each open set in $\mathbb{C}$ is a domain of holomorphy, and hence the theorem is true for $n = 1$. Let $n \geq 2$ and assume that the theorem is true for $n - 1$. Let U be a pseudoconvex open set in $\mathbb{C}^n$ and assume that U is not a domain of holomorphy. Then there are open sets V and W in $\mathbb{C}^n$ such that:

(a) V is connected and $V \not\subset U$.

(b) W is a connected component on $U \cap V$.

(c) For each $f \in \mathcal{H}(U)$ there is $\tilde{f} \in \mathcal{H}(V)$ such that $\tilde{f} = f$ on W.

Choose $a \in V \cap \partial U \cap \partial W$, choose $r > 0$ such that $B(a;r) \subset V$, and choose $b \in W \cap B(a;r)$. Fix a coordinate system in $\mathbb{C}^n$ such that a and b both lie in $\mathbb{C}^{n-1}$. Let $U_1 = U \cap \mathbb{C}^{n-1}$, let V_1 be the connected component of $V \cap \mathbb{C}^{n-1}$ which contains b, and let W_1 be any connected open subset of $W \cap \mathbb{C}^{n-1}$ containing b. We claim that:

(a_1) $a \in V_1$, $a \notin U_1$.

(b_1) $b \in W_1 \subset U_1 \cap V_1$.

(c_1) For each $f_1 \in \mathcal{H}(U_1)$ there is $\tilde{f}_1 \in \mathcal{H}(V_1)$ such that $\tilde{f}_1 = f_1$ on W_1.

It follows from the definition of V_1 that $B(a;r) \cap \mathbb{C}^{n-1} \subset V_1$, proving ($a_1$). ($b_1$) follows at once from the definition of W_1. To prove (c_1) let $f_1 \in \mathcal{H}(U_1)$. By Lemma 43.2 there is $f \in \mathcal{H}(U)$ such that $f = f_1$ on U_1. By (c) there is $\tilde{f} \in \mathcal{H}(V)$ such that $\tilde{f} = f$ on W. Then $\tilde{f}_1 = \tilde{f} \mid V_1 \in \mathcal{H}(V_1)$ and $\tilde{f}_1 = f_1$ on W_1, proving (c_1). Thus U_1 is a pseudoconvex open set in $\mathbb{C}^{n-1}$ which is not a domain of holomorphy, contradicting the induction hypothesis.

Theorems 11.5 and 43.1, and Corollary 37.7, can be summarized as follows.

43.3. THEOREM. *For an open subset* U *of* $\mathbb{C}^n$ *the following conditions are equivalent:*

(a) U *is a domain of existence.*

(b) U *is a domain of holomorphy.*

(c) U *is holomorphically convex.*

(d) U *is pseudoconvex.*

44. HOLOMORPHIC APPROXIMATION IN $\mathbb{C}^n$

In this section we shall prove an approximation theorem that generalizes the Oka-Weil Theorem 24.12. The key is the following lemma.

44.1. LEMMA. *Let* U *be a pseudoconvex open set in* $\mathbb{C}^n$. *Let* $u \in C^\infty(U)$ *be a strictly plurisubharmonic function such that the set* $K_c = \{z \in U : u(z) \leq c\}$ *is compact for each* $c \in \mathbb{R}$. *Then for each* $f \in \mathcal{H}(K_o)$ *there is a sequence* (f_j) *in* $\mathcal{H}(U)$ *such that* $\int_{K_o} |f_j - f|^2 \, d\lambda \to 0$.

PROOF. We have to show that $\mathcal{H}(K_o)$ is contained in the closure of $\mathcal{H}(U)$ in $L^2(K_o)$. Thus it suffices to show that $\mathcal{H}(U)^\perp \subset \mathcal{H}(K_o)^\perp$. In other words, it is sufficient to show that each $f_o \in L^2(K_o)$ satisfying

$$(44.1) \qquad \int_{K_o} f_o \bar{f} \, d\lambda = 0 \quad \text{for every} \quad f \in \mathcal{H}(U),$$

also satisfies

$$(44.2) \qquad \int_{K_o} f_o \bar{f} \, d\lambda = 0 \quad \text{for every} \quad f \in \mathcal{H}(K_o).$$

Fix $f_o \in L^2(K_o)$ satisfying (44.1), and define $f_o = 0$ on

$U \setminus K_o$, so that $f_o \in L^2(U)$. Let $\alpha, \beta, \gamma \in C^\infty(U;\mathbb{R})$ be three weight functions, and let A, B be the corresponding operators introduced in Definition 40.3. By Theorem 42.5 each $f \in N_A$ belongs to $C^\infty(U)$ and hence $N_A \subset \mathcal{H}(U)$. Thus it follows from (44.1) that $f_o e^\alpha \in (N_A)^\perp$. If it happens that

$$(44.3) \qquad \|g\|_\beta^2 \leq \|A^*g\|_\alpha^2 + \|Bg\|_\gamma^2 \quad \text{for every} \quad g \in D_{A^*} \cap D_B,$$

then by Corollary 39.8 there exists $g \in D_{A^*}$ such that

$$(44.4) \qquad A^*g = f_o e^\alpha$$

and

$$(44.5) \qquad \|g\|_\beta \leq \|f_o e^\alpha\|_\alpha .$$

If we write $g = \sum_j g_j d\bar{z}_j$ then it follows from Lemma 40.5 that $f_o = e^{-\alpha}A^*g = -\sum_j \frac{\partial}{\partial z_j}(g_j e^{-\beta})$. If we set $ge^{-\beta} = h = \sum_j h_j d\bar{z}_j$ then $h \in L^2_{01}(U, -\beta)$ and (44.4) and (44.5) can be rewritten in the form

$$f_o = -\sum_j \frac{\partial h_j}{\partial z_j}$$

and

$$\int_U \sum_j |h_j|^2 e^\beta d\lambda \leq \int_U |f_o|^2 e^\alpha d\lambda.$$

Let (η_i) be a sequence in $\mathcal{D}(U)$ such that $0 \leq \eta_i \leq 1$ on U for every i, and $\eta_i \equiv 1$ on each compact subset of U for all sufficiently large i. By Lemma 41.4 there is a function $\psi \in C^\infty(U;\mathbb{R})$ such that $|\bar{\partial}\eta_i|^2 \leq e^\psi$ on U for every i. Then the proof of Theorem 41.3 shows the existence of a convex, increasing function $\chi \in C^\infty(\mathbb{R};\mathbb{R})$ such that

$$(44.6) \qquad \sum_{j,k} \frac{\partial^2(\chi \circ u)}{\partial z_j \partial \bar{z}_k}(a) b_j \bar{b}_k \geq 2(|\bar{\partial}\psi(a)|^2 + e^{\psi(a)}) \sum_k |b_k|^2$$

for every $a \in U$ and $b \in \mathbb{C}^n$. By Exercise 15.B there is an increasing sequence of convex, increasing functions $\theta_p \in C^\infty(\mathbb{R};\mathbb{R})$ such that $\theta_p(t) = 0$ for every $t \leq 0$ and $\lim_{p\to\infty} \theta_p(t) = \infty$ for every $t > 0$. If we define $\chi_p = \chi + \theta_p$ for every $p \in \mathbb{N}$ then it is clear that $\chi_p \in C^\infty(\mathbb{R};\mathbb{R})$, χ_p is convex and increasing, $\chi_p(t) = \chi(t)$ for every $t \leq 0$ and $\lim_{p\to\infty} \chi_p(t) = \infty$ for every $t > 0$. Moreover, it follows from Exercise 35.E and (44.6) that

$$\sum_{j,k} \frac{\partial^2(\chi_p \circ u)}{\partial z_j \, \partial \bar{z}_k}(a) b_j \bar{b}_k \geq \sum_{j,k} \frac{\partial^2(\chi \circ u)}{\partial z_j \, \partial \bar{z}_k}(a) b_j \bar{b}_k$$

$$\geq 2(|\bar{\partial}\psi(a)|^2 + e^{\psi(a)}) \sum_k |b_k|^2.$$

If we define $\varphi_p = \chi_p \circ u$, $\alpha_p = \varphi_p - 2\psi$, $\beta_p = \varphi_p - \psi$ and $\gamma_p = \varphi_p$ for every $p \in \mathbb{N}$, then Proposition 41.1 guarantees the validity of the inequality (44.3) for the weight functions α_p, β_p, γ_p. Then the preceding argument shows the existence of a sequence of differential forms $h^p = \sum_j h_j^p \, d\bar{z}_j \in L^2_{01}(U, -\beta_p)$ such that

$$f_o = -\sum_j \frac{\partial h_j^p}{\partial z_j} \tag{44.7}$$

and

$$\int_U \sum_j |h_j^p|^2 e^{\chi_p \circ u - \psi} \, d\lambda \leq \int_U |f_o|^2 e^{\chi_p \circ u - 2\psi} \, d\lambda. \tag{44.8}$$

for every $p \in \mathbb{N}$. Since $f_o = 0$ on $U \setminus K_o$ and since $\chi_p(t) = \chi(t)$ for every $t \leq 0$ and every $p \in \mathbb{N}$, it follows that

$$\int_U \sum_j |h_j^p|^2 e^{\chi_p \circ u - \psi} \, d\lambda \leq \int_{K_o} |f_o|^2 e^{\chi \circ u - 2\psi} \, d\lambda$$

for every $p \in \mathbb{N}$. Set $C = \int_{K_o} |f_o|^2 e^{\chi \circ u - 2\psi} d\lambda$. Since the sequence

(χ_p) is increasing we conclude that

$$(44.9) \qquad \int_U \sum_j |h_j^p|^2 e^{\chi_q \circ u - \psi} d\lambda \leq C \quad \text{whenever} \quad p \geq q.$$

This shows that the sequence $(h^p)_{p=q}^{\infty}$ is bounded in $L^2_{01}(U, \psi - \chi_q \circ u)$ for each $q \in \mathbb{N}$. Since each bounded sequence in a Hilbert space has a weakly convergent subsequence, an inductive argument yields a differential form $h = \sum_j h_j d\bar{z}_j$ in $\bigcap_{q=1}^{\infty} L^2_{01}(U, \psi - \chi_q \circ u)$ which is a weak cluster point of $(h^p)_{p=q}^{\infty}$ in $L^2_{01}(U, \psi - \chi_q \circ u)$ for each $q \in \mathbb{N}$. Then it follows from (44.9) that

$$\int_U \sum_j |h_j|^2 e^{\chi_q \circ u - \psi} d\lambda \leq C \quad \text{for each} \quad q \in \mathbb{N}.$$

Since the sequence $(\chi_q(t))$ increases to ∞ for each $t > 0$, an application of the Monotone Convergence Theorem shows that $h = 0$ almost everywhere in $U \setminus K_o$. On the other hand, it follows from (44.7) that

$$\int_U f_o \bar{f} d\lambda = - \int_U \sum_j \frac{\partial h_j^p}{\partial z_j} \bar{f} d\lambda = \int_U \sum_j h_j^p \overline{\frac{\partial f}{\partial \bar{z}_j}} d\lambda$$

for every $f \in \mathcal{D}(U)$ and $p \in \mathbb{N}$. Whence it follows that

$$(44.10) \qquad \int_U f_o \bar{f} d\lambda = \int_U \sum_j h_j \overline{\frac{\partial f}{\partial \bar{z}_j}} d\lambda \quad \text{for every} \quad f \in \mathcal{D}(U).$$

Since K_o is a compact subset of U, and $f_o = h = 0$ on $U \setminus K_o$, (44.10) implies that

$$\int_{K_o} f_o \bar{f} d\lambda = \int_{K_o} \sum_j h_j \overline{\frac{\partial f}{\partial \bar{z}_j}} d\lambda \quad \text{for every} \quad f \in C^{\infty}(K_o).$$

In particular

$$\int_{K_o} f_o \bar{f} d\lambda = \int_{K_o} \sum_j h_j \overline{\frac{\partial f}{\partial \bar{z}_j}} d\lambda = 0 \quad \text{for every} \quad f \in \mathcal{H}(K_o).$$

This shows (44.2) and the lemma.

L^p convergence and uniform convergence are connected by the following lemma.

44.2. LEMMA. *Let U be an open set in $\mathbb{C}^n$. There for each compact set K and each open set V with $K \subset V \subset U$, there is a constant $c > 0$ such that*

$$\sup_K |f| \le c \int_V |f| d\lambda \quad \textit{for every} \quad f \in \mathcal{H}(U).$$

PROOF. (a) Let U be an open set in $\mathbb{C}$ and let D_1 and D_2 be two relatively compact open discs such that $\overline{D}_1 \subset D_2 \subset \overline{D}_2 \subset U$. We claim that there is a constant $c > 0$ such that

$$\sup_{D_1} |f| \le c \int_{D_2} |f| d\lambda \quad \text{for every} \quad f \in \mathcal{H}(U).$$

Indeed, choose $\varphi \in \mathcal{D}(D_2)$ such that $\varphi \equiv 1$ on a neighborhood of $\overline{D}_1$. Then by applying Lemma 22.3 to the product φf we obtain the formula

$$f(a) = \frac{1}{2\pi i} \int_{D_2} \frac{f(z)}{z-a} \frac{\partial \varphi}{\partial \bar{z}}(z) dz \wedge d\bar{z}$$

for every $f \in \mathcal{H}(U)$ and $a \in D_1$. Since $\dfrac{\partial \varphi}{\partial \bar{z}} \equiv 0$ on a neighborhood of $\overline{D}_1$ there exists $\delta > 0$ such that $|z-a| \ge \delta$ for every $a \in D_1$ and $z \in \operatorname{supp} \dfrac{\partial \varphi}{\partial \bar{z}}$. If we set $c = (2\pi\delta)^{-1} \sup_{D_2} \left|\dfrac{\partial \varphi}{\partial \bar{z}}\right|$ then $|f(a)| \le c \int_{D_2} |f| d\lambda$ for every $f \in \mathcal{H}(U)$ and $a \in D_1$, as asserted.

(b) Let U be an open set in $\mathbb{C}^n$, and let D_1 and D_2 be two relatively compact open polydiscs such that $\overline{D}_1 \subset D_2 \subset \overline{D}_2 \subset U$. Then by repeated applications of (a) we can find a constant $c > 0$ such that

$$\sup_{D_1} |f| \le c \int_{D_2} |f| d\lambda \quad \text{for every} \quad f \in \mathcal{H}(U).$$

(c) Since each compact set in $\mathbb{C}^n$ can be covered by finitely many polydiscs, the lemma follows at once form (b).

44.3. THEOREM. *Let U be a pseudoconvex open set in $\mathbb{C}^n$. Let K be a compact subset of U such that $\hat{K}_{Ps(U)} = K$. Then for each $f \in \mathcal{H}(K)$ there is a sequence (f_j) in $\mathcal{H}(U)$ which converges to f uniformly on a suitable neighborhood of K.*

PROOF. Choose an open set V such that $K \subset V \subset U$ and $f \in \mathcal{H}(V)$. By Theorem 38.6 there is a strictly plurisubharmonic function $u \in C^\infty(U)$ such that:

(a) The set $K_c = \{z \in U : u(z) \leq c\}$ is compact for each $c \in \mathbb{R}$.

(b) $u(z) < 0$ for every $z \in K$.

(c) $u(z) > 0$ for every $z \in U \setminus V$.

Since $u(z) < 0$ for every $z \in K$, there is a constant $c < 0$ such that $u(z) < c$ for every $z \in K$. Thus $K \subset \overset{\circ}{K}_c \subset K_c \subset \overset{\circ}{K}_0 \subset K_0 \subset V$. By Lemma 44.1 there is a sequence (f_j) in $\mathcal{H}(U)$ such that $\int_{K_0} |f_j - f|^2 d\lambda \to 0$. But then Lemma 44.2 implies that $\sup_{K_c} |f_j - f|^2 \to 0$, and the theorem has been proved.

44.4. COROLLARY. *Let U be a pseudoconvex open set in $\mathbb{C}^n$. Let V be an open subset of U such that $\hat{K}_{Ps(U)} \subset V$ for each compact set $K \subset V$. Then $\mathcal{H}(U)$ is dense in $(\mathcal{H}(V), \tau_0)$.*

44.5. THEOREM. *Let U be a holomorphically convex open set in $\mathbb{C}^n$. Then for each open set $V \subset U$ the following conditions are equivalent:*

(a) *$\hat{K}_{\mathcal{H}(U)} \cap V$ is compact for each compact set $K \subset V$.*

(b) *$\hat{K}_{\mathcal{H}(U)} \subset V$ for each compact set $K \subset V$.*

(c) *V is holomorphically convex and $\mathcal{H}(U)$ is dense in*

$(\mathcal{H}(V), \tau_c)$.

PROOF. (a) ⇒ (b): Let K be a compact subset of V and suppose that $\hat{K}_{\mathcal{H}(U)} \not\subset V$. Then $\hat{K}_{\mathcal{H}(U)} = A \cup B$, where $A = \hat{K}_{\mathcal{H}(U)} \cap V$ and $B = \hat{K}_{\mathcal{H}(U)} \setminus V$. Then A and B are two disjoint compact sets. Let $f \in \mathcal{H}(\hat{K}_{\mathcal{H}(U)})$ be equal to zero on a neighborhood of A and equal to one on a neighborhood of B. By Theorem 44.3 there is a function $g \in \mathcal{H}(U)$ such that $|g - f| < 1/2$ on $\hat{K}_{\mathcal{H}(U)}$. Hence $|g| < 1/2$ on K and $|g| > 1/2$ on B. This is impossible, for $B \subset \hat{K}_{\mathcal{H}(U)}$.

(b) ⇒ (c): Clearly V is holomorphically convex. And it follows from Corollary 44.4 that $\mathcal{H}(U)$ is dense in $(\mathcal{H}(V), \tau_c)$.

(c) ⇒ (a): Let K be a compact subset of V. Since $\mathcal{H}(U)$ is dense in $(\mathcal{H}(V), \tau_c)$ one can readily see that $\hat{K}_{\mathcal{H}(V)} = \hat{K}_{\mathcal{H}(U)} \cap V$. Since V is holomorphically convex we conclude that $\hat{K}_{\mathcal{H}(U)} \cap V$ is compact, as asserted.

44.6. THEOREM. *If U is a pseudoconvex open set in $\mathbb{C}^n$ then $\hat{K}_{Ps(U)} = \hat{K}_{\mathcal{H}(U)}$ for each compact set $K \subset U$.*

PROOF. The inclusion $\hat{K}_{Ps(U)} \subset \hat{K}_{\mathcal{H}(U)}$ is clear. To show the reverse inclusion let V be an open neighborhood of $\hat{K}_{Ps(U)}$ in U. By Lemma 38.1 there is a function $u \in Psc(U)$ such that $u < 0$ on K and $u > 0$ on $U \setminus V$. Set $W = \{z \in U : u(z) < 0\}$. Then W is pseudoconvex by Proposition 38.3. Clearly $K \subset W \subset V$, and $\hat{L}_{Ps(U)} \subset W$ for each compact set $L \subset W$. Thus $\mathcal{H}(U)$ is dense in $(\mathcal{H}(W), \tau_c)$ by Corollary 44.4. Since U and W are both pseudoconvex, it follows from Theorem 43.3 that U and W are both holomorphically convex. Then an application of Theorem 44.5 shows that $\hat{L}_{\mathcal{H}(U)} \subset W$ for each compact set $L \subset W$, and in particular $\hat{K}_{\mathcal{H}(U)} \subset W \subset V$. Since V was an arbitrary open neighborhood of $\hat{K}_{Ps(U)}$ in U we conclude that $\hat{K}_{\mathcal{H}(U)} \subset \hat{K}_{Ps(U)}$ and the proof of the theorem is complete.

EXERCISE

44.A. Let U be a holomorphically convex open set in $\mathbb{C}^n$, and let K be a compact subset of U such that $\hat{K}_{\mathcal{H}(U)} = K$. Show that if L is an open and closed subset of K then $\hat{L}_{\mathcal{H}(U)} = L$.

44.B. Let U be a holomorphically convex open set in $\mathbb{C}^n$, and let K be a compact subset of U. Show that each connected component of $\hat{K}_{\mathcal{H}(U)}$ contains points of K.

44.C. Let U be a holomorphically convex open set in $\mathbb{C}^n$, and let V be an open subset of U such that $\hat{K}_{\mathcal{H}(U)} \subset V$ for each compact set $K \subset V$. Show that if W is an open and closed subset of V then $\hat{K}_{\mathcal{H}(U)} \subset W$ for each compact set $K \subset W$.

44.D. Let E be a Banach space with the approximation property, and let U be a pseudoconvex open subset of E which is finitely Runge. Show that $\hat{K}_{P_s(U)} = \hat{K}_{P(E)}$ for each compact subset K of U. In particular U is polynomially convex.

45. THE LEVI PROBLEM IN BANACH SPACES

In this section we solve Problem 37.10 in the case of separable Banach spaces with the bounded approximation property. In this and the next section all Banach spaces considered will be complex.

Corollary 44.4 motivates the following definition.

45.1. DEFINITION. Let U be a pseudoconvex open subset of E. An open subset V of U is said to be *U-pseudoconvex* if $\hat{K}_{P_s(U)} \subset V$ for each compact set $K \subset V$.

The following examples can be readily verified by the reader.

45.2. EXAMPLES. Let U be a pseudoconvex open subset of E.

(a) Each convex open set $V \subset U$ is *U-pseudoconvex.*

(b) For each $f \in Ps(U)$ the open set $V = \{x \in U : f(x) < 0\}$ is U-pseudoconvex.

(c) If $(V_i)_{i \in I}$ is a family of U-pseudoconvex open sets, then the open set $V = \operatorname{int} \bigcap_{i \in I} V_i$ is U-pseudoconvex as well.

45.3. LEMMA. *Let U be a pseudoconvex open subset of E, and let V be a U-pseudoconvex open set. Then:*

(a) *V is pseudoconvex.*

(b) *$\mathcal{H}(U \cap M)$ is dense in $(\mathcal{H}(V \cap M), \tau_c)$ for each finite dimensional subspace M of E.*

PROOF. (a) If K is a compact subset of U then $\hat{K}_{Ps(V)} \subset \hat{K}_{Ps(U)} \subset V$. By Proposition 38.2 the set $\hat{K}_{Ps(U)}$ is compact. Hence $\hat{K}_{Ps(V)}$ is relatively compact in V and V is therefore pseudoconvex.

(b) If K is a compact subset of $V \cap M$ then one can readily see that $\hat{K}_{Ps(U \cap M)} \subset \hat{K}_{Ps(U)} \cap M \subset V \cap M$. Then an application of Corollary 44.4 yields the desired conclusion.

Let E be a Banach space with a monotone Schauder basis (e_n). Let E_n denote the subspace generated by $e_1, \ldots, e_n$, and let $T_n \in \mathcal{L}(E; E_n)$ denote the canonical projection. Since $\| T_n x - x \| \leq 2 \| x \|$ for every $x \in E$ and $n \in \mathbb{N}$ one can readily see that the function

$$\omega_j(x) = \sup_{n \geq j} \| T_n x - x \|$$

is continuous on E for each $j \in \mathbb{N}$. Then we have the following result.

45.4. LEMMA. *Let E be a Banach space with a monotone Schauder* *basis* (e_n), *and let* U *be a pseudoconvex open subset of* E. *Let* A_j, B_j *and* C_j *be the following open sets:*

$$A_j = \{x \in U : \sup_{n \geq j} \|T_n x - x\| < d_U(x)\},$$

$$B_j = \{x \in A_j : \|x\| < j \quad \text{and} \quad d_{A_j}(x) > 1/j\},$$

$$C_j = \{x \in B_j : \sup_{n \geq j} \|T_n x - x\| < d_{B_j}(x)\}.$$

Then:

(a) $U = \bigcup_{j=1}^{\infty} A_j = \bigcup_{j=1}^{\infty} B_j = \bigcup_{j=1}^{\infty} C_j$.

(b) $T_n(A_j) \subset U \cap E_n$ *and* $T_n(C_j) \subset B_j \cap E_n$ *whenever* $n \geq j$.

(c) $B_j \cap E_n$ *is relatively compact in* $A_j \cap E_n$ *for each* $j, n \in \mathbb{N}$.

(d) *Each* A_j *is* U-*pseudoconvex. In particular* $\mathcal{H}(U \cap E_n)$ *is dense in* $(\mathcal{H}(A_j \cap E_n), \tau_c)$ *for each* $j, n \in \mathbb{N}$.

PROOF. Assertions (a), (b) and (c) are clear. To show (d) it suffices to observe that the function $\log \omega_j(x) = \sup_{n \geq j} \log \|T_n x - x\|$ is plurisubharmonic on E and apply Example 45.2(b) and Lemma 45.3(b).

45.5. **LEMMA.** *Let* E *be a Banach space with a monotone Schauder basis* (e_n), *and let* U *be a pseudoconvex open subset of* E. *Then for each* $f_n \in \mathcal{H}(U \cap E_n)$ *and* $\varepsilon > 0$ *there exists* $f \in \mathcal{H}(U)$ *such that* $|f - f_n \circ T_n| \leq \varepsilon$ *on* C_n.

PROOF. Since $T_n(A_n) \subset U$ we have that $f_n \circ T_n \in \mathcal{H}(A_n)$ and in particular $f_n \circ T_n \in \mathcal{H}(A_n \cap E_{n+1})$. Since $\mathcal{H}(U \cap E_{n+1})$ is dense in $(\mathcal{H}(A_n \cap E_{n+1}), \tau_c)$, we can find $f_{n+1} \in \mathcal{H}(U \cap E_{n+1})$ such that

$$|f_{n+1} - f_n \circ T_n| \leq \varepsilon/2 \quad \text{on} \quad B_n \cap E_{n+1}$$

and hence

$$|f_{n+1} \circ T_{n+1} - f_n \circ T_n| \leq \varepsilon/2 \quad \text{on} \quad C_n .$$

Proceeding inductively we can find a sequence (f_j), with $f_j \in \mathcal{H}(U \cap E_j)$, and such that

$$|f_{j+1} \circ T_{j+1} - f_j \circ T_j| \leq \varepsilon \cdot 2^{-j+n-1} \quad \text{on} \quad C_j$$

for every $j \geq n$. Whence it follows that

$$|f_k \circ T_k - f_j \circ T_j| \leq \sum_{r=j}^{\infty} \varepsilon \cdot 2^{-r+n-1} = \varepsilon \cdot 2^{-j+n} \quad \text{on} \quad C_j$$

for every $k \geq j \geq n$. Thus the sequence $(f_j \circ T_j)$ converges uniformly on each C_j to a function $f \in \mathcal{H}(U)$. Moreover,

$$|f - f_j \circ T_j| \leq \varepsilon \cdot 2^{-j+n} \quad \text{on} \quad C_j$$

for every $j \geq n$, completing the proof.

45.6. LEMMA. *Let E be a Banach space with a monotone Schauder basis (e_n), and let U be a pseudoconvex open subset of E. Then $(\hat{C}_j)_{\mathcal{H}(U)} \subset (\hat{B}_j)_{Psc(U)}$ and in particular $d_U((\hat{C}_j)_{\mathcal{H}(U)}) \geq 1/j$ for every $j \in \mathbb{N}$.*

PROOF. Let $x \in (\hat{C}_j)_{\mathcal{H}(U)}$ and choose $k \geq j$ such that $x \in C_k$. We claim that

(45.1) $\quad T_n x \in (B_j \cap E_n)\hat{}_{\mathcal{H}(U \cap E_n)} \quad$ for every $n \geq k$.

To prove (45.1) let $n \geq k$ and let $f_n \in \mathcal{H}(U \cap E_n)$. Given $\varepsilon > 0$, by Lemma 45.5 we can find $f \in \mathcal{H}(U)$ such that $|f - f_n \circ T_n| \leq \varepsilon$ on C_n. Since $C_j \cup \{x\} \subset C_k \subset C_n$ we get that

$$|f_n \circ T_n(x)| \leq |f(x)| + \varepsilon \leq \sup_{C_j}|f| + \varepsilon$$

$$\leq \sup_{C_j}|f_n \circ T_n| + 2\varepsilon \leq \sup_{B_j \cap E_n} |f_n| + 2\varepsilon.$$

Since $\varepsilon > 0$ was arbitrary we conclude that $|f_n \circ T_n(x)| \leq \sup_{B_j \cap E_n} |f_n|$ and (45.1) follows. Thus by Theorem 44.6 we have that

$$T_n x \in (B_j \cap E_n)\hat{}_{\mathcal{H}(U \cap E_n)} = (B_j \cap E_n)\hat{}_{Psc(U \cap E_n)} \subset (\hat{B}_j)_{Psc(U)}$$

for every $n \geq k$. Letting $n \to \infty$ we get the desired conclusion.

The following theorem improves Theorem 28.7 and provides partial solutions to Problems 10.11, 11.6 and 37.10.

45.7. THEOREM. *Let E be a separable Banach space with the bounded approximation property. Then each pseudoconvex open set in E is a domain of existence.*

PROOF. (a) Assume first that E has a monotone Schauder basis, and let U be a pseudoconvex open subset of E. By Lemma 45.6 there is an increasing sequence of open sets C_j such that $U = \bigcup_{j=1}^{\infty} C_j$ and $d_U((\hat{C}_j)_{\mathcal{H}(U)}) > 0$ for every j. By Theorem 11.4 we may conclude that U is a domain of existence.

(b) If E is a separable Banach space with the bounded approximation property then by Pelczynski's Theorem 27.4 we may assume that E is a complemented subspace of a Banach space G which has a monotone Schauder basis. Let π be a continuous projection from G onto E, and let U be a pseudoconvex open subset of E. Then $\pi^{-1}(U)$ is a pseudoconvex open subset of G by Proposition 37.8, and thus $\pi^{-1}(U)$ is a domain of existence in G, by part (a). But then $U = \pi^{-1}(U) \cap E$ is a domain of existence in E by Proposition 11.7, and the proof of the theorem is complete.

Theorems 11.4 and 45.7, and Corollary 37.7, can by summarized as follows:

45.8. THEOREM. *Let E be a separable Banach space with the* *bounded approximation property. Then for each open subset U of E the following conditions are equivalent:*

(a) *U is a domain of existence.*

(b) *U is a domain of holomorphy.*

(c) *U is holomorphically convex.*

(d) *U is pseudoconvex.*

EXERCISES

45.A. Let E be a Banach space with a monotone Schauder basis, and let U be a pseudoconvex open set in E.

(a) By adapting the proofs of Lemmas 43.2 and 45.5 show that the restriction mapping $\mathcal{H}(U) \to \mathcal{H}(U \cap E_n)$ is surjective for each $n \in \mathbb{N}$.

(b) Using Exercise 26.A show that the restriction mapping $\mathcal{H}(U) \to \mathcal{H}(U \cap M)$ is surjective for each finite dimensional subspace M of E.

45.B. Let E be a separable Banach space with the bounded approximation property, let U be a pseudoconvex open subset of E, and let M be a finite dimensional subspace of E.

(a) Show that the restriction mapping $\mathcal{H}(U) \to \mathcal{H}(U \cap M)$ is surjective.

(b) Show that $\hat{K}_{\mathcal{H}(U)} = \hat{K}_{\mathcal{H}(U \cap M)}$ for each compact set $K \subset U \cap M$.

46. HOLOMORPHIC APPROXIMATION IN BANACH SPACES

In this section we obtain infinite dimensional versions of the approximation theorems from Section 44. We begin by extending Corollary 44.4.

46.1. THEOREM. *Let E be a separable Banach space with the bounded approximation property. Let U be a pseudoconvex open set in E, and let V be a U-pseudoconvex open set. Then $\mathcal{H}(U)$ is sequentially dense in $(\mathcal{H}(V), \tau_c)$.*

PROOF. (a) First assume that E has a monotone Schauder basis. Let U be a pseudoconvex open set in E, and let A_j, B_j and C_j be the open sets constructed in the preceding section. Let V be a U-pseudoconvex open set, and let B'_j and C'_j be the following open sets:

$$B'_j = \{x \in B_j \cap V : d_V(x) > 1/j\},$$

$$C'_j = \{x \in C_j \cap B'_j : \sup_{n \geq j} \| T_n x - x \| < d_{B'_j}(x)\}.$$

Then it is clear that $V = \bigcup_{j=1}^{\infty} B'_j = \bigcup_{j=1}^{\infty} C'_j$, $B'_j \cap E_n$ is relatively compact in $V \cap E_n$ for every $j, n \in \mathbb{N}$, and $T_n(C'_j) \subset B'_j$ whenever $n \geq j$. Let $f \in \mathcal{H}(V)$ be given. For each $n \in \mathbb{N}$ $\mathcal{H}(U \cap E_n)$ is dense in $(\mathcal{H}(V \cap E_n), \tau_c)$. Hence we can find $g_n \in \mathcal{H}(U \cap E_n)$ such that $|g_n - f| \leq 1/n$ on $B'_n \cap E_n$ and therefore

$$|g_n \circ T_n - f \circ T_n| \leq 1/n \quad \text{on} \quad C'_n .$$

By Lemma 45.5 for each $n \in \mathbb{N}$ we can find $f_n \in \mathcal{H}(U)$ such that

$$|f_n - g_n \circ T_n| \leq 1/n \quad \text{on} \quad C_n .$$

We claim that (f_n) converges to f in $(\mathcal{H}(V), \tau_c)$. Indeed, given a compact set $K \subset V$ and $\varepsilon > 0$ we first choose $\delta > 0$ such that $K + B(0;\delta) \subset V$ and $|f(y) - f(x)| \leq \varepsilon$ for all $x \in K$ and $y \in B(x;\delta)$. Next choose $n_o \in \mathbb{N}$ such that $1/n_o \leq \varepsilon, K \subset C'_{n_o}$ and $\| T_n x - x \| < \delta$ for every $x \in K$ and $n \geq n_o$. Then for every $x \in K$ and $n \geq n_o$ we have that

$$|f_n(x) - f(x)| \leq |f_n(x) - g_n \circ T_n(x)| + |g \circ T_n(x) - f \circ T_n(x)|$$

$$+ |f \circ T_n(x) - f(x)| \leq 3\varepsilon.$$

This completes the proof when E has a monotone Schauder basis.

(b) If E is a separable Banach space with the bounded approximation property then by Theorem 27.4 we may assume that E is a complemented subspace of a Banach space G which has a monotone Schauder basis. Let U be a pseudoconvex open set in E and let V be a U-pseudoconvex open set. Let π be a continuous projection from G onto E and let $U' = \pi^{-1}(U)$, $V' = \pi^{-1}(V)$. Then U' is a pseudoconvex open set in G, and V' is U'-pseudoconvex, so that $\mathcal{H}(U')$ is sequentially dense in $(\mathcal{H}(V'), \tau_c)$, by part (a). Now, let $f \in \mathcal{H}(V)$ be given. Then $f \circ \pi \in \mathcal{H}(V')$ and hence there is a sequence (g_n) in $\mathcal{H}(U')$ which converges to $f \circ \pi$ in $(\mathcal{H}(V'), \tau_c)$. Since $U = U' \cap E$ and $V = V' \cap E$ we may define $f_n = g_n | U \in \mathcal{H}(U)$ for every n, and then (f_n) converges to f in $(\mathcal{H}(V), \tau_c)$.

46.2. THEOREM. *Let U be a pseudoconvex open subset of E, and let K be a compact subset of U such that $\hat{K}_{Ps(U)} = K$. Then for each open set V with $K \subset V \subset U$ there is a U-pseudoconvex open set W such that $K \subset W \subset V$.*

PROOF. Let $0 < r < d_U(K)$ and set

$$L = \overline{co}(K) \cap \{x \in U : d_U(x) \geq r\}.$$

Then L is compact and $K \subset L \subset U$. Since $K = \hat{K}_{Ps(U)} = \hat{K}_{Psc(U)}$, for each point $a \in L \setminus V$ there is a function $f_a \in Psc(U)$ such that $\sup_K f_a < 0 < f_a(a)$. Since $L \setminus V$ is compact we can find functions $f_1, \ldots, f_m \in Psc(U)$ such that $\sup_K f_j < 0$ for $j = 1, \ldots, m$ and

$$L \setminus V \subset \bigcup_{j=1}^{m} \{x \in U : f_j(x) > 0\}.$$

If we define $f = sup\{-\log d_U + \log r, f_1,\ldots,f_m\}$ then $f \in Psc(U)$, $f < 0$ on K and

$$\overline{co}(K) \cap \{x \in U : f(x) \leq 0\} \subset V.$$

Proceeding as in the proof of Theorem 28.2 we can show the existence of $\delta > 0$ such that

$$(\overline{co}(K) + B(0;\delta)) \cap \{x \in U : f(x) \leq 0\} \subset V.$$

Then the open set $W = (\overline{co}(K) + B(0;\delta)) \cap \{x \in U : f(x) < 0\}$ has the required properties.

By combining Theorems 46.1 and 46.2 we obtain the following theorem, which can be regarded as a generalization of Theorem 44.3.

46.3. THEOREM. *Let E be a separable Banach space with the bounded approximation property. Let U be a pseudoconvex open subset of E, and let K be a compact subset of U such that $\hat{K}_{Ps(U)} = K$. Then for each $f \in \mathcal{H}(K)$ there are an open set V with $K \subset V \subset U$, and a sequence (f_j) in $\mathcal{H}(U)$ such that $f \in \mathcal{H}(V)$ and (f_j) converges to f in $(\mathcal{H}(V), \tau_c)$.*

46.4. THEOREM. *Let E be a separable Banach space with the bounded approximation property, and let U be a holomorphically convex open set in E. Then for each open set $V \subset U$ the following conditions are equivalent:*

(a) $\hat{K}_{\mathcal{H}(U)} \cap V$ *is compact for each compact set* $K \subset V$.

(b) $\hat{K}_{\mathcal{H}(U)} \subset V$ *for each compact set* $K \subset V$.

(c) V *is holomorphically convex and* $\mathcal{H}(U)$ *is sequentially dense in* $(\mathcal{H}(V), \tau_c)$.

(d) V *is holomorphically convex and* $\mathcal{H}(U)$ *is dense in* $(\mathcal{H}(V), \tau_c)$.

PROOF. In view of Theorems 46.1 and 46.3, the proof of Theorem 44.5 applies.

46.5. THEOREM. *Let E be a separable Banach space with the bounded approximation property, and let U be a pseudoconvex open set in E. Then $\hat{K}_{P_s(U)} = \hat{K}_{\mathcal{H}(U)}$ for each compact subset K of U.*

PROOF. Let V be any open set such that $\hat{K}_{P_s(U)} \subset V \subset U$. By Theorem 46.2 there is a U-pseudoconvex open set W such that $\hat{K}_{P_s(U)} \subset W \subset V$. By Theorem 46.1 $\mathcal{H}(U)$ is dense in $(\mathcal{H}(W), \tau_o)$. Since by Theorem 45.7 both U and W are holomorphically convex, we may apply Theorem 46.4 to conclude that $\hat{K}_{\mathcal{H}(U)} \subset W \subset V$. Since V was arbitrary we conclude that $\hat{K}_{\mathcal{H}(U)} \subset \hat{K}_{P_s(U)}$. Since the opposite inclusion is clear, the proof is complete.

EXERCISES

46.A. Let E be a separable Banach space with the bounded approximation property, let U be a pseudoconvex open subset of E, and let M be a complemented subspace of E.

(a) Show that the image of the restriction mapping $\mathcal{H}(U) \to \mathcal{H}(U \cap M)$ is dense in $\mathcal{H}(U \cap M)$ for the compact-open topology.

(b) Show that $\hat{K}_{\mathcal{H}(U)} = \hat{K}_{\mathcal{H}(U \cap M)}$ for each compact set $K \subset U \cap M$.

NOTES AND COMMENTS

Theorem 43.1 was obtained by K. Oka [3] for $n = 2$ in 1942, and by K. Oka [4], H. Bremermann [1] and F. Norguet [1] for arbitrary n in 1953 - 1954, thus solving a problem posed by E. Levi [1] in 1911. The proof of Theorem 43.1 given here is due to L. Hörmander [3]. Theorem 44.3, due to L. Hörmander [2], improves an approximation theorem in domains of holomorphy obtained by K. Oka [2]. Theorem 45.7 is due to L. Gruman [1] in the case of separable Hilbert spaces, to L. Gruman and C. Kiselman [1] in the case of Banach spaces with a Schauder basis, and to P. Noverraz [3] in the case of separable Banach spaces with the bounded approximation property. Theorems 46.1 and 46.2 are due to J. Mujica [4]. Theorems 46.3, 46.4 and 46.5 sharpen results of P. Noverraz [4]. The result in Exercise 46.A is also due to P. Noverraz [4].

CHAPTER XI

RIEMANN DOMAINS

47. RIEMANN DOMAINS

In this section we establish the basic properties of Riemann domains over Banach spaces. Throughout this chapter all Banach spaces considered will be complex.

We recall that if X and Y are two topological spaces then a mapping $f : X \to Y$ is said to be a *local homeomorphism* if for each $x \in X$ there is an open set U in X containing x such that $f(U)$ is open in Y and $f|U : U \to f(U)$ is a homeomorphism. Clearly every local homeomorphism is continuous and open.

47.1. DEFINITION. A *Riemann domain over* E is a pair (X,ξ) such that X is a Hausdorff topological space and $\xi : X \to E$ is a local homeomorphism. A *chart* in X is a connected open set $U \subset X$ such that $\xi|U : U \to \xi(U)$ is a homeomorphism. An *atlas* on X is a collection $(U_i)_{i \in I}$ of charts which cover X.

If (X,ξ) is a Riemann domain over E then X inherits many of the topological properties of E. In particular it is clear that X is a first countable space, and therefore a k-space. It is also clear that X is locally pathwise connected, and hence each connected open subset of X is pathwise connected. Furthermore, X is locally compact if and only if E is finite dimensional.

47.2. DEFINITIONS. (a) Let (X,ξ) and (Y,η) be two Riemann domains over E. A continuous mapping $\tau : X \to Y$ is said to be a *morphism* if $\eta \circ \tau = \xi$. A mapping $\tau : X \to Y$ is said to be an *isomorphism* if τ is a bijection and both τ and τ^{-1} are morphisms.

(b) More generally, let (X,ξ) be a Riemann domain over E, let (Y,η) be a Riemann domain over F, and let $T \in \mathcal{L}(E;F)$. Then a continuous mapping $\tau : X \to Y$ is said to be a *T-morphism* if $\eta \circ \tau = T \circ \xi$.

The next two lemmas will be very useful.

47.3. LEMMA. *Let (X,ξ) be a Riemann domain over E. Let A, B be subsets of X such that A is open, $A \cap B$ is nonvoid, $\xi(A) \cap \xi(B)$ is connected, and both $\xi | A : A \to \xi(A)$ and $\xi | B : B \to \xi(B)$ are homeomorphisms. Then $\xi | A \cup B$ is injective.*

PROOF. To show that $\xi | A \cup B$ is injective it clearly suffices to prove that $\xi(A \cap B) = \xi(A) \cap \xi(B)$, and to prove this it suffices to show that $\xi(A \cap B)$ is open and closed in $\xi(A) \cap \xi(B)$. Since A is open in X, the intersection $A \cap B$ is open in B. Hence $\xi(A \cap B)$ is open in $\xi(B)$, and therefore open in $\xi(A) \cap \xi(B)$.

To show that $\xi(A \cap B)$ is closed in $\xi(A) \cap \xi(B)$, let (x_j) be a sequence in $A \cap B$ such that $(\xi(x_j))$ converges to a point y in $\xi(A) \cap \xi(B)$. Then $(\xi | A)^{-1}(y) = \lim x_j = (\xi | B)^{-1}(y)$. Hence $y \in \xi(A \cap B)$ and the proof is complete.

47.4. LEMMA. *Let (X,ξ) be a Riemann domain over E. Let A be a subset of X such that $\xi(A)$ is a convex open subset of E and $\xi | A : A \to \xi(A)$ is a homeomorphism. Then A is a connected component of $\xi^{-1}(\xi(A))$.*

PROOF. It is clearly sufficient to show that A is open and closed in $\xi^{-1}(\xi(A))$. Given $a \in A$ we can find an open set U in X containing a such that $\xi(U)$ is an open ball in E and $\xi | U : U \to \xi(U)$ is a homeomorphism. Since $\xi(U) \cap \xi(A)$ is convex, and therefore connected, Lemma 47.3 implies that $\xi | U \cup A$ is injective. Whence it follows that $U \cap A = U \cap \xi^{-1}(\xi(A))$, proving that A is open in $\xi^{-1}(\xi(A))$.

To show that A is closed in $\xi^{-1}(\xi(A))$, let (x_j) be a sequence in A which converges to a point x in $\xi^{-1}(\xi(A))$. Hence

$\xi(x) = \lim \xi(x_j)$, and since $\xi \mid A : A \to \xi(A)$ is a homeomorphism it follows that $(\xi \mid A)^{-1}(\xi(x)) = \lim(\xi \mid A)^{-1}(\xi(x_j)) = \lim x_j = x$. Thus $x = (\xi \mid A)^{-1}(\xi(x)) \in A$ and the proof is complete.

47.5. **DEFINITION.** Let (X,ξ) be a Riemann domain over E, and let $x \in X$. We shall denote by $d_X(x)$ the supremum of all $r > 0$ for which there exists a set U in X containing x such that $\xi \mid U : U \to B_E(\xi(x);r)$ is a homeomorphism. For each r with $0 < r \leq d_X(x)$ we shall denote by $B_X(x;r)$ the connected component of $\xi^{-1}(B_E(\xi(x);r))$ which contains x.

47.6. **PROPOSITION.** *Let (X,ξ) be a Riemann domain over E.*

(a) *$\xi \mid B_X(x;r) : B_X(x;r) \to B_E(\xi(x);r)$ is a homeomorphism for each $x \in X$ and $0 < r \leq d_X(x)$.*

(b) *If X is connected and $d_X(a) < \infty$ for some $a \in X$ then $d_X(x) < \infty$ for every $x \in X$ and the function $d_X : X \to \mathbb{R}$ is continuous.*

(c) *If X is connected and $d_X(a) = \infty$ for some $a \in X$ then $d_X(x) = \infty$ for every $x \in X$ and $\xi : X \to E$ is a homeomorphism.*

PROOF. (a) Let $x \in X$. If $0 < r < d_X(x)$ then it follows from the definition of $d_X(x)$ that there is a set U_r in X containing x such that $\xi \mid U_r : U_r \to B_E(\xi(x);r)$ is a homeomorphism. It follows then from Lemma 47.4 that U_r is the connected component of $\xi^{-1}(B_E(\xi(x);r))$ which contains x, that is $U_r = B_X(x;r)$. If we set $U = \cup \{B_X(x;r) : 0 < r < d_X(x)\}$ then it follows from Lemma 47.3 that $\xi \mid U$ is injective. Since U is open we conclude that $\xi \mid U : U \to B_E(\xi(x);d_X(x))$ is a homeomorphism. Then another application of Lemma 47.4 shows that $U = B_X(x;d_X(x))$. Thus (a) has been proved. To establish (b) and (c) we shall first prove that if $a \in X$ then we have the inequalities

(47.1) $d_X(x) \geq d_X(a) - \|\xi(x) - \xi(a)\|$ for every $x \in B_X(a;d_X(a))$,

(47.2) $d_X(x) \leq d_X(a) - \|\xi(x) - \xi(a)\|$ for every $x \in B_X(a;\frac{1}{2}\, d_X(a))$.

Indeed, if $x \in B_X(a;d_X(a))$ then $B_E(\xi(x);d_X(a) - \|\xi(x) - \xi(a)\|)$ $\subset B_E(\xi(a);d_X(a))$ and (47.1) follows. On the other hand, if $x \in B_X(a;\frac{1}{2}\, d_X(a))$ then it follows from (47.1) that $d_X(x) > \frac{1}{2} d_X(a)$ $> \|\xi(x) - \xi(a)\|$. Hence $B_E(\xi(a);\, d_X(a) - \|\xi(x) - \xi(a)\|) \subset B_E(\xi(x);d_X(x))$ and (47.2) follows. (47.1) implies that the set $\{x \in X : d_X(x) = \infty\}$ is open, whereas (47.2) implies that the set $\{x \in X : d_X(x) < \infty\}$ is open. All the assertions in (b) and (c) are then clear.

47.7. PROPOSITION. *Each connected Riemann domain over a separable Banach space is second countable, and in particular separable and Lindelöf.*

PROOF. Fix $x \in X$ and $0 < r < d_X(x)$. For each $n \in \mathbb{N}$ let $A_m = \{y \in X : d_X(y) > r/m\}$ and let B_m be the connected component of A_m which contains x. Since X is pathwise connected it is clear that $X = \bigcup_{m=1}^{\infty} B_m$. For each $m, n \in \mathbb{N}$ let B_{mn} be the set of all $y \in B_m$ for which there are points $x_o,\ldots,x_n \in B_m$ such that $x_o = x$, $x_n = y$ and $x_j \in B_X(x_{j-1};r/2m)$ for $j = 1,\ldots,n$. We claim that $B_m = \bigcup_{n=1}^{\infty} B_{mn}$ for each $m \in \mathbb{N}$. Indeed, the set $\bigcup_{n=1}^{\infty} B_{mn}$ is clearly open, and to establish our claim it suffices to prove that $\bigcup_{n=1}^{\infty} B_{mn}$ is closed in B_m. Let y belong to the closure of $\bigcup_{n=1}^{\infty} B_{mn}$ in B_m. Then the intersection $B_X(y;r/2m) \cap B_{mn}$ is nonempty for some $n \in \mathbb{N}$ and whence it follows that $y \in B_{m,n+1}$. Thus we have shown that $B_m = \bigcup_{n=1}^{\infty} B_{mn}$ for each $m \in \mathbb{N}$ and hence $X = \bigcup_{m,n=1}^{\infty} B_{mn}$. Thus to show that X is second countable it is sufficient to prove that each B_{mn}

is second countable for the induced topology. To prove that each B_{mn} is second countable fix $m \in \mathbb{N}$. Then B_{m1} is second countable, for $B_{m1} \subset B_X(x;r/2m)$ and E is second countable. If B_{mn} is second countable for some n then B_{mn} is in particular Lindelöf and hence we can find a sequence (a_k) in B_{mn} such that $B_{mn} \subset \bigcup_{k=1}^{\infty} B_X(a_k;r/2m)$. Hence it follows that $B_{m,n+1} \subset \bigcup_{y \in B_{mn}} B_X(y;r/2m) \subset \bigcup_{k=1}^{\infty} B_X(a_k;r/m)$, and since each $B_X(a_k;r/m)$ is second countable, so is $B_{m,n+1}$. Thus each B_{mn} is second countable, and the proof of the proposition is complete.

47.8. **COROLLARY.** *Each connected Riemann domain over* $\mathbb{C}^n$ *is hemicompact.*

PROOF. Each locally compact Lindelöf space is hemicompact.

47.9. **DEFINITION.** Let (X,ξ) be a Riemann domain over E.

(a) By a *line segment* $[a,b]$ in X we mean a set in X containing the points a and b and homeomorphic under ξ to the line segment $[\xi(a),\xi(b)]$ in E.

(b) By a *polygonal line* $[x_o,x_1,\ldots,x_n]$ in X we mean a finite union of line segments of the form $[x_{j-1}, x_j]$ with $j = 1,\ldots,n$.

47.10. **PROPOSITION.** *Let* (X,ξ) *be a connected Riemann domain over* E. *For* $x, y \in X$ *let* $\rho_X(x,y)$ *be defined by*

$$\rho_X(x,y) = \inf \sum_{j=1}^{n} \|\xi(x_j) - \xi(x_{j-1})\|,$$

where the infimum is taken over all polygonal lines $[x_o,x_1,\ldots,x_n]$ *in* X *such that* $x_o = x$ *and* $x_n = y$. *Then* ρ_X *is a metric on* X *which generates the topology of* X. $\rho_X(x,y)$ *is called the geodesic distance between* x *and* y.

PROOF. Given $x \in X$ let A_m, B_m and B_{mn} be the sets introduced in the proof of Proposition 47.7. Then each $y \in X$ belongs to some B_{mn}. Hence there are points $x_o, \ldots, x_n \in X$ such that $x_o = x$, $x_n = y$ and $x_j \in B_X(x_{j-1}, d_X(x_{j-1}))$ for $j = 1, \ldots, n$. Hence $[x_o, \ldots, x_n]$ is a polygonal line in X joining x and y. This shows that $\rho_X(x,y) < \infty$ for every $x, y \in X$. We next show that if $\rho_X(x,y) = 0$ then $x = y$. Let $0 < \varepsilon < d_X(x)$. Then there is a polygonal line $[x_o, \ldots, x_n]$ in X such that $x_o = x$, $x_n = y$ and $\sum_{j=1}^{n} \|\xi(x_j) - \xi(x_{j-1})\| < \varepsilon$. Whence it follows that $\|\xi(x_k) - \xi(x)\| < \varepsilon$ for each $k = 1, \ldots, n$, and we conclude that $[\xi(x_{k-1}), \xi(x_k)] \subset B_E(\xi(x); \varepsilon)$ for each $k = 1, \ldots, n$. Then by repeated applications of Lemma 47.3 we obtain that $[x_{k-1}, x_k] \subset B_X(x;\varepsilon)$ for each $k = 1, \ldots, n$, and in particular $y \in B_E(x;\varepsilon)$. Since $\varepsilon > 0$ can be taken arbitrarily small we conclude that $y = x$. Since the properties $\rho_X(x,y) = \rho_X(y,x)$ and $\rho_X(x,z) \le \rho_X(x,y) + \rho_X(y,z)$ can be readily verified, it follows that ρ_X is a metric on X. And since it is clear that $\rho_X(x,y) = \|\xi(x) - \xi(y)\|$ if x and y both lie in some chart U in X then we conclude that ρ_X generates to topology of X. This completes the proof.

47.11. DEFINITION. Let (X,ξ) be a Riemann domain over E. A mapping $f : X \to F$ is said to be *holomorphic* (resp. of class C^k) if there is an atlas $(U_i)_{i \in I}$ on X such that $f \circ (\xi|U_i)^{-1} : \xi(U_i) \to F$ is holomorphic (resp. of class C^k) for each $i \in I$. We shall denote by $\mathcal{H}(X;F)$ (resp. $C^k(X;F)$ the vector space of all mappings $f : X \to F$ which are holomorphic (resp. of class C^k).

The set $Ps(X)$ of all plurisubharmonic functions $f : X \to [-\infty,\infty)$ can be defined in a similar manner. We can likewise define the vector space $\Omega_{pq}(X;F)$ of all F-valued differential forms of bidegree (p,q) on X, and the vector subspace $C^k_{pq}(X;F)$ of all members of $\Omega_{pq}(X;F)$ which are of class C^k. As usual we shall write $\mathcal{H}(X;\mathbb{C}) = \mathcal{H}(X)$, $C^k(X;\mathbb{C}) = C^k(X)$, etc.

47.12. DEFINITION. Let (X,ξ) be a Riemann domain over E, and let (Y,η) be a Riemann domain over F. A mapping $f : X \to Y$ is said to be *holomorphic* (resp. of class C^k) if f is continuous and $\eta \circ f : X \to F$ is holomorphic (resp. of class C^k). We shall denote by $\mathcal{H}(X;Y)$ (resp. $C^k(X;Y)$) the set of all mappings $f : X \to Y$ which are holomorphic (resp. of class C^k).

EXERCISES

47.A. Let (X,ξ) and (Y,η) be Riemann domains over E and F, respectively. Let $T \in \mathcal{L}(E;F)$ and let $\sigma : X \to Y$ and $\tau : X \to Y$ be two T-morphisms. Show that if $\sigma(a) = \tau(a)$ for some $a \in X$, and X is connected, then $\sigma(x) = \tau(x)$ for every $x \in X$.

47.B. Let (X,ξ) be a Riemann domain over E. Let A be a subset of X such that $\xi(A)$ is a convex subset of E and $\xi|A : A \to \xi(A)$ is a homeomorphism. Show the existence of an open set U in X containing A such that $\xi | U : U \to \xi(U)$ is a homeomorphism.

47.C. Let (X,ξ) be a connected Riemann domain over a separable Banach space E.

(a) Show that $\xi^{-1}(a)$ is a countable discrete subset of X for each $a \in E$.

(b) Show that if D is a countable dense subset of E then $\xi^{-1}(D)$ is a countable dense subset of X.

47.D. Let (X,ξ) and (Y,η) be Riemann domains over E and F, respectively, and let $f, g \in \mathcal{H}(X;Y)$. Show that if $f \equiv g$ on some nonvoid open subset of X, and X is connected, then $f \equiv g$ on all of X. This is the *Identity Principle*.

47.E. Let (X,ξ) be a connected Riemann domain over E, and let $f \in \mathcal{H}(X)$. Show that if f is not constant then f is open, that is, f maps open sets onto open sets. This is the *Open Mapping Principle*,

47.F. Let (X,ξ) be a connected Riemann domain over E, and let

$f \in \mathcal{H}(X)$. Show that if there is $a \in X$ such that $|f(x)| \le |f(a)|$ for every $x \in X$ then f is constant on X. This is the *Maximum Principle.*

47.G. Let (X,ξ), (Y,η) and (Z,ζ) be Riemann domains over the Banach spaces E, F and G, respectively. Show that:

(a) If $f \in \mathcal{H}(X;Y)$ and $g \in \mathcal{H}(Y;Z)$ then $g \circ f \in \mathcal{H}(X;Z)$.

(b) If $f \in C^k(X;Y)$ and $g \in C^k(Y;Z)$ then $g \circ f \in C^k(X;Z)$.

(c) If $f \in \mathcal{H}(X;Y)$ and $g \in Ps(Y)$ then $g \circ f \in Ps(X)$.

47.H. Let (X,ξ) be a Riemann domain over E, let $f \in \mathcal{H}(X;F)$ and let $m \in \mathbb{N}_0$. For each $x \in X$ let $P^m f(x) \in P(^mE;F)$ be defined by $P^m f(x) = P^m[f \circ (\xi|U)^{-1}](\xi(x))$, where U is any chart in X containing x. Show that $P^m f \in \mathcal{H}(X;P(^mE;F))$. Conclude that $P^m_t f \in \mathcal{H}(X;F)$ for each $t \in E$, where $P^m_t f(x) = P^m f(x)(t)$.

47.I. Let (X,ξ) and (Y,η) be Riemann domains over E. Given a morphism $\tau : X \to Y$ let $\tau^* : \mathcal{H}(Y) \to \mathcal{H}(X)$ be defined by $\tau^* g = g \circ \tau$ for every $g \in \mathcal{H}(Y)$. Show that $\tau^*(P^m_t g) = P^m_t(\tau^* g)$ for every $g \in \mathcal{H}(Y)$, $m \in \mathbb{N}_0$ and $t \in E$.

47.J. Let (X,ξ) be a Riemann domain over E. For each $f \in C^\infty(X;F)$, $x \in X$ and $j \in \mathbb{N}_0$ let $D^j f(x) \in \mathcal{L}^s(^jE_{\mathbb{R}};F_{\mathbb{R}})$ be defined by $D^j f(x) = D^j[f \circ (\xi \mid U)^{-1}](\xi(x))$, where U is any chart in U containing x. Endow $C^\infty(X;F)$ with the locally convex topology generated by the seminorms $f \to \sup_{k \in K} \| D^j f(x) \|$, with $j \in \mathbb{N}_0$ and $K \subset X$ compact. Show that $C^\infty(X;F)$ is complete. Show that if $E = \mathbb{C}^n$ then the topology of $C^\infty(X;F)$ is also generated by the seminorms $f \to \sup_{k \in K} \left\| \frac{\partial^\alpha f}{\partial x^\alpha}(x) \right\|$, with $\alpha \in \mathbb{N}^n_0$ and $K \subset X$ compact, where $\frac{\partial^\alpha f}{\partial x^\alpha}(x)$ is defined in the obvious way.

48. DISTRIBUTIONS ON RIEMANN DOMAINS

In this section we shall extend to the case of Riemann domains several results from Sections 16 and 17. A key tool is the following theorem on existence of partitions of unity.

48.1. THEOREM. *Let (X,ξ) be a Riemann domain over a separable Hilbert space E. Then for each open cover $(U_i)_{i\in I}$ of X there is a C^∞ partition of unity $(\varphi_i)_{i\in I}$ on X which is subordinated to $(U_i)_{i\in I}$.*

PROOF. By restricting our attention to each connected component of X we may assume that X itself is connected. But then X is a Lindelöf space and the proof of Theorem 15.4 can be directly adapted.

If X is a topological space then we shall denote by $K(X)$ the subspace of all $f \in C(X)$ which have compact support.

48.2. THEOREM. *Let (X,ξ) be a Riemann domain over $\mathbb{C}^n$. Then there exists a unique regular Borel measure μ_X on X such that*

$$\int_U f\,d\mu_X = \int_{\xi(U)} f\circ(\xi\mid U)^{-1}d\lambda \tag{48.1}$$

for each chart U in X and each $f \in K(U)$.

PROOF. Let $f \in K(X)$, let $(U_i)_{i\in I}$ be an atlas on X, and let $(\varphi_i)_{i\in I}$ be a partition of unity on X subordinated to $(U_i)_{i\in I}$. Then $\varphi_i f \equiv 0$ for all but finitely many indices. If we define

$$Tf = \sum_{i\in I}\int_{\xi(U_i)} (\varphi_i f)\circ(\xi\mid U_i)^{-1}d\lambda$$

then one can readily verify that the definition of Tf is independent from the atlas and partition of unity chosen. And clearly T is a positive linear functional on $K(X)$, that is

$Tf \geq 0$ whenever $f \geq 0$. Thus, by the Riesz Representation Theorem there is a regular Borel measure μ_X on X such that $Tf = \int_X f d\mu_X$ for every $f \in K(X)$. Clearly μ_X verifies (48.1). And uniqueness of μ_X follows at once from the uniqueness in the Riesz Representation Theorem.

If (X,ξ) is a Riemann domain over $\mathbb{C}^n$ and $1 \leq p < \infty$ then we shall denote by $L^p(X)$ the Banach space of all equivalent classes of Borel measurable functions f on X such that $\int_X |f|^p d\mu_X < \infty$. The vector space $L^p(X, loc)$ is defined similarly. Since the measure μ_X is regular one can readily obtain the following result, whose straightforward proof is left to the reader as an exercise.

48.3. PROPOSITION. *Let (X,ξ) be a Riemann domain over $\mathbb{C}^n$, and let U be a chart in X. Then:*

(a) $\mu_X(A) = \lambda(\xi(A))$ *for each Borel set* $A \subset U$.

(b) $\int_U f d\mu_X = \int_{\xi(U)} f \circ (\xi \mid U)^{-1} d\lambda$ *for each* $f \in L^1(X)$.

Let (X,ξ) be a Riemann domain over $\mathbb{C}^n$. We shall denote by $\mathcal{D}(X)$ the subspace of all $f \in C^\infty(X)$ with compact support. If K is a compact subset of X then we shall denote by $\mathcal{D}(K)$ the subspace of all $f \in \mathcal{D}(X)$ such that $supp\, f \subset K$. The members of $\mathcal{D}(X)$ are called *test functions* on X.

48.4. DEFINITION. Let (X,ξ) be a Riemann domain over $\mathbb{C}^n$, let $\delta > 0$, and let $X_\delta = \{x \in X : d_U(x) > \delta\}$. Given $f \in L^1(X, loc)$ and $\varphi \in \mathcal{D}(\overline{B}(0;\delta))$ we define their *convolution* $f * \varphi : X_\delta \to \mathbb{C}$ by

$$(f * \varphi)(x) = \int_{\overline{B}(0;\delta)} f \circ (\xi \mid U_x)^{-1}(\xi(x) - t)\varphi(t)d\lambda(t)$$

$$= \int_{\overline{B}(\xi(x);\delta)} f \circ (\xi \mid U_x)^{-1}(t)\varphi(\xi(x) - t)d\lambda(t)$$

where $U_x = B_X(x; d_X(x))$.

48.5. PROPOSITION. *Let* (X,ξ) *be a Riemann domain over* $\mathbb{C}^n$, *and let* $f \in L^1(X, loc)$ *and* $\varphi \in \mathcal{D}(\overline{B}(0;\delta))$. *Then:*

(a) $f * \varphi \in C^\infty(X_\delta)$.

(b) $\dfrac{\partial^\alpha}{\partial x^\alpha}(f * \varphi) = f * \dfrac{\partial^\alpha \varphi}{\partial x^\alpha}$ *for each multi-index* α.

(c) *If* f *has compact support and* $\delta > 0$ *is sufficiently small then* $f * \varphi \in \mathcal{D}(X)$.

PROOF. Fix $a \in U_\delta$. Then for each $x \in B_X(a; d_X(a) - \delta)$ we have that $\overline{B}(\xi(x);\delta) \subset B(\xi(a); d_X(a)) = \xi(U_a)$, and therefore

$$(f * \varphi)(x) = \int_{\xi(U_a)} f \circ (\xi \mid U_a)^{-1}(t)\varphi(\xi(x) - t)d\lambda(t).$$

Then (a) and (b) follow by differentiation under the integral sign. To show (c) let (U_i) be an atlas on X and let (ψ_i) be a C^∞ partition of unity on X subordinated to (U_i). Then $f = \Sigma\, \psi_i f$ and $\psi_i f \equiv 0$ for all but finitely many indices. Since $\psi_i f \in \mathcal{D}(U_i)$ for each i it follows from Proposition 16.4 that $f * \varphi = \Sigma\, (\psi_i f) * \varphi \in \mathcal{D}(X)$ if $\delta > 0$ is sufficiently small.

48.6. PROPOSITION. *Let* (X,ξ) *be a Riemann domain over* $\mathbb{C}^n$, *let* $1 \leq p < \infty$, *and let* $f \in L^p(X)$ *be a function with compact support. Then* $\int_X |f * \rho_\delta - f|^p d\mu_X \to 0$ *when* $\delta \to 0$.

PROOF. Let (U_i) be an atlas on X and let (ψ_i) be a C^∞ partition of unity on X subordinated to (U_i). Then $f = \Sigma\psi_i f$ and $\psi_i f \equiv 0$ for all but finitely many indices. It follows from

Propositions 48.3 and 16.6, and Exercise 16.A, that

$$\lim_{\delta\to 0}\int_X |(\psi_i f) * \rho_\delta - \psi_i f|^p d\mu_X = \lim_{\delta\to 0}\int_{U_i} |(\psi_i f) * \rho_\delta - \psi_i f|^p d\mu_X = 0$$

for each i. Then an application of Minkowski's inequality yields the desired conclusion.

48.7. COROLLARY. *If (X,ξ) is a Riemann domain over $\mathbb{C}^n$ and $1 \leq p < \infty$ then $\mathcal{D}(X)$ is dense in $L^p(X)$.*

PROOF. Let $f \in L^p(X)$. Then $f = 0$ almost everywhere on X_i for all but countably many components X_i of X. Then the proof of Corollary 16.7 applies.

48.8. DEFINITION. Let (X,ξ) be a Riemann domain over $\mathbb{C}^n$.

(a) For each compact set $K \subset X$ the vector space $\mathcal{D}(K)$ will be endowed with the locally convex topology induced by $C^\infty(X)$. The vector space $\mathcal{D}(X)$ will be endowed with the finest locally convex topology such that the inclusion mapping $\mathcal{D}(K) \hookrightarrow \mathcal{D}(X)$ is continuous for each compact set $K \subset X$.

(b) Each continuous linear functional on $\mathcal{D}(X)$ is called a *distribution* on X. The vector space $\mathcal{D}'(X)$ of all distributions on X will be always endowed with the topology of pointwise convergence.

As in Section 17 the vector space $L^1(X, loc)$ can be canonically identified with a vector subspace of $\mathcal{D}'(X)$.

48.9. DEFINITION. Let (X,ξ) be a Riemann domain over $\mathbb{C}^n$ and let U be an open subset of X.

(a) If $T \in \mathcal{D}'(X)$ and $T \mid U$ will denote the restriction of T to $\mathcal{D}(U)$. Thus $T \mid U \in \mathcal{D}'(U)$.

(b) Given $S, T \in \mathcal{D}'(X)$ we shall say that $S = T$ on U if $S\varphi = T\varphi$ for every $\varphi \in \mathcal{D}(U)$.

48.10. PROPOSITION. *Let (X,ξ) be a Riemann domain over $\mathbb{C}^n$. Let*

$(U_i)_{i\in I}$ be an open cover of X. Suppose that for each $i \in I$ there exists $T_i \in \mathcal{D}'(U_i)$ such that $T_i = T_j$ on $U_i \cap U_j$ whenever $U_i \cap U_j$ is nonvoid. Then there is a unique $T \in \mathcal{D}'(X)$ such that $T \mid U_i = T_i$ for every $i \in I$.

PROOF. Let $\varphi \in \mathcal{D}(X)$. If $(\psi_i)_{i\in I}$ is a C^∞ partition of unity on X subordinated to $(U_i)_{i\in I}$ then $\psi_i\varphi \in \mathcal{D}(U_i)$ for every $i \in I$ and $\psi_i\varphi \equiv 0$ for all but finitely many indices. If we define $T\varphi = \sum_{i\in I} T_i(\psi_i\varphi)$ then one can readily verify that the definition of $T\varphi$ is independent from the partition of unity chosen. T is clearly linear. To show that T is continuous let K be a compact subset of X and let $J = \{i \in I : K \cap U_i \neq \phi\}$. Since $T\varphi = \sum_{i\in J} T_i(\psi_i\varphi)$ for every $\varphi \in \mathcal{D}(K)$, and since each of the mappings $\varphi \in \mathcal{D}(K) \to \psi_i\varphi \in \mathcal{D}(U_i)$ is continuous we conclude that the restriction of T to $\mathcal{D}(K)$ is continuous. This shows that $T \in \mathcal{D}'(X)$. If $\varphi \in \mathcal{D}(U_i)$ then $\psi_j\varphi \in \mathcal{D}(U_j \cap U_i)$ for every j and therefore

$$T\varphi = \sum_j T_j(\psi_j\varphi) = \sum_j T_i(\psi_j\varphi) = T_i(\sum_j \psi_j\varphi) = T_i\varphi,$$

proving that $T \mid U_i = T_i$ for every i. Finally, to show uniqueness let T be a distribution on X such that $T \mid U_i = T_i$ for every i. If $\varphi \in \mathcal{D}(X)$ then

$$T\varphi = T(\sum_i \psi_i\varphi) = \sum_i T(\psi_i\varphi) = \sum_i T_i(\psi_i\varphi)$$

and the proposition has been proved.

Let (X,ξ) be a Riemann domain over $\mathbb{C}^n$ and let $T \in \mathcal{D}'(X)$. Proceeding as in Section 17 we can show the existence at a largest open set $V \subset X$ (possibly empty) such that $T = 0$ on V. The set $X \setminus V$ is called the *support* of T and is denoted by *supp T*.

Before defining derivatives of distributions we establish

the *formula of integration by parts.*

48.11. PROPOSITION. *Let (X,ξ) be a Riemann domain over $\mathbb{C}^n$. Then*

$$\int_X \varphi \frac{\partial f}{\partial x_j} d\mu_X = - \int_X \frac{\partial \varphi}{\partial x_j} f d\mu_X$$

for every $f \in C^\infty(X)$, $\varphi \in \mathcal{D}(X)$ *and* $j = 1,\ldots,2n$.

PROOF. The formula is clearly true if the support of φ lies in a chart U in X. To establish the formula for an arbitrary $\varphi \in \mathcal{D}(X)$, let (U_i) be an atlas on X and let (ψ_i) be a C^∞ partition of unity on X subordinated to (U_i). Then

$$\int \varphi \frac{\partial f}{\partial x_j} d\mu_X = \int \sum_i \psi_i \varphi \frac{\partial f}{\partial x_j} d\mu_X = - \int \sum_i \frac{\partial}{\partial x_j}(\psi_i \varphi) f d\mu_X$$

$$= - \int \sum_i \frac{\partial \psi_i}{\partial x_j} \varphi f d\mu_X - \int \sum_i \psi_i \frac{\partial \varphi}{\partial x_j} f d\mu_X$$

$$= - \int \frac{\partial}{\partial x_j} (\sum_i \psi_i) \varphi f d\mu_X - \int \frac{\partial \varphi}{\partial x_j} f d\mu_X$$

$$= - \int \frac{\partial \varphi}{\partial x_j} f d\mu_X .$$

48.12. COROLLARY. *Let (X,ξ) be a Riemann domain over $\mathbb{C}^n$. Then*

$$\int_X \varphi \frac{\partial^\alpha f}{\partial x^\alpha} d\mu_X = (-1)^{|\alpha|} \int_X \frac{\partial^\alpha \varphi}{\partial x^\alpha} f d\mu_X$$

for every $f \in C^\infty(X)$, $\varphi \in \mathcal{D}(X)$ *and* $\alpha \in \mathbb{N}_o^{2n}$.

48.13. DEFINITION. Let (X,ξ) be a Riemann domain over $\mathbb{C}^n$. Given $T \in \mathcal{D}'(X)$ and $\alpha \in \mathbb{N}_o^{2n}$ we define $\frac{\partial^\alpha T}{\partial x^\alpha} \in \mathcal{D}'(X)$ by $\frac{\partial^\alpha T}{\partial x^\alpha}(\varphi) = (-1)^{|\alpha|} T(\frac{\partial^\alpha \varphi}{\partial x^\alpha})$ *for every* $\varphi \in \mathcal{D}(X)$.

The mapping $T \in \mathcal{D}'(X) \to \partial^\alpha T / \partial x^\alpha \in \mathcal{D}'(X)$ is clearly continuous. If $f \in C^{\infty}(X)$ then it follows from Corollary 48.12 that the definition of the derivative $\partial^{\alpha}f/\partial x^{\alpha}$ in the sense of distributions coincides with the classical definition.

48.14. DEFINITION. Let (X,ξ) be a Riemann domain over $\mathbb{C}^n$. Given $f \in C^{\infty}(X)$ and $T \in \mathcal{D}'(X)$, their product $fT \in \mathcal{D}'(X)$ is defined by $(fT)(\varphi) = T(f\varphi)$ for every $\varphi \in \mathcal{D}(U)$.

If $f \in C^{\infty}(X)$ and $T \in \mathcal{D}'(X)$ then the formula

$$\frac{\partial}{\partial x_j}(fT) = \frac{\partial f}{\partial x_j} T + f \frac{\partial T}{\partial x_j}$$

can be proved exactly as in Section 17.

EXERCISES

48.A. Let (X,ξ) be a Riemann domain over $\mathbb{C}^n$, and let $1 \le p < \infty$. Show that a Borel measurable function f on X belongs to $L^p(X,loc)$ if and only if $f \circ (\xi \mid U)^{-1}$ belongs to $L^p(\xi(U),loc)$ for each chart U in X.

48.B. Let (X,ξ) be a connected Riemann domain over $\mathbb{C}^n$, and let $1 \le p < \infty$. For each $\varphi \in C^{\infty}(X;\mathbb{R})$ let $L^p(X,\varphi)$ denote the Banach space of all equivalent classes of Borel measurable functions f on X such that $\int_X |f|^p e^{-\varphi} d\mu_X < \infty$. Show that $L^p(X,loc)$ is the union of the spaces $L^p(X,\varphi)$, with $\varphi \in C^{\infty}(X;\mathbb{R})$.

48.C. Let (X,ξ) be a Riemann domain over $\mathbb{C}^n$, let $(U_i)_{i\in I}$ be an atlas on X, and let $(\psi_i)_{i\in I}$ be a C^{∞} partition of unity on X subordinated to the atlas $(U_i)_{i\in I}$. Let $T \in \mathcal{D}'(X)$ be a distribution with compact support, and let $\varphi \in \mathcal{D}(\overline{B}(0;\delta))$.

(a) Show that T can be written as a finite sum $T = \sum_{i\in J} \psi_i T$. Show that each $\psi_i T$ has compact support and belongs to $\mathcal{D}'(U_i)$.

(b) Show that if $\delta > 0$ is sufficiently small then the convolution $(\psi_i T) * \varphi$ can be defined in a natural way and belongs to $\mathcal{D}(U_i)$ for each $i \in J$. Show that if we set $T * \varphi = \sum_{i \in J} (\psi_i T) * \varphi$ then $T * \varphi \in \mathcal{D}(X)$ and

$$\frac{\partial^\alpha}{\partial x^\alpha}(T * \varphi) = T * \frac{\partial^\alpha \varphi}{\partial x^\alpha} = \frac{\partial^\alpha T}{\partial x^\alpha} * \varphi \quad \text{for every} \quad \alpha.$$

(c) Show that $T * \rho_\delta \to T$ in $\mathcal{D}'(X)$ when $\delta \to 0$.

49. PSEUDOCONVEX RIEMANN DOMAINS

This section is devoted to the study of pseudoconvex Riemann domains over Banach spaces. The results in this section are natural generalizations of the results from Section 37.

49.1. DEFINITION. Let (X, ξ) be a Riemann domain over E, and let $(x,t) \in X \times E$. We shall denote by $\delta_X(x,t)$ the supremum of all $r > 0$ for which there is a set D in X containing x such that $\xi \mid D : D \to \xi(x) + \Delta rt$ is a homeomorphism. For each r with $0 < r < \delta_X(x,t)$ we shall denote by $\Delta_X(x;t;r)$ the connected component of $\xi^{-1}(\xi(x) + \Delta rt)$ which contains x.

49.2. PROPOSITION. *Let (X, ξ) be a Riemann domain over E.*

(a) *$\xi \mid \Delta_X(x;t;r) : \Delta_X(x;t;r) \to \xi(x) + \Delta rt$ is a homeomorphism for every $(x,t) \in X \times E$ and $0 < r < \delta_X(x,t)$.*

(b) *The function $\delta_X : X \times E \to (0, \infty]$ is lower semicontinuous.*

(c) *$d_X(x) = \inf\{\delta_X(x,t) : t \in E, \|t\| = 1\}$ for each $x \in X$.*

PROOF. (a) Given $(x,t) \in X \times E$ and $0 < r < \delta_X(x,t)$ be can find a set D in X containing x such that $\xi \mid D : D \to \xi(x) + \Delta rt$ is a homeomorphism. By Lemma 47.4 $D = \Delta_X(x;t;r)$ and (a) has been proved.

(b) Let $(x_o, t_o) \in X \times E$ and $0 < r < s < \delta_X(x_o, t_o)$. Then

$\xi \mid \bar{\Delta}_X(x_o;t_o;s) : \bar{\Delta}_X(x_o;t_o;s) \to \xi(x_o) + \bar{\Delta} s t_o$ is a homeomorphism. By Exercise 47.B there is an open set U in X containing $\bar{\Delta}_X(x_o;t_o;s)$ such that $\xi \mid U : U \to \xi(U)$ is a homeomorphism. Then $\xi(x_o) + \bar{\Delta} s t_o \subset \xi(U)$ and we can find a neighborhood V of x_o in X and a neighborhood W of t_o in E such that $\xi(V) + \bar{\Delta} s W \subset \xi(U)$. Whence it follows that $\delta(x,t) \geq s > r$ for every $(x,t) \in V \times W$ and (b) has been proved.

(c) Let $x \in X$. Since

$$B_E(\xi(x);r) = \cup \{\xi(x) + \Delta r t : t \in E, \|t\| = 1\}$$

for every $r > 0$, the inequality $d_X(x) \leq \inf\{\delta_X(x,t) : t \in E, \|t\| = 1\}$ is clear. To show the reverse inequality let $0 < r < \inf\{\delta_X(x,t) : t \in E, \|t\| = 1\}$. For each $t \in E$ with $\|t\| = 1$ $\xi \mid \bar{\Delta}_X(x;t;r) : \bar{\Delta}_X(x;t;r) \to \xi(x) + \bar{\Delta} r t$ is a homeomorphism, and then by Exercise 47.B there is an open set U_t in X containing $\bar{\Delta}_X(x;t;r)$ such that $\xi \mid U_t : U_t \to \xi(U_t)$ is a homeomorphism. Set $U = \cup \{U_t : t \in E, \|t\| = 1\}$. Since we may assume that $\xi(U_t)$ is convex for every t, an application of Lemma 47.3 shows that $\xi|U$ is injective, and hence $\xi|U : U \to \xi(U)$ is a homeomorphism. Since $\xi(U) \supset B_E(\xi(x);r)$ we conclude that $d_X \geq r$ and the proof of the proposition is complete.

49.3. DEFINITION. A Riemann domain (X,ξ) over E is said to be *pseudoconvex* if the function $-\log \delta_X$ is plurisubharmonic on $X \times E$.

49.4. DEFINITION. Let (X,ξ) be a Riemann domain over E. For each set $A \subset X$ we define

$$\hat{A}_{Ps(X)} = \{x \in X : f(x) \leq \sup_A f \quad \text{for all} \quad f \in Ps(X)\},$$

$$\hat{A}_{Psc(X)} = \{x \in X : f(x) \leq \sup_A f \quad \text{for all} \quad f \in Psc(X)\}.$$

If (X,ξ) is a Riemann domain over E then for each set $A \subset X$ we shall set $d_X(A) = \inf_{a \in A} d_X(a)$.

49.5. THEOREM. *For a Riemann domain (X,ξ) over E the following conditions are equivalent:*

(a) *X is pseudoconvex.*

(b) *The function $-\log \delta_X(\cdot, t)$ is plurisubharmonic on X for each $t \in E$.*

(c) *The function $-\log d_X$ is plurisubharmonic on X.*

(d) *$d_X(\hat{A}_{Psc(X)}) = d_X(A)$ for each set $A \subset X$.*

(e) *$d_X(\hat{K}_{Psc(X)}) > 0$ for each compact set $K \subset X$.*

(f) *$d_X(\hat{K}_{Ps(X)}) > 0$ for each compact set $K \subset X$.*

(g) *If (H,D) is a Hartogs figure in $\mathbb{C}^2$ then for each $f \in \mathcal{H}(H;X)$ there exists $\tilde{f} \in \mathcal{H}(D;X)$ such that $\tilde{f} \mid H = f$.*

PROOF. The implications (a) ⇒ (b), (b) ⇒ (c), (c) ⇒ (d), (d) ⇒ (e) and (e) ⇒ (f) should be clear.

(f) ⇒ (g): Consider the sets

$$D = \{(\lambda,\zeta) \in \mathbb{C}^2 : |\lambda| < R \quad \text{and} \quad |\zeta| < R\},$$

$$H = \{(\lambda,\zeta) \in D : |\lambda| > r \quad \text{or} \quad |\zeta| < s\},$$

where $0 < r < R$ and $0 < s < R$. Given $f \in \mathcal{H}(H;X)$ we want to find $\tilde{f} \in \mathcal{H}(D;X)$ such that $\tilde{f} \mid H = f$. Fix b with $r < b < R$. For each $t \in (0,R]$ let

$$D_t = \{(\lambda,\zeta)\ \mathbb{C}^2 : |\lambda| < b \quad \text{and} \quad |\lambda| < t\},$$

and set

$$T = \{t \in (0,R] : \text{there is } f_t \in \mathcal{H}(D_t;X) \text{ such that } f_t = f \text{ on } D_t \cap H\}.$$

We shall prove that $T = (0,R]$. This will imply (g), for the function $\tilde{f}$ defined by $\tilde{f} = f$ on H and $\tilde{f} = f_R$ on D_R will then be the required extension. Since $(0,s] \subset T$ the set T is nonvoid. Since $t_n \in T$ for each $n \in \mathbb{N}$ implies that $\sup t_n \in T$ the set T is closed in $(0,R]$. To show that T is open in $(0,R]$ fix $t \in T$ with $t < R$, fix a with $r < a < b$ and consider the set

$$J = \{(\lambda,\zeta) \in \mathbb{C}^2 : a \leq |\lambda| \leq b \quad \text{and} \quad |\zeta| \leq t\}.$$

Then J is a compact subset of H and $K = f(J)$ is a compact subset of X. If $u \in Ps(X)$ then $u \circ f_t \in Ps(D_t)$ by Exercise 47.G and hence $u \circ f_t(\lambda,\zeta)$ is a subharmonic function of λ for $|\lambda| < b$ for each fixed ζ with $|\zeta| < t$. If $a < c < b$ then

$$\sup_{|\lambda| \leq c} u \circ f_t(\lambda,\zeta) = \sup_{|\lambda|=c} u \circ f_t(\lambda,\zeta) = \sup_{|\lambda|=c} u \circ f(\lambda,\zeta) \leq \sup_K u.$$

Hence $f_t(\lambda,\zeta) \in \hat{K}_{Ps(X)}$ for every $(\lambda,\zeta) \in D_t$ and by (f) there exists $\varepsilon > 0$ such that $d_X(f_t(D_t)) > \varepsilon$. Set $g = \xi \circ f \in \mathcal{H}(H;E)$. By Example 10.2 and Exercise 8.I there exists $\tilde{g} \in \mathcal{H}(D;E)$ such that $\tilde{g} = g$ on H. Since $\tilde{g} = g = \xi \circ f = \xi \circ f_t$ on $H \cap D_t$ we see that $\tilde{g} = \xi \circ f_t$ on D_t. Since $\tilde{g}$ is uniformly continuous on each compact subset of D we can find $\delta > 0$ with $t + \delta < R$ such that

$$\|\tilde{g}(\lambda,\zeta) - \tilde{g}(\lambda',\zeta')\| < \varepsilon \tag{49.1}$$

for all $(\lambda,\zeta), (\lambda',\zeta') \in D_{t+\delta}$ such that $|\lambda - \lambda'| \leq 2\delta$, $|\zeta - \zeta'| < 2\delta$. Given $(\lambda,\zeta) \in D_t$ we claim that

$$f_t(\lambda',\zeta') \in B_X(f_t(\lambda,\zeta);\varepsilon) \tag{49.2}$$

for all $(\lambda',\zeta') \in D_t$ such that $|\lambda' - \lambda| < 2\delta$, $|\zeta' - \zeta| < 2\delta$. Indeed, consider the convex set

$$A = \{(\lambda',\zeta') \in D_t : |\lambda' - \lambda| < 2\delta, \ |\zeta' - \zeta| < 2\delta\}.$$

Then (49.1) implies that $\xi \circ f_t(A) \subset B_E(\xi \circ f_t(\lambda,\zeta);\varepsilon)$, and since $f_t(A)$ is connected and contains the point $f_t(\lambda,\zeta)$ we conclude that $f_t(A) \subset B_X(f_t(\lambda,\zeta);\varepsilon)$ and (49.2) has been proved. Next we define $f_{t+\delta} : D_{t+\delta} \to X$ by

$$f_{t+\delta}(\lambda,\zeta) = (\xi \mid B_X(f_t(\lambda',\zeta');\varepsilon))^{-1} \circ \tilde{g}(\lambda,\zeta)$$

if $(\lambda',\zeta') \in D_t$ satisfies $|\lambda' - \lambda| < \delta$, $|\zeta' - \zeta| < \delta$. Observe that (49.1) guarantees that $\tilde{g}(\lambda,\zeta) \in B_E(\xi \circ f_t(\lambda',\zeta');\varepsilon)$. We have to show that $f_{t+\delta}$ is well defined. Indeed, let $(\lambda,\zeta) \in D_{t+\delta}$ and let (λ',ζ'), $(\lambda'',\zeta'') \in D_t$ such that $|\lambda' - \lambda| < \delta$, $|\zeta' - \zeta| < \delta$, $|\lambda'' - \lambda| < \delta$, $|\zeta'' - \zeta| < \delta$. Then $|\lambda' - \lambda''| < 2\delta$, $|\zeta' - \zeta''| < 2\delta$ and (49.2) implies that $f_t(\lambda',\zeta') \in B_X(f_t(\lambda'',\zeta'');\varepsilon)$. Then Lemma 47.3 implies that $\xi \mid B_X(f_t(\lambda',\zeta');\varepsilon) \cup B_X(f_t(\lambda'',\zeta'');\varepsilon)$ is injective and hence $f_{t+\delta}$ is well defined. It is clear that $f_{t+\delta} \in \mathcal{H}(D_{t+\delta};X)$ and $f_{t+\delta} = f_t$ on D_t. This shows that T is open in $(0,R]$ and (g) follows.

(g) ⇒ (a): Assuming (g) we have to prove that the function $-\log \delta_X$ is plurisubharmonic on $X \times E$. It suffices to show that the function $-\log \delta_X((\xi \mid U_o)^{-1}(x),y)$ is plurisubharmonic on $\xi(U_o) \times E$ for each chart U_o in X such that $\xi(U_o)$ is convex. Fix $(a,b) \in \xi(U_o) \times E$ and $(s,t) \in E \times E$ such that

$$(49.3) \qquad a + \lambda s \in \xi(U_o) \quad \text{for all} \quad \lambda \in \bar{\Delta}.$$

Let $P \in \mathcal{P}(\mathbb{C})$ such that

$$-\log \delta_X((\xi \mid U_o)^{-1}(a + \lambda s), b + \lambda t) \leq \operatorname{Re} P(\lambda)$$

for every $\lambda \in \partial\Delta$, or equivalently

$$(49.4) \qquad \delta_X((\xi \mid U_o)^{-1}(a + \lambda s), e^{-P(\lambda)}(b + \lambda t)) \geq 1$$

for every $\lambda \in \partial\Delta$. We have to show that (49.4) holds for every $\lambda \in \bar{\Delta}$. It suffices to prove that

$$(49.5) \qquad \delta_X((\xi \mid U_o)^{-1}(a + \lambda s), \rho e^{-P(\lambda)}(b + \lambda t)) \geq 1$$

for every $\lambda \in \bar{\Delta}$ and $0 < \rho < 1$. Fix ρ with $0 < \rho < 1$ and consider the function $g \in \mathcal{H}(\mathbb{C}^2;E)$ defined by

$$g(\lambda,\zeta) = a + \lambda s + \rho\zeta e^{-P(\lambda)}(b + \lambda t).$$

(49.3) implies that $g(\bar{\Delta} \times \{0\}) \subset \xi(U_0)$, and a compactness argument yields an open disc D_0 centered at zero and containing $\bar{\Delta}$, and an open disc D_0' centered at zero, such that $g(D_0 \times D_0') \subset \xi(U_0)$. By (49.4) and Exercise 47.B for each $\lambda \in \partial\Delta$ we can find a chart U_λ in X containing $(\xi \mid U_0)^{-1}(a + \lambda s)$ such that $\xi(U_\lambda)$ is convex and contains $g(\{\lambda\} \times \bar{\Delta})$. Then for each $\lambda \in \partial\Delta$ we can find an open disc D_λ centered at λ, and an open disc D_λ' centered at zero and containing $\bar{\Delta}$, such that $g(D_\lambda \times D_\lambda') \subset \xi(U_\lambda)$. Set

$$G = (D_0 \times D_0') \cup \bigcup_{|\lambda|=1} (D_\lambda \times D_\lambda'), \quad U = U_0 \cup \bigcup_{|\lambda|=1} U_\lambda .$$

We want to find a function $f \in \mathcal{H}(G;X)$ such that $\xi \circ f = g$. Define $f : G \to X$ by $f = (\xi \mid U_0)^{-1} \circ g$ on $D_0 \times D_0'$ and $f = (\xi \mid U_\lambda)^{-1} \circ g$ on $D_\lambda \times D_\lambda'$ for each $\lambda \in \partial\Delta$. We must show that f is well defined. If $\lambda \in \partial\Delta$ then $(\xi \mid U_0)^{-1}(a + \lambda s) \in U_0 \cap U_\lambda$ and $\xi(U_0) \cap \xi(U_\lambda)$ is convex. Then Lemma 47.3 implies that $\xi | U_0 \cup U_\lambda$ is injective and therefore $(\xi \mid U_0)^{-1} \circ g = (\xi \mid U_\lambda)^{-1} \circ g$ on $(D_0 \times D_0') \cap (D_\lambda \times D_\lambda')$. Next consider $\lambda, \mu \in \partial\Delta$ such that $(D_\lambda \times D_\lambda') \cap (D_\mu \times D_\mu')$ is nonvoid. Then $D_\lambda \cap D_\mu$ is nonvoid too, and this clearly implies that $D_\lambda \cap D_\mu \cap D_0$ is nonvoid as well. If $p \in D_\lambda \cap D_\mu \cap D_0$ then

$$(p,0) \in (D_\lambda \times D_\lambda') \cap (D_\mu \times D_\mu') \cap (D_0 \times D_0').$$

Since

$$(\xi \mid U_\lambda)^{-1} \circ g = (\xi \mid U_0)^{-1} \circ g = (\xi \mid U_\mu)^{-1} \circ g$$

on $(D_\lambda \times D_\lambda') \cap (D_\mu \times D_\mu') \cap (D_0 \times D_0')$, the Identity Principle 47.D implies that $(\xi \mid U_\lambda)^{-1} \circ g = (\xi \mid U_\mu)^{-1} \circ g$ on $(D_\lambda \times D_\lambda') \cap (D_\mu \times D_\mu')$. Thus f is well defined. Clearly $f \in \mathcal{H}(G;X)$ and $\xi \circ f = g$. Observe that G contains an open set H of the form $H = H_1 \cup H_2$, where

$$H_1 = \{(\lambda,\zeta) \in \mathbb{C}^2 : |\lambda| < 1+\varepsilon \quad \text{and} \quad |\zeta| < \varepsilon\},$$

$$H_2 = \{(\lambda,\zeta) \in \mathbb{C}^2 : 1-\varepsilon < |\lambda| < 1+\varepsilon \quad \text{and} \quad |\zeta| < 1+\varepsilon\}.$$

Thus $f \in \mathcal{H}(H;X)$ and by (g) f has an extension $\tilde{f} \in \mathcal{H}(D;X)$, where

$$D = \{(\lambda,\zeta) \in \mathbb{C}^2 : |\lambda| < 1+\varepsilon \quad \text{and} \quad |\zeta| < 1+\varepsilon\}.$$

Since $\xi \circ f = g$ on H we see that $\xi \circ \tilde{f} = g$ on D. For each $\lambda \in \bar{\Delta}$ set $A_\lambda = \tilde{f}(\{\lambda\} \times \Delta)$. Note that

$$\xi(A_\lambda) = g(\{\lambda\} \times \Delta) = \{a + \lambda s + \rho\zeta e^{-P(\lambda)}(b + \lambda t) : |\zeta| < 1\}.$$

If $b + \lambda t = 0$ then (49.5) is obviously true. And if $b + \lambda t \neq 0$ then we immediately see that $g \mid \{\lambda\} \times \Delta : \{\lambda\} \times \Delta \to g(\{\lambda\} \times \Delta)$ is a homeomorphism. Whence it follows that $\xi \mid A_\lambda : A_\lambda \to \xi(A_\lambda)$ is a homeomorphism and (49.5) holds. This shows (a) and completes the proof of the theorem.

EXERCISES

49.A. Let X and Y be topological spaces and let $f : X \to Y$ be a local homeomorphism. Let N be a topological subspace of Y and let $M = f^{-1}(N)$. Show that $f \mid M : M \to N$ is a local homeomorphism.

49.B. Let (X,ξ) be a Riemann domain over E. Let E_o be a closed vector subspace of E, let $X_o = \xi^{-1}(E_o)$ and let $\xi_o = \xi \mid X_o$. Show that (X_o,ξ_o) is a Riemann domain over E_o.

49.C. Let (X,ξ) be a Riemann domain over E.

(a) Show that an upper semicontinuous function $f : X \to [-\infty,\infty)$ is plurisubharmonic if and only if its restriction

$f \mid \xi^{-1}(E_o)$ is plurisubharmonic for each finite dimensional subspace E_o of E.

(b) Show that X is pseudoconvex if and only if $\xi^{-1}(E_o)$ is pseudoconvex for each finite dimensional subspace E_o of E.

49.D. Let (X,ξ) and (Y,η) be Riemann domains over E and F respectively.

(a) Show that $(X \times Y, (\xi,\eta))$ is a Riemann domain over $E \times F$.

(b) Show that $\delta_{X\times Y}((x,y),(s,t)) = min\{\delta_X(x,s), \delta_X(y,t)\}$ for all $(x,y) \in X \times Y$ and $(s,t) \in E \times F$.

(c) Show that $X \times Y$ is pseudoconvex if and only if X and Y are pseudoconvex.

50. PLURISUBHARMONIC FUNCTIONS ON RIEMANN DOMAINS

In this section we extend Theorem 38.6 to the case of Riemann domains. This is a key step in the solution of the $\bar{\partial}$ equation in Riemann domains. We shall give the main result after a series of preparatory lemmas.

50.1. LEMMA. *Let (X,ξ) be a Riemann domain over $\mathbb{C}^n$, and let U be a connected component of $X_\delta = \{x \in X : d_X(x) > \delta\}$. If ρ_U denotes the geodesic distance in U then the set $\{x \in U : \rho_U(x_o,x) \le c\}$ is relatively compact in X for each $x_o \in U$ and $c \in \mathbb{R}$.*

PROOF. Choose ε with $0 < \varepsilon \delta/3$. Fix $x_o \in U$ and set $K_c = \{x \in U : \rho_U(x_o,x) \le c\}$ for each $c \in \mathbb{R}$. Since ρ_U is a metric the set K_c is obviously compact when $c \le 0$. To prove that K_c is relatively compact for every $c \in \mathbb{R}$ it clearly suffices to prove that if K_c is relatively compact for a certain $c \ge 0$ then $K_{c+\varepsilon}$ is relatively compact as well. Before proving this we show that

(50.1) $$K_{c+\varepsilon} \subset \bigcup_{x \in K_c} B_X(x;2\varepsilon).$$

Indeed, given $y \in K_{c+\varepsilon}$ there is a polygonal line $[x_o,\ldots,x_m]$ in U, with $x_m = y$, such that

(50.2) $$\sum_{j=1}^{m} \|\xi(x_j) - \xi(x_{j-1})\| < c + 2\varepsilon$$

If $\sum_{j=1}^{m} \|\xi(x_j) - \xi(x_{j-1})\| \leq c$ then $\rho_U(x_o,y) \leq c$ and $y \in K_c$. Thus we may assume that $c < \sum_{j=1}^{m} \|\xi(x_j) - \xi(x_{j-1})\| < c + 2\varepsilon$. Then we can find k with $1 \leq k \leq m$ and $x \in [x_{k-1},x_k]$ such that

(50.3) $$\sum_{j=1}^{k-1} \|\xi(x_j) - \xi(x_{j-1})\| + \|\xi(x) - \xi(x_{k-1})\| = c,$$

and therefore

(50.4) $$\|\xi(x_k) - \xi(x)\| + \sum_{j=k+1}^{m} \|\xi(x_j) - \xi(x_{j-1})\| < 2\varepsilon.$$

By (50.3) $\rho_U(x_o,x) \leq c$ and hence $x \in K_c$. By (50.4) $\|\xi(x_j) - \xi(x)\| < 2\varepsilon$ for $j = k, k+1,\ldots,m$, and it follows from repeated applications of Lemma 47.3 that the points $x_k, x_{k+1},\ldots,x_m$ all lie in $B_X(x;2\varepsilon)$. In particular $y = x_m \in B_X(x;2\varepsilon)$ and (50.1) has been proved. If K_c is relatively compact for a certain $c \geq 0$ then there are points $a_1,\ldots,a_m \in K_c$ such that $K_c \subset \bigcup_{j=1}^{m} B_X(a_j;\varepsilon)$. Then it follows from (50.1) that $K_{c+\varepsilon} \subset \bigcup_{j=1}^{m} B_X(a_j;3\varepsilon)$ and $K_{c+\varepsilon}$ is relatively compact set as well. This completes the proof of the lemma.

50.2. **LEMMA.** *Let (X,ξ) be a pseudoconvex Riemann domain over $\mathbb{C}^n$, and let U be a connected component of X_δ. Then there is a function $f \in Psc(U)$ such that the set $\{x \in U : f(x) \leq c\}$ is compact for each $c \in \mathbb{R}$. In particular the set $\hat{K}_{Psc(U)}$ is compact for each compact set $K \subset U$.*

PROOF. Fix ε with $0 < 2\varepsilon < \delta$ and let V be the connected component of X_ε which contains U. Observe that $U \subset V_\varepsilon$. Fix $x_o \in V$ and set $\sigma(x) = \rho_V(x_o, x)$ for every $x \in V$. It is easy to see that

$$(50.5) \qquad |\sigma(x) - \sigma(y)| \leq \| \xi(x) - \xi(y) \|$$

if x and y both lie in a same chart. Then it follows from Proposition 17.18 that $\partial\sigma/\partial x_j \in L^1(V, loc)$ and $|\partial\sigma / \partial x_j| \leq 1$ almost everywhere in V for every j. Then it follows from Proposition 48.5 that $\sigma * \rho_\varepsilon \in C^\infty(U)$ and for each $x \in U$ and $j,k = 1,\dots,n$ we can write

$$\frac{\partial^2(\sigma * \rho_\varepsilon)}{\partial z_j \partial \bar{z}_k}(x) = \left(\frac{\partial\sigma}{\partial z_j} * \frac{\partial\rho_\varepsilon}{\partial \bar{z}_k}\right)(x)$$

$$= \int_{\bar{B}(\xi(x);\varepsilon)} \frac{\partial\sigma}{\partial z_j} \circ (\xi \mid U_x)^{-1}(t)\rho_\varepsilon(\xi(x) - t)d\lambda(t),$$

where $U_x = B_X(x; d_X(x))$. Hence the functions $\partial^2(\sigma * \rho_\varepsilon) / \partial z_j \partial \bar{z}_k$ are uniformly bounded on U by a constant which depends only on ε. Hence we can find a constant $c_\varepsilon > 0$ such that the function

$$u_\varepsilon(x) = (\sigma * \rho_\varepsilon)(x) + c_\varepsilon \| \xi(x) \|^2$$

is plurisubharmonic on U. It follows from (50.5) that

$$\sigma(x) - \sigma * \rho_\varepsilon(x) = \int_{\bar{B}(0;\varepsilon)} [\sigma(x) - \sigma \circ (\xi|U_x)^{-1}(\xi(x) - t)] \rho_\varepsilon(t)\, d\lambda(t) \leq \varepsilon$$

for every $x \in U$. Then by Lemma 50.1 we may conclude that the set $\{x \in U : u_\varepsilon(x) \leq c\}$ is relatively compact in X for each $c \in \mathbb{R}$.

We next consider the function

$$v(x) = -1/(-\log d_X(x) + \log\delta).$$

Since the function $\varphi(t) = -1/t$ is convex and increasing for $t < 0$ we see that $v \in Psc(X_\delta)$. Set $f = \max\{u_\varepsilon, v\}$. Then $f \in Psc(U)$, and since $v(x_j) \to \infty$ when (x_j) approaches the boundary of U we conclude that the set $\{x \in U : f(x) \leq c\}$ is compact for each $c \in \mathbb{R}$. The proof of the lemma is complete.

50.3. LEMMA. *Let (X,ξ) be a pseudoconvex Riemann domain over $\mathbb{C}^n$, and let U be a connected component of X_δ. Let K be a compact subset of U and let V be an open neighborhood of $\hat{K}_{Ps(U)}$ in U. Then there is a function $f \in Psc(U)$ such that:*

(a) *The set $\{x \in U : f(x) \leq c\}$ is compact for each $c \in \mathbb{R}$.*

(b) *$f(x) < 0$ for every $x \in K$.*

(c) *$f(x) > 0$ for every $x \in U \setminus V$.*

PROOF. In view of Lemma 50.2, the proof of Lemma 38.1 applies.

50.4. LEMMA. *Let (X,ξ) be a connected pseudoconvex Riemann domain over $\mathbb{C}^n$, and let $r \geq s > 0$. Then $\hat{K}_{Psc(X_s)}$ is a compact subset of X_r for each compact set $K \subset X_r$.*

PROOF. Let K be a compact subset of X_r. Since X is connected we can find δ with $s > \delta > 0$ such that K is contained in a connected component U of X_δ. By Lemma 50.2 $\hat{K}_{Psc(U)}$ is a compact subset of U. Let $X_s(K)$ denote the union of those connected components of X_s which intersect K. Then $X_s(K) \subset U$ and $\hat{K}_{Psc(X_s)} = \hat{K}_{Psc(X_s(K))} \subset \hat{K}_{Psc(U)}$. This shows that $\hat{K}_{Psc(X_s)}$ is a compact subset of U. But on the other hand, since $K \subset X_r$ there is $\varepsilon > 0$ such that $d_X \geq r + \varepsilon$ on K, and therefore $d_X \geq r + \varepsilon$ on $\hat{K}_{Psc(X)}$. Hence $\hat{K}_{Psc(X_s)} \subset \hat{K}_{Psc(X)} \subset X_r$ and $\hat{K}_{Psc(X_s)}$ is a compact subset of X_r, as asserted.

50.5. LEMMA. *Let (X,ξ) be a connected pseudoconvex Riemann domain over $\mathbb{C}^n$. Then there is a function $f \in Psc(X)$ such that the set $\{x \in X : f(x) \leq c\}$ is compact for each $c \in \mathbb{R}$. In particular* *the set* $\hat{K}_{Psc(X)}$ *is compact for each compact set* $K \subset X$.

PROOF. By Corollary 47.8 X is hemicompact. Let (A_j) be an increasing sequence of compact subsets of X such that each compact subset of X is contained in some A_j. Let (r_j) be a sequence of strictly positive numbers decreasing to zero such that A_j is contained in a connected component U_j of X_{r_j}. Set $K_j = (\hat{A}_j)_{Psc(U_{j+1})}$ for each j. Then K_j is a compact subset of $U_{j+1} \cap X_{r_j}$ by Lemmas 50.2 and 50.4. For each j let V_j denote the union of those connected components of X_{r_j} which intersect K_j. Observe that $U_j \subset V_j \subset U_{j+1}$ and V_j has only finitely many connected components. By Lemma 50.2 there is a function $v_1 \in Psc(V_1)$ such that the set $\{x \in V_1 : v_1(x) \leq c\}$ is compact for each $c \in \mathbb{R}$. Choose $c_1 > 1$ such that $v_1 < c_1$ on K_1 and set $L_1 = \{x \in V_1 : v_1(x) \leq c_1\}$. Since L_1 is contained in some A_j we may assume without loss of generality that $L_1 \subset A_2$. Since $(\hat{A}_1)_{Psc(V_2)} = (\hat{A}_1)_{Psc(U_2)} = K_1 \subset V_1$, an application of Lemma 50.3 yields a function $v_2 \in Psc(V_2)$ such that the set $\{x \in V_2 : v_2(x) \leq c\}$ is compact for each $c \in \mathbb{R}$, $v_2 < 0$ on A_1 and $v_2 > 2$ on $V_2 \setminus V_1$. Choose $c_2 > 2$ such that $v_2 < c_2$ on K_2 and set $L_2 = \{x \in V_2 : v_2(x) \leq c_2\}$. Proceeding inductively we can find a sequence of functions $(v_j)_{j=1}^{\infty}$ with $v_j \in Psc(V_j)$, and a sequence of constants $(c_j)_{j=1}^{\infty}$, with $c_j > j$, such that if we set

$$L_j = \{x \in V_j : v_j(x) \leq c_j\},$$

then $A_j \subset L_j \subset A_{j+1}$ for every $j \geq 1$, and $v_j < 0$ on A_{j-1} and $v_j > j$ on $V_j \setminus V_{j-1}$ for every $j \geq 2$. Let $\chi \in C^{\infty}(\mathbb{R};\mathbb{R})$ be a convex increasing function such that $\chi(t) = 0$ for $t \leq 0$ and $\chi(t) \geq t$ for $t \geq 1$. If we define

$$f_j = \sum_{i=1}^{j} \chi \circ v_i$$

for each $j \in \mathbb{N}$, then $f_j = f_k$ on A_i whenever $j, k \geq i$ and hence the limit $f = \lim f_j$ exists and belongs to $Psc(X)$. To complete the proof of the lemma we shall show that

$$(50.6) \qquad \{x \in X : f(x) \leq j\} \subset L_j \quad \text{for each} \quad j \in \mathbb{N}.$$

Let $x \in X \setminus L_j$. Then there is $k \geq j$ such that $x \notin L_k$ and $x \in L_{k+1}$. We distinguish two cases. If $x \in V_k$ then $v_k(x) > k$ and therefore $\chi \circ v_k(x) > k$. If $x \notin V_k$ then $v_{k+1}(x) > k + 1$ and therefore $\chi \circ v_{k+1}(x) > k + 1$. In each case $f(x) > k \geq j$, proving (50.6) and the lemma.

50.6. THEOREM. *Let (X, ξ) be a connected pseudoconvex Riemann domain over $\mathbb{C}^n$. Let K be a compact subset of X and let V be an open neighborhood of $\hat{K}_{Ps(X)}$. Then there exists a strictly plurisubharmonic function $f \in C^\infty(X)$ such that:*

(a) *The set $\{x \in X : f(x) \leq c\}$ is compact for each $c \in \mathbb{R}$.*

(b) *$f(x) < 0$ for every $x \in K$.*

(c) *$f(x) > 0$ for every $x \in X \setminus V$.*

PROOF. In view of Lemma 50.5, the proofs of Lemma 38.1 and Theorem 38.6 can be directly adapted.

EXERCISES

50.A. Let (X, ξ) be a pseudoconvex Riemann domain over $\mathbb{C}^n$. Show that $\hat{K}_{Ps(X)} = \hat{K}_{Psc(X)}$ for each compact set $K \subset X$.

50.B. Let (X, ξ) be a pseudoconvex Riemann domain over E, let $f \in Ps(X)$ and let $U = \{x \in X : f(x) < 0\}$. Show that U is also pseudoconvex.

50.C. Let (X, ξ) be a connected pseudoconvex Riemann domain over $\mathbb{C}^n$. Show that for each $\delta > 0$ there exists a strictly plurisubharmonic function $f_\delta \in C^\infty(X_\delta)$ such that the set $\{x \in X_\delta : f_\delta(x) \leq c\}$ is compact for each $c \in \mathbb{R}$.

51. THE $\bar\partial$ EQUATION IN RIEMANN DOMAINS

The definitions and properties of differential forms and the $\bar\partial$ operator admit straightforward extensions to the case of Riemann domains. The necessary ingredients from differentiation and integration in Riemann domains were introduced in Section 48 so as to make these extensions clear. In particular the spaces $L^2_{pq}(X)$, $L^2_{pq}(X,loc)$ and $L^2_{pq}(X,\varphi)$ have the expected meanings, and the operator $\bar\partial$ induces operators

$$\bar\partial : D_o \subset L^2(X,loc) \to L^2_{01}(X,loc)$$

and

$$\bar\partial : D_1 \subset L^2_{01}(X,loc) \to L^2_{02}(X,loc)$$

similar to those introduced in Section 40. Likewise, if α, β $\gamma \in C^\infty(X;\mathbb{R})$ are three weight functions then the operator $\bar\partial$ induces operators

$$A : D_A \subset L^2(X,\alpha) \to L^2_{01}(X,\beta)$$

and

$$B : D_B \subset L^2_{01}(X,\beta) \to L^2_{02}(X,\gamma)$$

similar to those introduced in Section 40. Then the results from Sections 40, 41 and 42 admit straightforward extensions to the case of Riemann domains. An examination of the corresonding proofs shows this. We restrict ourselves to state the main results in the form they will be most convenient to us.

51.1. THEOREM. *Let (X,ξ) be a connected pseudoconvex Riemann domain over $\mathbb{C}^n$, and let $g_o \in L^2_{01}(X,loc)$ with $\bar\partial g_o = 0$. Then:*

(a) *There exist weight functions $\alpha, \beta, \gamma \in C^\infty(X;\mathbb{R})$ such* *that* $g_o \in L^2_{01}(X,\beta)$ *and*

$$\|g\|^2_\beta \leq \|A^*g\|^2_\alpha + \|Bg\|^2_\gamma \quad \textit{for every} \quad g \in D_{A^*} \cap D_B .$$

(b) *For each* $g \in L^2_{01}(X,\beta)$ *such that* $\overline{\partial} g = 0$ *there exists* $f \in L^2(X,\alpha)$ *such that* $\overline{\partial} f = g$ *and* $\|f\|_\alpha \leq \|g\|_\beta$.

51.2. **THEOREM.** *Let* (X,ξ) *be a pseudoconvex Riemann domain over* $\mathbb{C}^n$, *and let* $g \in C^\infty_{01}(X)$ *with* $\overline{\partial} g = 0$. *Then each solution* $f \in L^2(X,loc)$ *of the equation* $\overline{\partial} f = g$ *belongs to* $C^\infty(X)$.

NOTES AND COMMENTS

The results in Section 47 can be found in M. Schottenloher [1] or G. Coeuré [1]. Our presentation in Section 49 follows M. Schottenloher [3], whereas in Section 50 we have essentially followed the book of L. Hörmander [3]. The results in Section 51 are special cases of results obtained by L. Hörmander [2] [3].

CHAPTER XII

THE LEVI PROBLEM IN RIEMANN DOMAINS

52. THE CARTAN-THULLEN THEOREM IN RIEMANN DOMAINS

In this section we extend to the case of Riemann domains the notions of domain of holomorphy, domain of existence and holomorphically convex domain, and study the relationships between these notions.

52.1. DEFINITION. Let (X,ξ) be a Riemann domain over E and let $\mathcal{F} \subset \mathcal{H}(X)$.

(a) If (Y,η) is a Riemann domain over E then a morphism $\tau : X \to Y$ is said to be an $\mathcal{F}$-*extension* of X if for each $f \in \mathcal{F}$ there is a unique $\tilde{f} \in \mathcal{H}(Y)$ such that $\tilde{f} \circ \tau = f$. A morphism $\tau : X \to Y$ is said to be a *holomorphic extension* of X if τ is an $\mathcal{H}(X)$-extension of X.

(b) X is said to be an $\mathcal{F}$-*domain of holomorphy* if each $\mathcal{F}$-*extension* of X is an isomorphism. X is said to be a *domain of holomorphy* if X is an $\mathcal{H}(X)$-*domain of holomorphy*. X is said to be the *domain of existence* of a function $f \in \mathcal{H}(X)$ if X is an $\{f\}$-*domain of holomorphy*.

If $\tau : X \to Y$ is an $\mathcal{F}$-extension of X then it is clear that each connected component of Y intersects $\tau(X)$. We observe that a morphism $\tau : X \to Y$ is a holomorphic extension of X if and only if the mapping $\tau^* : g \in \mathcal{H}(Y) \to g \circ \tau \in \mathcal{H}(X)$ is an algebra isomorphism.

52.2. DEFINITION. Let (X,ξ) be a Riemann domain over E and let $\mathcal{F} \subset \mathcal{H}(X)$.

(a) For each set $A \subset X$ we define

$$\hat{A}_F = \{x \in X : |f(x)| \leq \sup_A |f| \quad \text{for all} \quad f \in F\}$$

(b) X is said to be F-*convex* if the set $\hat{K}_F$ is compact for each compact set $K \subset X$. X is said to be *holomorphically convex* if X is $\mathcal{H}(X)$-*convex*.

(c) X is said to be *metrically* F-*convex* if $d_X(\hat{K}_F) > 0$ for each compact set $K \subset X$. X is said to be *metrically holomorphically convex* if X is metrically $\mathcal{H}(X)$-convex.

(d) X is said to be F-*separated* if for each $x, y \in X$ such that $x \neq y$ and $\xi(x) = \xi(y)$ there exists an $f \in F$ such that $f(x) \neq f(y)$. X is said to be *holomorphically separated* if X is $\mathcal{H}(X)$-separated.

52.3. **LEMMA.** *Let (X,ξ) be a Riemann donain over E. Then X is holomorphically separated if an only if each holomorphic extension of X is injective.*

PROOF. We first assume that X is holomorphically separated. Let (Y,η) be a Riemann domain over E, let $\tau : X \to Y$ be a holomorphic extension of X, and let $x, y \in X$ with $x \neq y$. If $\xi(x) = \xi(y)$ then by hypothesis there exists an $f \in \mathcal{H}(X)$ such that $f(x) \neq f(y)$. Hence $\tilde{f} \circ \tau(x) \neq \tilde{f} \circ \tau(y)$ and therefore $\tau(x) \neq \tau(y)$. If $\xi(x) \neq \xi(y)$ then $\eta \circ \tau(x) \neq \eta \circ \tau(y)$ and $\tau(x) \neq \tau(y)$ too. Thus τ is injective.

Conversely assume that X is not holomorphically separated. Define an equivalence relation on X as follows. Let $x \sim y$ if $\xi(x) = \xi(y)$ and $f(x) = f(y)$ for every $f \in \mathcal{H}(X)$. Let $Y = X/\sim$ be the quotient space and let $\pi : X \to Y$ be the quotient mappping. Clearly ξ induces a unique mapping $\eta : Y \to E$ such that $\eta \circ \pi = \xi$ and each $f \in \mathcal{H}(X)$ induces a unique function $\tilde{f} : Y \to \mathbb{C}$ such that $\tilde{f} \circ \pi = f$. Since Y has the quotient topology, the mapping η is continuous, and so is each $\tilde{f}$. Let $x, y \in X$ with $\pi(x) \neq \pi(y)$. Then either $\eta \circ \pi(x) = \xi(x) \neq \xi(y) = \eta \circ \pi(x)$, or else there is an $f \in \mathcal{H}(X)$ such that $\tilde{f} \circ \pi(x) = f(x) \neq f(y) = \tilde{f} \circ \pi(y)$. Whence it follows that Y is a Hausdorff space. We next show that the mapping π is open. Let A be an open subset of X. To show that $\pi(A)$ is open in Y it suffices to prove that $\pi^{-1}(\pi(A))$ is open in X, by the definition of the quotient topology. Let $x \in \pi^{-1}(\pi(A))$. Choose $a \in A$ such that $x \sim a$, and choose $r > 0$ such that $r < d_X(x)$, $r < d_X(a)$ and $B_X(a;r) \subset A$. We claim that $B_X(x;r) \subset \pi^{-1}(\pi(A))$. Indeed, since $x \sim a$ then for each $f \in \mathcal{H}(X)$ and $t \in B_E(0;r)$ we have that

$$f \circ \xi_x^{-1}(\xi(x) + t) = \sum_{m=0}^{\infty} P_t^m f(x) = \sum_{m=0}^{\infty} P_t^m f(a) = f \circ \xi_a^{-1}(\xi(a) + t)$$

where $\xi_z = \xi \mid B_X(z;d_X(z))$ for each $z \in X$. Whence it follows that $\xi_x^{-1}(\xi(x) + t) \sim \xi_a^{-1}(\xi(a) + t)$ for each $t \in B_E(0;r)$. Thus $B_X(x;r) \subset \pi^{-1}(\pi(A))$ and π is an open mapping, as asserted. Now, let $x \in X$ and choose an open set U in X containing x such that $\xi \mid U : U \to \xi(U)$ is a homeomorphism. Since π is continuous and open, it follows that both mappings $\pi \mid U : U \to \pi(U)$ and $\eta \mid \pi(U) : \pi(U) \to \xi(U)$ are homeomorphisms. Thus (Y,η) is a Riemann domain over E and $\pi : X \to Y$ is a morphism. Since $\tilde{f} \circ \pi = f$, it is clear that $\tilde{f} \in \mathcal{H}(Y)$ for each $f \in \mathcal{H}(X)$. Thus π is a holomorphic extension of X which is not injective, and the proof of the lemma is complete.

52.4. DEFINITION. Let (X,ξ) be a Riemann domain over E, and let A be a subset of X such that $d_X(A) = \rho > 0$. Then for each set $B \subset B_E(0;\rho)$ we shall set

$$A + B = \bigcup_{x \in A} \xi_x^{-1}(\xi(x) + B),$$

where $\xi_x = \xi \mid B_X(x;d_X(x))$ for each $x \in X$.

52.5. LEMMA. *Let (X,ξ) be a Riemann domain over E, let K be a compact subset of X, and let $\rho = d_X(K)$. Then $K + L$ is a compact subset of X for each compact set $L \subset B_E(0;\rho)$.*

PROOF. Choose $0 < r < s < \rho$ such that $L \subset B_E(0;r)$. Since K

is compact there are points $x_1,\ldots,x_m \in K$ such that $K \subset \bigcup_{j=1}^{m} B_X(x_j;s-r)$. Then one can easily see that

$$K + L = \bigcup_{j=1}^{m} \xi_{x_j}^{-1}[\xi(K \cap \bar{B}_X(x_j;s-r)) + L],$$

and the desired conclusion follows.

The next results parallels Theorem 11.4.

52.6. THEOREM. *Let (X,ξ) be a Riemann domain over E. Consider the following conditions:*

(a) *X is a domain of existence.*

(b) *X is holomorphically separated and X is the union of an increasing sequence of open sets A_j such that $d_X((\hat{A}_j)_{\mathcal{H}(X)}) > 0$ for every j.*

(c) *X is holomorphically separated and for each sequence (x_j) in X such that $d_X(x_j) \to 0$ there is a function $f \in \mathcal{H}(X)$ such that $\sup|f(x_j)| = \infty$.*

(d) *X is a domain of holomorphy.*

(e) *X is holomorphically separated and $d_X(\hat{K}_{\mathcal{H}(X)}) = d_X(K)$ for each compact set $K \subset X$.*

(f) *X is holomorphically separated and metrically holomorphically convex.*

(g) *X is holomorphically separated and pseudoconvex.*

Then (a) ⇒ (b) ⇒ (c) ⇒ (d) ⇒ (e) ⇒ (f) ⇒ (g).

PROOF. (a) ⇒ (b): Suppose X is the domain of existence of a function $f \in \mathcal{H}(X)$. Then X is holomorphically separated by Lemma 52.3. Now, consider the open sets

$$B_j = \{x \in X : |f(x)| < j\}, \quad A_j = \{x \in B_j : d_{B_j}(x) > 1/j\}.$$

Clearly $X = \bigcup_{j=1}^{\infty} A_j$ and $A_j \subset A_{j+1}$ for every j. We shall prove that $d_X((\hat{A}_j)_{\mathcal{H}(X)}) \geq 1/j$ for every j. Indeed, let $z \in (\hat{A}_j)_{\mathcal{H}(X)}$. Then for each $t \in B_E(0;1/j)$ and $m \in \mathbb{N}$ we have that

$$|P^m f(z)(t)| \leq \sup_{x \in A_j} |P^m f(x)(t)| \leq \sup_{B_j} |f| \leq j.$$

Hence $\|P^m f(z)\| \leq j^{m+1}$ for every $m \in \mathbb{N}$ and it follows from the Cauchy-Hadamard Formula that the series $\sum_{m=0}^{\infty} P^m f(z)(t)$ defines a function f_z which is holomorphic on the ball $B_E(\xi(z);1/j)$ and satisfies $f_z \circ \xi = f$ on a neighborhood of z. Let X_o be the disjoint union of X and $B_E(\xi(z);1/j)$, with its natural topology, and let $\xi_o : X_o \to E$ be defined $\xi_o = \xi$ on X and $\xi_o = identity$ on $B_E(\xi(z);1/j)$. Clearly (X_o, ξ_o) is a Riemann domain over E and the inclusion mapping $X \hookrightarrow X_o$ is a morphism. Let $f_o \in \mathcal{H}(X_o)$ be defined by $f_o = f$ on X and $f_o = f_z$ on $B_E(\xi(z);1/j)$. Define an equivalence relation on X_o as follows. Let $x \sim y$ if $\xi_o(x) = \xi_o(y)$ and $P^m_t f_o(x) = P^m_t f_o(y)$ for every $m \in \mathbb{N}$ and $t \in E$. Let $Y = X_o/\sim$ be the quotient space and let $\pi : X_o \to Y$ be the quotient mapping. There is a unique mapping $\eta : Y \to E$ such that $\eta \circ \pi = \xi_o$ and there is a unique function $\tilde{f} : Y \to \mathbb{C}$ such that $\tilde{f} \circ \pi = f_o$. Proceeding as in the proof of Lemma 52.3 we can show (Y, η) is a Riemann domain over E. If follows that $\tilde{f} \in \mathcal{H}(Y)$ and the mapping $\tau = \pi | X : X \to Y$ is an $\{f\}$-extension of X. Since X is the domain of existence of f, τ is an isomorphism. Since clearly $d_Y(\tau(z)) \geq 1/j$, we conclude that $d_X(z) \geq 1/j$ too, and (b) has been proved.

(b) ⇒ (c): The proof of the implication (b) ⇒ (c) in Theorem 11.4 applies.

(c) ⇒ (d): Suppose X satisfies (c). Let (Y, η) be a Riemann domain over E, and let $\tau : X \to Y$ be a holomorphic extension of X. By Lemma 52.3 the mapping τ is injective. Since every morphism is a local homeomorphism, we conclude that $\tau : X \to \tau(X)$ is a homeomorphism. If $\tau : X \to Y$ is not an isomorphism then $\tau(X) \neq Y$. Since each connected component of Y intersects $\tau(X)$, there is a point Y in the boundary of $\tau(X)$ in Y. Since $\tau : X \to \tau(X)$ is a homeomorphism there is a sequence (x_j) in X such that $\tau(x_j) \to y$. It follows that $d_X(x_j) \to 0$ and by (c) there is a function $f \in \mathcal{H}(X)$ such that $sup|f(x_j)| = \infty$. Let $\tilde{f} \in \mathcal{H}(Y)$ such that $\tilde{f} \circ \tau = f$. Then $f(x_j) = \tilde{f} \circ \tau(x_j) \to \tilde{f}(y)$, a contradiction. Thus τ is an isomorphism and (d) follows.

(d) ⇒ (e): Suppose X is a domain of holomorphy. Then X is holomorphically separated by Lemma 52.3. Now, let K be a compact subset of X, let $r = d_X(K)$ and let $z \in \hat{K}_{\mathcal{H}(X)}$. We shall first prove that for each $f \in \mathcal{H}(X)$ there is function f_z which is holomorphic on the ball $B_E(\xi(z);r)$ and satisfies $f_o \circ \xi = f$ on a neighborhood of z. Indeed, let $f \in \mathcal{H}(X)$. Given $t \in B_E(0;r)$ choose $\rho > 1$ such that $\rho t \in B_E(0;r)$. Then $K + \bar{\Delta}\rho t$ is a compact subset of X by Lemma 52.5. Choose $\varepsilon > 0$ such that $d_X(K + \bar{\Delta}\rho t) > \rho\varepsilon$ and f is bounded, by c say, on the set $K + \bar{\Delta}\rho t + B_E(0;\rho\varepsilon)$. Then

$$|P^m f(z)(t + h)| \leq \sup_{x \in K} |P^m f(x)(t + h)| \leq c\rho^{-m}$$

for every $h \in B_E(0;\varepsilon)$ and $m \in \mathbb{N}$. Hence the series $\sum_{m=0}^{\infty} P^m f(z)(t)$ defines a function f_z which is holomorphic on the ball $B_E(\xi(z);r)$ and satisfies $f_z \circ \xi = f$ on a neighborhood of z, as we wanted. Proceeding as in the proof of the implication (a) ⇒ (b) we can find a Riemann domain (Y,η) over E, and a holomorphic extension $\tau : X \to Y$ of X such that $d_Y(\tau(z)) \geq r$. Since X is a domain of holomorphic, τ must be an isomorphism. Hence $d_X(x) \geq r$ and (e) has been proved.

Since the implications (e) ⇒ (f) and (f) ⇒ (g) are obvious, the proof of the theorem is complete.

If E is separable and X is an open subset of E then Theorem 11.4 asserts that (b) ⇒ (a). But it is unknown whether this implication holds in the case of Riemann domains.

52.7. **DEFINITION.** Let (X,ξ) be a Riemann domain over E. An increasing sequence $\mathcal{U} = (U_j)_{j=1}^{\infty}$ of open subsets of X is said to be a *regular cover* of X if $X = \bigcup_{j=1}^{\infty} U_j$ and $d_{U_{j+1}}(U_j) > 0$ for every j. We shall denote by $\mathcal{H}^{\infty}(\mathcal{U})$ the algebra

$$\mathcal{H}^{\infty}(\mathcal{U}) = \{f \in \mathcal{H}(X) : \sup_{U_j}|f| < \infty \quad \text{for every} \quad j\},$$

endowed with the topology generated by the seminorms $f \to \sup_{U_j}|f|$.

$\mathcal{H}^{\infty}(\mathcal{U})$ is clearly a Fréchet algebra. If $f \in \mathcal{H}^{\infty}(\mathcal{U})$ then it is easy to see that $P_t^m f \in \mathcal{H}^{\infty}(\mathcal{U})$ for every $m \in \mathbb{N}$ and $t \in E$. These two properties of $\mathcal{H}^{\infty}(\mathcal{U})$ play a role in the next two lemmas.

52.8. **LEMMA.** *Let (X,ξ) be a Riemann domain over E and let $\mathcal{F} \subset \mathcal{H}(X)$.*

(a) *If X is $\mathcal{F}$-separated then each $\mathcal{F}$-extension of X is injective.*

(b) *If $P_t^m f \in \mathcal{F}$ for every $f \in \mathcal{F}$, $m \in \mathbb{N}_0$ and $t \in E$, and if each $\mathcal{F}$-extension of X is injective, then X is $\mathcal{F}$-separated.*

PROOF. Just examine the proof of Lemma 52.3.

52.9. **LEMMA.** *Let (X,ξ) be a Riemann domain over E. Suppose there is a regular cover $\mathcal{U}$ of X such that X is $\mathcal{H}^{\infty}(\mathcal{U})$ separated. Then for each countable set $P \subset X$ there exists a function $g \in \mathcal{H}^{\infty}(\mathcal{U})$ such that $g(x) \neq g(y)$ for all $x, y \in P$ such that $x \neq y$ and $\xi(x) = \xi(y)$.*

PROOF. Consider the countable set

$$Q = \{(x,y) \in P \times P : x \neq y \quad \text{and} \quad \xi(x) \neq \xi(y)\}.$$

Since X is $\mathcal{H}^{\infty}(\mathcal{U})$-separated, the set

$$S_{xy} = \{g \in \mathcal{H}^{\infty}(\mathcal{U}) : g(x) \neq g(y)\}$$

is nonvoid for each $(x,y) \in Q$. The set S_{xy} is clearly open in $\mathcal{H}^\infty(\mathfrak{U})$. We claim that S_{xy} is dense in $\mathcal{H}^\infty(\mathfrak{U})$. Indeed, given $f \in \mathcal{H}^\infty(\mathfrak{U})$ with $f \notin S_{xy}$, choose $g \in S_{xy}$ and set $g_n = f + \frac{1}{n} g$. Clearly $g_n \in S_{xy}$ for every n and $g_n \to f$ in $\mathcal{H}^\infty(\mathfrak{U})$. Since $\mathcal{H}^\infty(\mathfrak{U})$ is a Baire space the set

$$S = \cap \{S_{xy} : (x,y) \in Q\}$$

is dense in $\mathcal{H}^\infty(\mathfrak{U})$, an in particular nonvoid. This completes the proof.

52.10. **THEOREM.** *Let (X,ξ) be a Riemann domain over E. Consider the following conditions:*

(a) *X is a domain of existence.*

(b) *There is a regular cover $\mathfrak{U}$ of X such that X is an $\mathcal{H}^\infty(\mathfrak{U})$-domain of holomorphy.*

(c) *There is a regular cover $\mathfrak{U}$ of X such that X is $\mathcal{H}^\infty(\mathfrak{U})$-separated and $d_X(\hat{U}_{\mathcal{H}^\infty(\mathfrak{U})}) > 0$ for every $U \in \mathfrak{U}$.*

(d) *There is a regular cover $\mathfrak{U}$ of X such that X is $\mathcal{H}^\infty(\mathfrak{U})$-separated and $d_X(\hat{U}_{\mathcal{H}(X)}) > 0$ for every $U \in \mathfrak{U}$.*

Then (a) $\Rightarrow$ (b) $\Rightarrow$ (c) $\Rightarrow$ (d). If E is separable then these conditions are equivalent.

PROOF. (a) $\Rightarrow$ (b): Suppose X is the domain of existence of a function $f \in \mathcal{H}(X)$. Consider the open sets

$$V_j = \{x \in X : |f(x)| < j\}, \quad U_j = \{x \in V_j : d_{V_j}(x) > 2^{-j}\}.$$

Since $d_{U_{j+1}}(U_j) > 2^{-j-1}$ for every j, it is clear that the sequence $\mathfrak{U} = (U_j)_{j=1}^\infty$ is a regular cover of X. Since $f \in \mathcal{H}^\infty(\mathfrak{U})$, (b) follows.

(b) ⇒ (c): Let $\mathcal{U} = (U_j)_{j=1}^{\infty}$ be a regular cover of X such that X is an $\mathcal{H}^{\infty}(\mathcal{U})$-domain of holomorphy. Then X is $\mathcal{H}^{\infty}(\mathcal{U})$-separated by Lemma 52.8. If we set $r_j = d_{U_{j+1}}(U_j)$ then the proof of the implication (a) ⇒ (b) in Theorem 52.6 shows that $d_X((\hat{U}_j)_{\mathcal{H}^{\infty}(\mathcal{U})}) \geq r_j$ for every j. This shows (c).

Since the implication (c) ⇒ (d) is obvious, it only remains to show that (d) ⇒ (a) when E is separable. Thus let $\mathcal{U} = (U_j)_{j=1}^{\infty}$ be a regular cover of X such that X is $\mathcal{H}^{\infty}(\mathcal{U})$-separated and $d_X((\hat{U}_j)_{\mathcal{H}(X)}) > 0$ for every j. Let D be a countable dense subset of E, let $P = \xi^{-1}(D)$ and let

$$Q = \{(x,y) \in P \times P : x \neq y \quad \text{and} \quad \xi(x) = \xi(y)\}.$$

By restricting our attention to each connected component of X we may assume without loss of generality that X itself is connected. Then P is a countable dense subset of X by Exercise 47.C, and by Lemma 52.9 there exists a function $g \in \mathcal{H}^{\infty}(\mathcal{U})$ such that $g(x) \neq g(y)$ for every $(x,y) \in Q$. Now, let (x_j) be a sequence in P with the property that each point of P appears in the sequence (x_j) infinitely many times. Set $B_X(x) = B_X(x; d_X(x))$ for each $x \in X$ and set $V_j = (\hat{U}_j)_{\mathcal{H}(X)}$ for each $j \in \mathbb{N}$. Then $B_X(x) \not\subset V_j$ for each $x \in X$ and $j \in \mathbb{N}$, and after replacing the sequence (V_j) by a subsequence, if necessary, we can find a sequence (y_j) in X such that $y_j \in B_X(x_j)$, $y_j \notin V_j$ and $y_j \in V_{j+1}$ for every $j \in \mathbb{N}$. Hence we can inductively find a sequence (f_j) in $\mathcal{H}(X)$ such that

$$\sup_{V_j} |f_j| < 2^{-j} \quad \text{and} \quad |f_j(y_j)| > |g(y_j)| + j + 1 + \left|\sum_{i=1}^{j-1} f_i(y_j)\right|$$

for every $j \in \mathbb{N}$. Hence the series $\sum_{j=1}^{\infty} f_j$ converges uniformly on each V_j to a function $f \in \mathcal{H}(X)$ such that $|f(y_j)| > |g(y_j)| + j$ for every $j \in \mathbb{N}$. Since the set of quotients $(f(x) - f(y)/(g(x) - g(y))$, with $(x,y) \in Q$, is countable, there exists $\theta \in (0,1)$ such that $f(x) - f(y) \neq \theta(g(x) - g(y))$ for every $(x,y) \in Q$. If we set

$h = f - \theta g$ then $h \in \mathcal{H}(X)$, $h(x) \neq h(y)$ for every $(x,y) \in Q$ and $|h(y_j)| > j$ for every $j \in \mathbb{N}$. We shall prove that X is the domain of existence of h. Indeed, let (Y,η) be a Riemann domain over E, let $\tau : X \to Y$ be an $\{h\}$-extension of X, and let $\tilde{h} \in \mathcal{H}(Y)$ with $\tilde{h} \circ \tau = h$. To prove that τ is injective let $a, b \in X$ with $\tau(a) = \tau(b)$. Then for every $m \in \mathbb{N}_0$ and $t \in E$ we have that $P_t^m \tilde{h}(\tau(a)) = P_t^m \tilde{h}(\tau(b))$, and therefore $P_t^m h(a) = P_t^m h(b)$ by Exercise 47.I. If $a \neq b$ then we can find $r > 0$ such that $r < d_X(a)$, $r < d_X(b)$ and $B_X(a;r) \cap B_X(b;r) = \emptyset$. Then for each $t \in B_E(0;r)$ we have that

$$h \circ \xi_a^{-1}(\xi(a) + t) = \sum_{m=0}^{\infty} P_t^m h(b) = \sum_{m=0}^{\infty} P_t^m h(b) = h \circ \xi_b^{-1}(\xi(b) + t).$$

Since $\xi(a) = \xi(b)$, and since D is dense in E, there exists $t \in B_E(0;r)$ such that $\xi(a) + t = \xi(b) + t \in D$. If we set $x = \xi_a^{-1}(\xi(a) + t)$ and $y = \xi_b^{-1}(\xi(b) + t)$ then $(x,y) \in Q$ but $h(x) = h(y)$, a contradiction. Thus τ is injective, and whence it follows that $\tau : X \to \tau(X)$ is a homeomorphism. If $\tau(X) \neq Y$ then there is a point b in the boundary of $\tau(X)$ in Y. We shall prove that $\tilde{h}$ is unbounded on $B_Y(b;2r)$ whenever $0 < 2r < d_Y(b)$. Indeed, if $0 < 2r < d_Y(b)$ then we can find $a \in P$ such that $\tau(a) \in B_Y(b;r)$. Hence $d_X(a) < r$ and it follows that

$$\tau(B_X(a)) = B_Y(\tau(a);d_X(a)) \subset B_Y(\tau(a);r) \subset B_Y(b;2r).$$

By the definition of (x_j) there is a strictly increasing sequence (j_k) in $\mathbb{N}$ such that $x_{j_k} = a$ for every k. Hence $y_{j_k} \in B_X(a)$ and therefore $\tau(y_{j_k}) \in B_Y(b;2r)$ for every k. Since $|\tilde{h} \circ \tau(y_{j_k})| = |h(y_{j_k})| > j_k$ for every k, we conclude that $\tilde{h}$ is unbounded on $B_Y(b;2r)$, as asserted. This contradiction completes the proof of the theorem.

52.11. THEOREM. *Let (X,ξ) be a Riemann domain over $\mathbb{C}^n$. Then the following conditions are equivalent:*

(a) *X is a domain of existence.*

(b) *X is a domain of holomorphy.*

(c) *X is holomorphically separated and metrically holomorphically convex.*

PROOF. Without loss of generality we may assume that X is connected. Then there is an increasing sequence (V_j) of relatively compact open sets such that $X = \bigcup_{j=1}^{\infty} V_j$. If we define

$$U_j = \{x \in V_j : d_{V_j}(x) > 2^{-j}\}$$

for each $j \in \mathbb{N}$, then $\mathcal{U} = (U_j)_{j=1}^{\infty}$ is a regular cover of X and $\mathcal{H}(X) = \mathcal{H}^{\infty}(\mathcal{U})$. Then the theorem is a direct consequence of Theorem 52.10

EXERCISES

52.A. Let U be an open subset of E. Show that U is a domain of holomorphy (resp. a domain of existence) in the sense of Definition 52.1 if and only if U is a domain of holomorphy (resp. a domain of existence) in the sense of Definition 10.4 (resp. Definition 10.8).

52.B. Let (X,ξ) be a Riemann domain over E. Show that X is a domain of holomorphy (resp. a domain of existence) if and only if each connected component of X is a domain of holomorphy (resp. a domain of existence).

52.C. Let (X,ξ) be a Riemann domain over E. Show that X is holomorphically convex (resp. metrically holomorphically convex) if and only if each connected component of X is holomorphically convex (resp. metrically holomorphically convex).

52.D. Let (X,ξ) be a Riemann domain over E. Show that X is holomorphically separated if and only if each connected component of X is holomorphically separated.

52.E. Let (X,ξ) be a Riemann domain over E. Show that if X is holomorphically convex (resp. metrically holomorphically convex) then $\xi^{-1}(E_o)$ is holomorphically convex (resp. metrically holomorphically convex) for each closed vector subspace E_o of E.

52.F. Let (X,ξ) be a Riemann domain over E. Show that if X is holomorphically separated then $\xi^{-1}(E_o)$ is holomorphically separated for each closed vector subspace E_o of E.

53. THE LEVI PROBLEM IN FINITE DIMENSIONAL RIEMANN DOMAINS

In this section we show that each pseudoconvex Riemann domain over $\mathbb{C}^n$ is holomorphically convex and holomorphically separated, and therefore a domain of existence.

To begin with we observe that the results on holomorphic approximation from Section 44 admit straightforward extensions to the case of Riemann domains.

53.1. LEMMA. *Let (X,ξ) be a connected pseudoconvex Riemann domain over $\mathbb{C}^n$. Let $u \in C^\infty(X)$ be a strictly plurisubharmonic function such that the set $K_c = \{x \in X : u(x) \leq c\}$ is compact for each $c \in \mathbb{R}$. Then for each $f \in \mathcal{H}(K_o)$ there is a sequence (f_j) in $\mathcal{H}(X)$ such that $\int_{K_o} |f_j - f|^2 d\mu_X \to 0$.*

53.2. LEMMA. *Let (X,ξ) be a Riemann domain over $\mathbb{C}^n$. Then for each compact set K and each open set V with $K \subset V \subset X$ there is a constant $c > 0$ such that*

$$\sup_K |f| \leq c \int_V |f| d\mu_X \quad \text{for every} \quad f \in \mathcal{H}(X).$$

53.3. THEOREM. *Let (X,ξ) be a pseudoconvex Riemann domain over $\mathbb{C}^n$. Let K be a compact subset of X such that $\hat{K}_{Ps(X)} = K$. Then for each $f \in \mathcal{H}(K)$ there is a sequence (f_j) in $\mathcal{H}(X)$ which converges to f uniformly on a suitable neighborhood of K.*

53.4. COROLLARY. *Let (X,ξ) be a pseudoconvex Riemann domain over*

$\mathbb{C}^n$. *Let* U *be an open subset of* X *such that* $\hat{K}_{P\mathcal{S}(X)} \subset U$ *for each compact set* $K \subset U$. *Then* $\mathcal{H}(X)$ *is dense in* $(\mathcal{H}(U), \tau_c)$.

53.5. THEOREM. *Let* (X, ξ) *be a holomorphically convex Riemann domain over* $\mathbb{C}^n$, *and let* U *be an open subset of* X. *Then the following conditions are equivalent:*

(a) $\hat{K}_{\mathcal{H}(X)} \subset U$ *is compact for each compact set* $K \subset X$.

(b) $\hat{K}_{\mathcal{H}(X)} \subset U$ *for each compact set* $K \subset U$.

(c) U *is holomorphically convex and* $\mathcal{H}(X)$ *is dense in* $(\mathcal{H}(U), \tau_c)$.

We cannot yet extend Theorem 44.6 to the case of Riemann domains, for we first have to prove that every pseudoconvex Riemann domain over $\mathbb{C}^n$ is holomorphically convex. To prove this result we need the following lemma.

53.6. LEMMA. *Let* (X, ξ) *be a Riemann domain over* $\mathbb{C}^n$, *let* $u \in C^\infty(X)$ *be a strictly plurisubharmonic function on* X *and let* $a \in X$. *Then there are an open neighborhood* V *of* a *and a function* $v \in \mathcal{H}(V)$ *such that* $v(a) = 0$ *and* $\operatorname{Re} v(z) < u(z) - u(a)$ *for every* $z \in V \setminus \{a\}$.

PROOF. Let U be a chart in X containing a. If $z_1, \ldots, z_n$ denote the complex coordinates of $\xi(z)$ for each $z \in U$ then the Taylor expansion of u can be written in the form

$$u(z) - u(a) = \sum_j \frac{\partial u}{\partial z_j}(a)(z_j - a_j) + \sum_j \frac{\partial u}{\partial \bar{z}_j}(a)(\bar{z}_j - \bar{a}_j)$$

$$+ \frac{1}{2} \sum_{j,k} \frac{\partial^2 u}{\partial z_j \partial z_k}(a)(z_j - a_j)(z_k - a_k)$$

$$+ \frac{1}{2} \sum_{j,k} \frac{\partial^2 u}{\partial \bar{z}_j \partial \bar{z}_k}(a)(\bar{z}_j - \bar{a}_j)(\bar{z}_k - \bar{a}_k)$$

$$+ \sum_{j,k} \frac{\partial^2 u}{\partial z_j \partial \bar{z}_k}(a)(z_j - a_j)(\bar{z}_k - \bar{a}_k) + \varphi(z),$$

where $\lim_{z\to a} (\varphi(z)/\|\xi(z) - \xi(a)\|^2) = 0$. If we set

$$v(z) = 2\sum_j \frac{\partial u}{\partial z_j}(a)(z_j - a_j) + \sum_{j,k} \frac{\partial^2 u}{\partial z_j \partial z_k}(a)(z_j - a_j)(z_k - a_k)$$

then $v \in \mathcal{H}(U)$ and

$$u(z) - u(a) = Re\, v(z) + \sum_{j,k} \frac{\partial^2 u}{\partial z_j \partial \bar{z}_k}(a)(z_j - a_j)(\bar{z}_k - \bar{a}_k) + \varphi(z).$$

Since the Hermitian form $\sum_{j,k} \frac{\partial^2 u}{\partial z_j \partial \bar{z}_k}(a) t_j \bar{t}_k$ is strictly positive, there is a neighborhood V of a such that $u(z) - u(z) > Re\, v(z)$ for every $z \in V \setminus \{a\}$.

53.7. THEOREM. *Each pseudoconvex Riemann domain over* $\mathbb{C}^n$ *is holomorphically convex and holomorphically separated.*

PROOF. Let (X, ξ) be a pseudoconvex Riemann domain over $\mathbb{C}^n$. Without loss of generality we may assume that X is connected. By Theorem 50.6 there is a strictly plurisubharmonic function $u \in C^\infty(X)$ such that the set $K_r = \{x \in X : u(x) \leq r\}$ is compact for each $r \in \mathbb{R}$. To show that X is holomorphically convex it suffices to prove that $(\hat{K}_r)_{\mathcal{H}(X)} = K_r$ for each $r \in \mathbb{R}$. Now, fix $r \in \mathbb{R}$ and let $a \in X \setminus K_r$. By Lemma 53.6 there are an open neighborhood V of a and a function $v \in \mathcal{H}(V)$ such that $v(a) = 0$ and

(53.1) $\quad Re\, v(x) < u(x) - u(a) \quad$ for every $\quad x \in V \setminus \{a\}$.

Choose $\tau \in \mathcal{D}(V)$ such that $\tau \equiv 1$ on a neighborhood of a. Since $supp\, \bar{\partial}\tau$ is a compact subset of $V \setminus \{a\}$ there exists $\varepsilon > 0$ such that

$$Re\, v(x) < u(x) - u(a) - 2\varepsilon \quad \text{for every} \quad x \in supp\, \bar{\partial}\tau .$$

We shall set $U_t = \{x \in X : u(x) < t\}$ for each $t \in \mathbb{R}$, and if $t > u(a)$ then we shall denote by $\tilde{U}_t$ the union of those connected components of U_t which intersect $K_r \cup \{a\}$. Note that if $t > u(a)$ then $\tilde{U}_t$ is a pseudoconvex Riemann domain with finitely many connected components. Set $s = u(a) + \varepsilon$. Then

(53.2) $\quad Re\, v(x) < -\varepsilon \quad$ for every $\quad x \in U_s \cap supp\, \bar{\partial}\tau.$

For each $k \in \mathbb{N}$ set

(53.3) $$w_k = e^{kv}\tau - f_k,$$

where $f_k \in C^\infty(\tilde{U}_s)$ will be chosen so that $w_k \in \mathcal{H}(\tilde{U}_s)$. The equation $\bar{\partial} w_k = 0$ is equivalent to the equation $\bar{\partial} f_k = g_k$, where

(53.4) $$g_k = e^{kv}\bar{\partial}\tau.$$

By Theorem 51.1 there are weight functions $\alpha, \beta, \gamma \in C^\infty(\tilde{U}_s; \mathbb{R})$ such that $\|g\|_\beta^2 \le \|A^*g\|_\alpha^2 + \|Bg\|_\gamma^2$ for every $g \in D_{A^*} \cap D_B$. Since each g_k belongs to $\mathcal{D}(V \setminus \{a\})$ it follows from Theorems 51.1 and 51.2 that we can find a sequence (f_k) in $C^\infty(\tilde{U}_s)$ such that $\bar{\partial} f_k = g_k$ and $\|f_k\|_\alpha \le \|g_k\|_\beta$ for every $k \in \mathbb{N}$. Using (53.4) and (53.2) we can then find a constant $c > 0$ such that

(53.5) $$\|f_k\|_\alpha \le \|g_k\|_\beta \le ce^{-k\varepsilon}$$

for every $k \in \mathbb{N}$. Since each f_k is holomorphic on a neighborhood of a, an application of Lemma 53.2 shows that $\lim_{k\to\infty} f_k(a) = 0$, and then a glance at (53.3) shows that

(53.6) $$\lim_{k\to\infty} w_k(a) = 1.$$

On the other hand, it follows from (53.3), (53.1) and (53.5) that

$$\lim_{k\to\infty}\int_{\tilde{U}_t} |w_k|^2 d\mu_X = 0$$

for each $t < u(a)$, and then another application of Lemma 53.2 shows that

(53.7) $$\lim_{k\to\infty} \sup_{K_r} |w_k| = 0.$$

By (53.6) and (53.7) we can find $k \in \mathbb{N}$ such that $\sup_{K_r} |w_k| < |w_k(a)|$. Since $w_k \in \mathcal{H}(\tilde{U}_s)$ an application of Corollary 53.4 yields a function $\varphi \in \mathcal{H}(X)$ such that $\sup_{K_r} |\varphi| < |\varphi(a)|$. Thus $(\hat{K}_r)_{\mathcal{H}(X)} = K_r$, as asserted.

To prove that X is holomorphically separated, let a, b be two distinct points of X. Without loss of generality we may assume that $u(b) \leq u(a)$. If V is the neighborhood of a from the first part of the proof, then we may assume that $b \notin V$. We still set $s = u(a) + \varepsilon$ and $U_s = \{x \in X : u(x) < s\}$, but we now denote by $\tilde{U}_s$ the union of those connected components of U_s which contain a or b. Then the first part of the proof still applies, and in particular it follows from (53.3), (53.5) and Lemma 53.3 that

$$\lim_{k\to\infty} w_k(b) = - \lim_{k\to\infty} f_k(b) = 0.$$

Since $\lim_{k\to\infty} w_k(a) = 1$ there exists $k \in \mathbb{N}$ such that $w_k(a) \neq w_k(b)$. Then another application of Corollary 53.4 yields a function $\psi \in \mathcal{H}(X)$ such that $\psi(a) \neq \psi(b)$. The proof of the theorem is now complete.

Theorems 52.6, 52.11 and 53.7 can be summarized as follows.

53.8. THEOREM. *Let (X, ξ) be a Riemann domain ove $\mathbb{C}^n$. Then the following conditions are equivalent*

(a) *X is a domain of existence.*

(b) *X is a domain of holomorphy.*

(c) *X is holomorphically convex.*

(d) *X is metrically holomorphically convex.*

(e) *X is pseudoconvex.*

Now we can repeat the proof of Theorem 44.6 to prove the following theorem.

53.9. THEOREM. *If* (X,ξ) *is a pseudoconvex Riemann domain over* $\mathbb{C}^n$ *then* $\hat{K}_{Ps(X)} = \hat{K}_{\mathcal{H}(X)}$ *for each compact set* $K \subset X$.

For later use we give the following separation theorem.

53.10. THEOREM. *Let* (X,ξ) *be a connected, holomorphically separated Riemann domain over* $\mathbb{C}^n$. *Then there is an injective mapping* $f \in \mathcal{H}(X;\mathbb{C}^N)$ *for a suitable* $N \in \mathbb{N}$.

To prove this theorem we need three auxiliary lemmas.

53.11. LEMMA. *Let* (X,ξ) *be a holomorphically separated Riemann domain over* $\mathbb{C}^n$, *and let* K *be a compact subset of* X. *Then there is a mapping* $f \in \mathcal{H}(X;\mathbb{C}^N)$ *which is injective on* K.

PROOF. Each $a \in K$ has an open neighborhood U_a such that $\xi \mid U_a$ is injective. Thus we can find open sets $U_1,\dots,U_p$ in X such that $K \subset U_1 \cap \dots \cap U_p$ and $\xi \mid U_j$ is injective for $j = 1,\dots,p$. If we set

$$U = (U_1 \times U_1) \cup \dots \cup (U_p \times U_p)$$

then the diagonal of $K \times K$ is contained in U and $\xi(x) \neq \xi(y)$ for every $(x,y) \in U$ with $x \neq y$. On the other hand, for each $(a,b) \in (K \times K) \setminus U$ there exists $\varphi \in \mathcal{H}(X)$ such that $\varphi(a) \neq \varphi(b)$, and hence $\varphi(x) \neq \varphi(y)$ for all (x,y) in a suitable neighborhood of (a,b). Thus we can find open sets $V_1,\dots,V_q$ in $X \times X$ and functions $\varphi_1,\dots,\varphi_q \in \mathcal{H}(X)$ such that

$$(K \times K) \setminus U \subset V_1 \cup \dots \cup V_q$$

and $\varphi_j(x) \neq \varphi_j(y)$ for all $(x,y) \in V_j$ and $j = 1,\dots,q$. Since $K \times K \subset U \cup V_1 \cup \dots \cup V_q$ it is clear that the mapping $f = (\xi,\varphi_1,\dots,\varphi_q) \in \mathcal{H}(X;\mathbb{C}^{n+q})$ is injective on K.

53.12. LEMMA. *Let (X,ξ) be a Riemann domain over $\mathbb{C}^n$, let K be a compact subset of X, and let $f \in \mathcal{H}(X;\mathbb{C}^N)$. If $N > n$ then the compact set $f(K)$ has Lebesgue measure zero in $\mathbb{C}^n$.*

PROOF. Without loss of generality we may assume that K is contained in a chart U in X. Let $0 < \delta < d_U(X)$. If we set $f = (f_1,\dots,f_N)$ then it follows from the Mean Value Theorem that there is a constant $c > 0$ such that

$$(53.8) \qquad |f_j(x+t) - f_j(x)| \leq c\,\|t\|$$

for all $x \in K$, $t \in B(0;\delta)$ and $j = 1,\dots,N$. Let $0 < \varepsilon < \delta$, and let C be a subset of X which intersects K and which is homeomorphic under ξ to a cube in $\mathbb{C}^n$ with side ε. Then $C \subset U$ and it follows from (53.8) and Proposition 48.3 that

$$(53.9) \qquad \lambda(f(C)) \leq c\varepsilon^{2N} = c\mu_X(C)\varepsilon^{2N-2n}.$$

There are finitely many such "cubes" $C_1,\dots,C_p$ in U such that $K \subset C_1 \cup \dots \cup C_p$ and $\sum_{j=1}^{p} \mu_X(C_j) < \mu_X(K) + 1$. Then it follows from (53.9) that $\lambda(f(K)) \leq c(\mu_X(K) + 1)\varepsilon^{2N-2n}$. Since ε can be taken arbitrarily small we conclude that $\lambda(f(K)) = 0$, as we wanted.

53.13. LEMMA. *Let (X,ξ) be a connected Riemann domain over $\mathbb{C}^n$, let K be a compact subset of X, and let $(f_1,\dots,f_{N+1}) \in \mathcal{H}(X;\mathbb{C}^{N+1})$ be a mapping which is injective on K. If $N > 2n + 1$ then for each $\varepsilon > 0$ we can find $(\zeta_1,\dots,\zeta_N) \in \mathbb{C}^N$, with $|\zeta_j| < \varepsilon$ for $j = 1,\dots,N$, and such that the mapping*

$$(f_1 - \zeta_1 f_{N+1}, \dots, f_N - \zeta_N f_{N+1}) \in \mathcal{H}(X;\mathbb{C}^N)$$

is injective on K.

PROOF. Let $g \in \mathcal{H}(X^2 \times \mathbb{C}) \to \mathbb{C}^N$ be defined by

$$g(x,y,\lambda) = \lambda(f_1(x) - f_1(y),\dots,f_N(x) - f_N(y)).$$

Since $X^2 \times \mathbb{C}$ is hemicompact, it follows from Lemma 53.12 that the set $g(X^2 \times \mathbb{C})$ has Lebesgue measure zero in $\mathbb{C}^N$. Thus to complete the proof of the lemma it suffices to prove that the mapping

$$(f_1 - \zeta_1 f_{N+1}, \ldots, f_N - \zeta_N f_{N+1}) \in \mathcal{H}(X;\mathbb{C}^N)$$

is injective on K for each $\zeta = (\zeta_1,\ldots,\zeta_N) \in \mathbb{C}^N \setminus g(X^2 \times \mathbb{C})$. Now, let x,y be two distinct points of K. If $f_{N+1}(x) = f_{N+1}(y)$ then there exists j with $1 \leq j \leq N$ such that $f_j(x) \neq f_j(y)$, and therefore $f_j(x) - \zeta_j f_{N+1}(x) \neq f_j(y) - \zeta_j f_{N+1}(y)$. If $f_{N+1}(x) \neq f_{N+1}(y)$, then since $\zeta \notin g(X^2 \times \mathbb{C})$, there exists j with $1 \leq j \leq N$ such that

$$\zeta_j \neq (f_{N+1}(x) - f_{N+1}(y))^{-1}(f_j(x) - f_j(y)),$$

and therefore $f_j(x) - \zeta_j f_{N+1}(x) \neq f_j(y) - \zeta_j f_{N+1}(y)$. This completes the proof.

PROOF OF THEOREM 53.10. Let $(K_m)_{m=1}^{\infty}$ be an increasing sequence of compact subsets of X such that each compact subset of X is contained in some K_m. Let F be the set of all $f \in \mathcal{H}(X;\mathbb{C}^{2n+2})$ such that the mapping $(\xi,f) \in \mathcal{H}(X;\mathbb{C}^{3n+2})$ is not injective. Likewise, for each $m \in \mathbb{N}$ let F_m be the set of all $f \in \mathcal{H}(X;\mathbb{C}^{2n+2})$ such that the mapping $(\xi,f) \in \mathcal{H}(X;\mathbb{C}^{3n+2})$ is not injective on K_m. Certainly $F = \bigcup_{m=1}^{\infty} F_m$. Since $\mathcal{H}(X;\mathbb{C}^{2n+2})$ is a Fréchet space for the compact-open topology, to prove the theorem it is sufficient to show that each F_m is nowhere dense. We first show that F_m is closed in $\mathcal{H}(X;\mathbb{C}^{2n+2})$. Indeed, let (f_j) be a sequence in F_m which converges to a mapping $f \in \mathcal{H}(X;\mathbb{C}^{2n+2})$ for the compact-open topology. For each $j \in \mathbb{N}$ we can find $x_j \neq y_j$ in K_m such that $\xi(x_j) = \xi(y_j)$ and $f_j(x_j) = f_j(y_j)$. Without loss of generality we may assume that (x_j) converges to a point x, whereas (y_j) converges to a point y.

Then clearly $\xi(x) = \xi(y)$, and it follows from the inequality

$$|f(x) - f(y)| \leq |f(x) - f(x_j)| + |f(x_j) - f_j(x_j)|$$
$$+ |f_j(y_j) - f(y_j)| + |f(y_j) - f(y)|$$

that $f(x) = f(y)$. We claim that $x \neq y$. Otherwise let U be a chart in X containing $x = y$. Then $x_j \in U$ and $y_j \in U$ for sufficiently large j. But this is impossible, for $x_j \neq y_j$ and $\xi(x_j) = \xi(y_j)$. Thus $x \neq y$ and $f \in F_m$. Thus F_m is closed in $\mathcal{H}(X;\mathbb{C}^{2n+2})$, and to complete the proof we show that F_m has empty interior. Indeed, let $f = (f_1,\ldots,f_{2n+2}) \in F_m$. By Lemma 53.11 there exists a mapping $g = (g_1,\ldots,g_p) \in \mathcal{H}(X;\mathbb{C}^p)$ which is injective on K_m. The clearly the mapping $(f,g) \in \mathcal{H}(X;\mathbb{C}^{2n+2+p})$ is also injective on K_m. Then by repeated applications of Lemma 53.13 we can for each $\varepsilon > 0$ find complex numbers ζ_{jk} with $|\zeta_{jk}| < \varepsilon$ such that the mapping

$$h_\varepsilon = (f_1 - \sum_{k=1}^{p} \zeta_{1k} g_k, \ldots, f_{2n+2} - \sum_{k=1}^{p} \zeta_{2n+2,k} g_k) \in \mathcal{H}(X;\mathbb{C}^{2n+2})$$

is injective on K_m. Obviously $(\xi, h_\varepsilon) \in \mathcal{H}(X;\mathbb{C}^{3n+2})$ is also injective on K_m. Thus $h_\varepsilon \notin F_m$ and since (h_ε) converges to f when $\varepsilon \to 0$ we conclude that F_m has empty interior. This completes the proof of the theorem.

54. THE LEVI PROBLEM IN INFINITE DIMENSIONAL RIEMANN DOMAINS

In this section we prove that each pseudoconvex Riemann domain over a separable Banach space with the bounded approximation property is a domain of existence. We also prove that such domains are holomorphically convex.

54.1. DEFINITION. Let (X,ξ) be a pseudoconvex Riemann domain over E. An open set $U \subset X$ is said to be *X-pseudoconvex* if $\hat{K}_{Ps(X)} \subset U$ for each compact set $K \subset U$.

54.2. EXAMPLES. Let (X,ξ) be a pseudoconvex Riemann domain over E.

(a) If V is a convex open set in E then the open set $U = \xi^{-1}(V)$ is X-pseudoconvex.

(b) If $f \in Ps(X)$ then the open set $U = \{x \in X : f(x) < 0\}$ is X-pseudoconvex.

(c) If $(U_i)_{i\in I}$ is a family of X-pseudoconvex open sets then the open set $U = int \bigcap_{i\in I} U_i$ is X-pseudoconvex as well.

54.3. LEMMA. *Let (X,ξ) be a pseudoconvex Riemann domain over E, and let U be an X-pseudoconvex open set. Then:*

(a) *U is pseudoconvex.*

(b) *$\mathcal{H}(\xi^{-1}(E_o))$ is dense in $(\mathcal{H}(U \cap \xi^{-1}(E_o)), \tau_c)$ for each finite dimensional subspace E_o of E.*

The proofs of these statements are left as exercises to the reader.

If E is a Banach space with a Schauder basis $(e_n)_{n=1}^{\infty}$ then $(z_n)_{n=1}^{\infty}$ will denote the sequence of coordinate functionals, E_n will denote the subspace generated by $e_1,\dots,e_n$ and T_n will denote the canonical projection from E onto E_n. If (X,ξ) is a Riemann domain over E then we shall set $X_n = \xi^{-1}(E_n)$. Then X_n is a Riemann domain over E_n.

54.4. LEMMA. *Let E be a Banach space with a monotone Schauder basis $(e_n)_{n=1}^{\infty}$. Let (X,ξ) be a pseudoconvex Riemann domain over E. For each $j \in \mathbb{N}$ let*

$$A_j = \{x \in X : \sup_{n \geq j} \| T_n \circ \xi(x) - \xi(x) \| < d_X(x)\},$$

and let $\tau_j \in \mathcal{H}(A_j; X_j)$ be defined by

$$\tau_j(x) = \xi_x^{-1} \circ T_j \circ \xi(x).$$

Then:

(a) $(A_j)_{j=1}^{\infty}$ *is an increasing sequence of open sets which cover* X. *Each* A_j *is* X-*pseudoconvex and contains* X_j. $\mathcal{H}(X_n)$ *is dense in* $(\mathcal{H}(A_j \cap X_n), \tau_c)$ *for each* $j, n \in \mathbb{N}$.

(b) $\xi \circ \tau_j = T_j \circ \xi$ *on* A_j, τ_j = *identity on* X_j *and* $\tau_j \circ \tau_{j+1} = \tau_{j+1} \circ \tau_j = \tau_j$ *on* A_j *for every* $j \in \mathbb{N}$.

(c) *For each compact set* $K \subset X$ *and* $0 < \varepsilon < d_X(K)$ *there exists* $j \in \mathbb{N}$ *such that* $K \subset A_j$ *and* $\tau_n(x) \in B_X(x;\varepsilon)$ *for every* $x \in K$ *and* $n \geq j$.

PROOF. Since the function $w_j(t) = \sup_{n \geq j} \| T_n t - t \|$ is continuous on E for each $j \in \mathbb{N}$, all the assertions in the lemma can be readily verified.

54.5. LEMMA. *Let* E *be a Banach space with a monotone Schauder basis* $(e_n)_{n=1}^{\infty}$. *Let* (X,ξ) *be a connected pseudoconvex Riemann domain over* E. *Then there are two increasing sequences of open sets* $(B_j)_{j=1}^{\infty}$ *and* $(C_j)_{j=1}^{\infty}$ *such that:*

(a) $C_j \subset B_j \subset A_j$ *for every* $j \in \mathbb{N}$. $X = \bigcup_{j=1}^{\infty} B_j = \bigcup_{j=1}^{\infty} C_j$.

(b) $d_{A_j}(B_j) \geq 2^{-j}$ *and* $B_j \cap X_n$ *is relatively compact in* $A_j \cap X_n$ *for every* $j, n \in \mathbb{N}$.

(c) $d_{C_{j+1}}(C_j) \geq 2^{-j-1}$ *and* $\tau_n(C_j) \subset B_j \cap X_n$ *for every* $j \in \mathbb{N}$ *and every* $n \geq j$.

PROOF. Consider the open sets

$$P_j = \{x \in A_j : d_X(x) > 2^{-j}\},$$

$$Q_j = \{x \in P_j : \sup_{n \geq j} \| T_n \circ \xi(x) - \xi(x) \| < d_{P_j}(x)\}.$$

Then each of the sequences $(P_j)_{j=1}^{\infty}$ and $(Q_j)_{j=1}^{\infty}$ is increasing and covers X. We also have that $\tau_n(Q_j) \subset P_j \cap X_n$ whenever $n \geq j$ and in particular

(54.1) $\quad \tau_j(Q_j) \subset P_j \cap X_j = Q_j \cap X_j \quad$ for every $\quad j \in \mathbb{N}$.

Without loss of generality we may assume that $Q_1 \cap X_1$ is nonvoid. Fix a point $x_o \in Q_1 \cap X_1$, let Q_j' be the connected component of Q_j which contains the point x_o, and define

$$B_j = \{x \in Q_j' : \rho_{Q_j'}(x_o, x) < j\},$$

where $\rho_{Q_j'}$ denotes the geodesic distance in Q_j'. Since the geodesic distance is a continuous function, each B_j is open. And since X is pathwise connected, it follows that $X = \bigcup_{j=1}^{\infty} Q_j' = \bigcup_{j=1}^{\infty} B_j$. If L is a polygonal line in Q_j' joining x_o to a point $x \in Q_j' \cap X_j$ then it follows from (54.1) that $\tau_j(L)$ is a polygonal line in $Q_j' \cap X_j$ joining x_o to x. In particular $Q_j' \cap X_j$ is connected. We claim that

(54.2) $\quad \rho_{Q_j'}(x_o, x) = \rho_{Q_j' \cap X_j}(x_o, x) \quad$ for every $\quad x \in Q_j' \cap X_j$.

Indeed, the inequality $\rho_{Q_j'}(x_o, x) \leq \rho_{Q_j' \cap X_j}(x_o, x)$ is obvious. To show the opposite inequality let $\varepsilon > 0$ be given and let $L = [x_o, \ldots, x_m]$ be a polygonal line in Q_j' joining x_o to x and such that

$$\sum_{k=1}^{m} \|\xi(x_k) - \xi(x_{k-1})\| \leq \rho_{Q_j'}(x_o, x) + \varepsilon.$$

Then $\tau_j(L)$ is a polygonal line in $Q_j' \cap X_j$ joining x_o to x and

$$\rho_{Q_j' \cap X_j}(x_o, x) \leq \sum_{k=1}^{m} \|\xi \circ \tau_j(x_k) - \xi \circ \tau_j(x_{k-1})\|$$

$$= \sum_{k=1}^{m} \|T_j \circ \xi(x_k) - T_j \circ \xi(x_{k-1})\| \leq \sum_{k=1}^{m} \|\xi(x_k) - \xi(x_{k-1})\| \leq \rho_{Q_j'}(x_o, x) + \varepsilon.$$

Since $\varepsilon > 0$ was arbitrary, (54.2) follows. From (54.2) and Lemma 50.1 it follows that the set

$$B_j \cap X_j = \{x \in Q_j' \cap X_j : \rho_{Q_j' \cap X_j}(x_o, x) < j\}$$

is relatively compact in $A_j \cap X_j$ for each $j \in \mathbb{N}$. Whence it follows that $B_j \cap X_n$ is relatively compact in $A_j \cap E_n$ for all $j, n \in \mathbb{N}$. If we define

$$R_j = \{x \in B_j : \sup_{n \geq j} \| T_n \circ \xi(x) - \xi(x) \| < d_{B_j}(x)\},$$

$$C_j = \{x \in R_j : d_{R_j}(x) > 2^{-j}\},$$

then the remaining assertions in the lemma follow at once.

Observe that the sequence $C = (C_j)_{j=1}^{\infty}$ constructed in Lemma 54.5 is a regular cover of X.

54.6. LEMMA. *Let E be a Banach space with a monotone Schauder basis $(e_n)_{n=1}^{\infty}$. Let (X, ξ) be a connected pseudoconvex Riemann domain over E. Then for each $f_n \in \mathcal{H}(X_n)$ and $\varepsilon > 0$ there exists $f \in \mathcal{H}^{\infty}(C)$ such that $f = f_n$ on X_n and $|f - f_n \circ \tau_n| \leq \varepsilon$ on C_n.*

PROOF. Consider the sets

$$P = X_n \cup (B_n \cap X_{n+1})\hat{}_{\mathcal{Ps}(X_{n+1})}, \quad Q = X_{n+1} \setminus A_n .$$

If follows from Lemma 54.4 and 54.5 that $(B_n \cap X_{n+1})\hat{}_{\mathcal{Ps}(X_{n+1})}$ is a compact subset of $A_n \cap X_{n+1}$. Whence it follows that P and Q are two disjoint closed subsets of X_{n+1}. Choose $\varphi \in C^{\infty}(X_{n+1})$ such that $\varphi \equiv 1$ on a neighborhood of P and $\varphi \equiv 0$ on a neighborhood of Q. Set

$$g = \varphi(f_n \circ \tau_n) - (z_{n+1} \circ \xi)u,$$

where $u \in C^{\infty}(X_{n+1})$ will be chosen so that $g \in \mathcal{H}(X_{n+1})$.

The equation $\bar{\partial} g = 0$ is equivalent to the equation $\bar{\partial} u = v$, where $v \in C^{\infty}_{01}(X_{n+1})$ is defined by

$$v = \frac{f_n \circ \tau_n}{z_{n+1} \circ \xi} \bar{\partial}\varphi .$$

Since $\bar{\partial} v = 0$ on X_{n+1} then equation $\bar{\partial} u = v$ has a solution $u \in C^{\infty}(X_{n+1})$ by Theorem 51.2. Thus $g \in \mathcal{H}(X_{n+1})$ and it is clear that $g = f_n$ on X_n. Since $\bar{\partial} u = v = 0$ on a neighborhood of the compact set $(B_n \cap X_{n+1})\hat{}_{P\mathcal{S}(X_{n+1})}$, an application of Theorem 53.3 yields a function $h \in \mathcal{H}(X_{n+1})$ such that

$$|h - u| \leq \varepsilon/2M \quad \text{on} \quad B_n \cap X_{n+1},$$

where $M = \sup_{B_n \cap X_{n+1}} |z_{n+1} \circ \xi|$. If we define

$$f_{n+1} = g + (z_{n+1} \circ \xi)h = \varphi(f_n \circ \tau_n) + (z_{n+1} \circ \xi)(h - u),$$

then $f_{n+1} = g = f_n$ on X_n and $|f_{n+1} - f_n \circ \tau_n| \leq \varepsilon / 2$ on $B_n \cap X_{n+1}$. Since $\tau_{n+1}(C_n) \subset B_n \cap X_{n+1}$ we obtain that

$$|f_{n+1} \circ \tau_{n+1} - f_n \circ \tau_n| \leq \varepsilon/2 \quad \text{on} \quad C_n .$$

Proceeding inductively we obtain a sequence $(f_j)_{j=n}^{\infty}$ such that $f_j \in \mathcal{H}(X_j)$, $f_{j+1} = f_j$ on X_j and

$$|f_{j+1} \circ \tau_{j+1} - f_j \circ \tau_j| \leq \varepsilon \cdot 2^{n-j-1} \quad \text{on} \quad C_j$$

for every $j \geq n$. If follows that

$$|f_k \circ \tau_k - f_j \circ \tau_j| \leq \varepsilon \cdot 2^{n-j} \quad \text{on} \quad C_j$$

for every $k \geq j \geq n$, and hence the sequence $(f_j \circ \tau_j)$ converges uniformly on each C_i to a function $f \in \mathcal{H}(X)$. Clearly $f = f_j$ on X_j and

$$|f - f_j \circ \tau_j| \leq \varepsilon \cdot 2^{n-j} \quad \text{on} \quad C_j$$

for every $j \geq n$. From this estimate we also conclude that f is bounded on each C_j, for $\tau_j(C_j) \subset B_j \cap X_j$ and this set is relatively compact in X_j.

Now we are ready to extend Theorem 45.7 to the case of Riemann domains.

54.7. THEOREM. *Let E be a separable Banach space with the bounded approximation property. Then each pseudoconvex Riemann domain over E is a domain of existence.*

PROOF. (a) We first assume that E has a monotone Schauder basis. Let (X,ξ) be a pseudoconvex Riemann domain over E. Without loss of generality we may assume that X is connected. We first show that X is $\mathcal{H}^\infty(C)$-separated. Indeed, let x,y be two distinct points in X. First choose $j \in \mathbb{N}$ such that $x,y \in C_j$ and next choose $n \geq j$ such that $\tau_n(x) \neq \tau_n(y)$ and $\tau_n(x)$ and $\tau_n(y)$ both lie in C_j. By Theorem 53.7 X_n is holomorphically separated and hence there exists $f_n \in \mathcal{H}(X_n)$ such that $|f_n \circ \tau_n(x) - f_n \circ \tau_n(y)| = 3\varepsilon > 0$. By Lemma 54.6 there exists $f \in \mathcal{H}^\infty(C)$ such that $|f - f_n \circ \tau_n| \leq \varepsilon$ on C_n. Hence it follows that $|f(x) - f(y)| \geq \varepsilon$ and X is $\mathcal{H}^\infty(C)$-separated, as asserted.

Next we shall prove that

$$(54.3) \qquad (\hat{C}_j)_{\mathcal{H}^\infty(C)} \subset (\hat{B}_j)_{Psc(X)} \quad \text{for every} \quad j \in \mathbb{N},$$

and therefore $d_X((\hat{C}_j)_{\mathcal{H}^\infty(C)}) \geq 2^{-j}$ for every $j \in \mathbb{N}$. To prove this let $x \in (\hat{C}_j)_{\mathcal{H}^\infty(C)}$ and choose $k \geq j$ such that $x \in C_k$. We claim that

$$(54.4) \qquad \tau_n(x) \in (B_j \cap X_n)\hat{}_{\mathcal{H}(X_n)} \quad \text{for every} \quad n \geq k.$$

Indeed, given $f_n \in \mathcal{H}(X_n)$ and $\varepsilon > 0$, by Lemma 54.6 we can find $f \in \mathcal{H}^\infty(C)$ such that $|f - f_n \circ \tau_n| \leq \varepsilon$ on C_n. Whence it follows that

$$|f_n \circ \tau_n(x)| \le |f(x)| + \varepsilon \le \sup_{C_j}|f| + \varepsilon$$

$$\le \sup_{C_j}|f_n \circ \tau_n| + 2\varepsilon \le \sup_{B_j \cap X_n} |f_n| + 2\varepsilon,$$

and since $\varepsilon > 0$ was arbitrary, (54.4) follows. Then, using Theorem 53.9 we obtain that

$$\tau_n(x) \in (B_j \cap X_n)\hat{}_{\mathcal{H}(X_n)} = (B_j \cap X_n)\hat{}_{Psc(X_n)} \subset (\hat{B}_j)_{Psc(X)}$$

for every $n \ge k$. Letting $n \to \infty$ we get that $x \in (\hat{B}_j)_{Psc(X)}$ and (54.3) has been proved.

We have shown that X is $\mathcal{H}^\infty(C)$-separated and $d_X((\hat{C}_j)_{\mathcal{H}^\infty(C)}) > 0$ for every $j \in \mathbb{N}$. Thus X is a domain of existence by Theorem 52.10.

(b) If E is a separable Banach space with the bounded approximation property then by Theorem 27.4 we may assume that E is a complemented subspace of a Banach space G which has a monotone Schauder basis. Thus we may assume that $G = E \times F$ for a suitable Banach space F. Then $(X \times F, (\xi, id))$ is a pseudoconvex Riemann domain over G by Exercise 49.D. Hence $X \times F$ is a domain of existence by part (a). Hence X is a domain of existence by Exercise 54.B.

54.8. **LEMMA.** *Let (X,ξ) be a Riemann domain over E. Let $\mathcal{F}$ be a subalgebra of $\mathcal{H}(X)$ such that $P_t^m f \in \mathcal{F}$ for every $f \in \mathcal{F}$, $m \in \mathbb{N}$ and $t \in E$. Let A be a subset of X such that $d_X(\hat{A}_{\mathcal{F}}) = r > 0$. Then*

$$\hat{A}_{\mathcal{F}} + B \subset (A + B)\hat{}_{\mathcal{F}}$$

for each balanced set $B \subset B_E(0;r)$.

PROOF. Let $y \in \hat{A}_{\mathcal{F}}$, $t \in B$ and $0 < \theta < 1$. Then for each $f \in \mathcal{F}$ we have that

$$|f(y + \theta t)| \le \sum_{m=0}^{\infty} \theta^m |P_t^m f(y)| \le \sum_{m=0}^{\infty} \theta^m \sup_{x \in A} |P_t^m f(x)| \le (1 - \theta)^{-1} \sup_{A+B} |f|.$$

After applying the preceding inequality to f^n, taking nth root and letting $n \to \infty$ we get that

$$|f(y + \theta t)| \leq \sup_{A+B} |f|.$$

Whence it follows that $y + \theta t \in (A + B)\hat{}_F$. Letting $\theta \to 1$ we get the desired conclusion.

54.9. LEMMA. *Let E be a Banach space with a monotone Schauder basis $(e_n)_{n=1}^{\infty}$. Let (X, ξ) be a connected pseudoconvex Riemann domain over E. Then:*

(a) *The set $(\hat{C}_j)_{\mathcal{H}^\infty(C)} \cap X_n$ is compact for all $j, n \in \mathbb{N}$.*

(b) *The set $(\hat{C}_j)_{\mathcal{H}^\infty(C)} \cap \xi^{-1}(K)$ is compact for each $j \in \mathbb{N}$ and each compact set $K \subset E$.*

PROOF. To prove (a) it certainly suffices to prove that

$$(54.5) \qquad (\hat{C}_j)_{\mathcal{H}^\infty(C)} \cap X_n \subset (B_j \cap X_n)\hat{}_{\mathcal{H}(X_n)} \quad \text{for every } n \geq j.$$

Let $x \in (\hat{C}_j)_{\mathcal{H}^\infty(C)} \cap X_n$ and $f_n \in \mathcal{H}(X_n)$. By Lemma 54.6 for each $\varepsilon > 0$ there exists $f \in \mathcal{H}^\infty(C)$ such that $f = f_n$ on X_n and $|f - f_n \circ \tau_n| \leq \varepsilon$ on C_n. Hence

$$|f_n(x)| = |f(x)| \leq \sup_{C_j}|f| \leq \sup_{C_j}|f_n \circ \tau_n| + \varepsilon \leq \sup_{B_j \cap X_n} |f_n| + \varepsilon,$$

and since $\varepsilon > 0$ was arbitrary, (54.5) follows.

To prove (b) fix $j \in \mathbb{N}$ and choose $\varepsilon > 0$ such that $3\varepsilon < d_X((\hat{C}_j)_{\mathcal{H}^\infty(C)})$ and $3\varepsilon < d_{C_{j+1}}(C_j)$. Then it follows from Lemma 54.8 that

$$(54.6) \qquad (\hat{C}_j)_{\mathcal{H}^\infty(C)} + B_E(0; 3\varepsilon) \subset (\hat{C}_{j+1})_{\mathcal{H}^\infty(C)}.$$

Let (x_i) be a sequence in $(\hat{C}_j)_{\mathcal{H}^\infty(C)} \cap \xi^{-1}(K)$. Then $(\xi(x_i)) \subset K$

and after replacing (x_i) by a subsequence, if necessary, we may assume that $(\xi(x_i))$ converges to a point $a \in K$. Hence we can find $i_o \in \mathbb{N}$ such that $\|\xi(x_i) - a\| < \varepsilon$ for every $i \geq i_o$. On the other hand, since $T_n a \to a$ we can find $n \in \mathbb{N}$ and $b \in E_n$ such that $\|b - a\| < \varepsilon$. Thus $\|b - \xi(x_i)\| < 2\varepsilon$ for every $i \geq i_o$, and hence for each $i \geq i_o$ there exists a unique point $y_i \in B_X(x_i;2\varepsilon)$ such that $\xi(y_i) = b$. Then a glance at (54.6) shows that

$$y_i \in (\hat{C}_j)_{\mathcal{H}^\infty(C)} + B_E(0;2\varepsilon) \subset (\hat{C}_{j+1})_{\mathcal{H}^\infty(C)}$$

for every $i \geq i_o$. Since $\xi(y_i) = b \in E_n$ we conclude that $y_i \in (\hat{C}_{j+1})_{\mathcal{H}^\infty(C)} \cap X_n$ for every $i \geq i_o$. By (a) the set $(\hat{C}_{j+1})_{\mathcal{H}^\infty(C)} \cap X_n$ is compact and therefore, after replacing (y_i) by a subsequence, if necessary, we may assume that (y_i) converges to a point y. Observe that $d_X(y) \geq \varepsilon$. Since $\xi(y_i) = b$ for every $i \geq i_o$ we see that $\xi(y) = b$ too. And since $\|a - \xi(y)\| = \|a - b\| < \varepsilon$, there is a unique point $x \in B_X(y;\varepsilon)$ such that $\xi(x) = a$. Choose $i_1 \geq i_o$ such that $y_i \in B_X(y;\varepsilon)$ for every $i \geq i_1$. Since $\xi(x_i) \to a = \xi(x)$, we conclude that $x_i \to x$. This shows (b) and the lemma.

54.10. DEFINITION. Let (X,ξ) be a Riemann domain over E, and let $F \in \mathcal{H}(X)$. A subset B of X is said to be F-*bounding* if each $f \in F$ is bounded on B. We shall say that B is a *bounding* subset of X if B is $\mathcal{H}(X)$-bounding.

54.11. THEOREM. *Let E be a separable Banach space with the bounded approximation property. Let (X,ξ) be a pseudoconvex Riemann domain over E. Then each bounding subset of X is relatively compact. In particular the set $\hat{K}_{\mathcal{H}(X)}$ is compact for each compact set $K \subset X$.*

PROOF. (a) We first assume that E has a monotone Schauder basis. Let (X,ξ) be a pseudoconvex Riemann domain over E, and let B be a bounding subset of X. Clearly B is contained in the union of finitely many connected components of X, and hence we may assume from the outset that X is connected. Then the proof of Proposition 12.3 shows the existence of $j \in \mathbb{N}$ such that $B \subset (\hat{C}_j)_{\mathcal{H}(X)}$. On the other hand it is clear that $\xi(B)$ is a bounding subset of E, and is therefore relatively compact by Theorem 12.5. Thus $B \subset (\hat{C}_j)_{\mathcal{H}(X)} \cap \xi^{-1}\overline{(\xi(B))}$ and B is relatively compact by Lemma 54.9.

(b) The general case can be reduced to case (a) as in the proof of Theorem 54.7. The details are left to the reader as an exercise.

Theorems 52.6, 54.7 and 54.11 can be summarized as follows.

54.12. THEOREM. *Let E be a separable Banach space with the bounded approximation property. Let (X,ξ) be a Riemann domain over E. Then the following conditions are equivalent:*

(a) *X is a domain of existence.*

(b) *X is a domain of holomorphy.*

(c) *X is a holomorphically convex.*

(d) *X is metrically holomorphically convex.*

(e) *X is pseudoconvex.*

EXERCISES

54.A. Let (X,ξ) be a Riemann domain over E, and let U be an X-pseudoconvex open set. Let X' be a connected component of X which intersects U, and let $U' = U \cap X'$. Show that U' is X'-pseudoconvex.

54.B. Let (X,ξ) be a Riemann domain over E. Let E_o be a closed vector subspace of E, let $X_o = \xi^{-1}(E_o)$ and let $\xi_o = \xi \mid X_o$. Show the following:

(a) If $\mathcal{U} = (U_j)_{j=1}^{\infty}$ is a regular cover of X then

$\mathcal{U}_o = (U_j \cap X_o)_{j=1}^{\infty}$ is a regular cover of X_o. If X is $\mathcal{H}^{\infty}(\mathcal{U})$-separated then X_o is $\mathcal{H}^{\infty}(\mathcal{U}_o)$-separated. If $d_X((\hat{U}_j)_{\mathcal{H}^{\infty}(\mathcal{U})}) > 0$ for every $j \in \mathbb{N}$ then $d_X((U_j \cap X_o)\hat{}_{\mathcal{H}^{\infty}(\mathcal{U}_o)}) > 0$ for every $j \in \mathbb{N}$.

(b) If X is a domain of existence then X_o is a domain of holomorphy. If X is a domain of existence and E_o is separable then X_o is also a domain of existence.

(c) If U is X-pseudoconvex then $U \cap X_o$ is X_o-pseudoconvex.

55. HOLOMORPHIC APPROXIMATION IN INFINITE DIMENSIONAL RIEMANN DOMAINS

In this section we extend to the case of Riemann domains the approximation theorems from Section 46.

55.1. THEOREM. *Let E be a separable Banach space with the bounded approximation property. Let (X,ξ) be a pseudoconvex Riemann domain over E, and let U be an X-pseudoconvex open set. Then $\mathcal{H}(X)$ is sequentially dense in $(\mathcal{H}(U),\tau_c)$.*

PROOF. In view of Exercise 54.A, we may restrict our attention to those connected components of X which intersect U, and hence we may assume that X itself is connected. If E has a monotone Schauder basis then the proof can proceed parallel to the proof of the corresponding case in Theorem 46.1. And the case where E is a separable Banach space with the bounded approximation property can be reduced to the preceding case in the usual manner using Exercise 54.B. The details are left to the reader as an exercise.

55.2. THEOREM. *Let E be a separable Banach space with the bounded approximation property. Let (X,ξ) be a pseudoconvex Riemann domain over E. Let K be a compact subset of X such that $\hat{K}_{Psc(X)} = K$. Then for each open neighborhood U of K there is* *an X-pseudoconvex open set* V *such that* $K \subset V \subset U$.

PROOF. (a) We first assume that E has a monotone Schauder basis. Without loss of generality we may assume that the Riemann domain X is connected. Choose $j \in \mathbb{N}$ such that $K \subset C_j$. Then the set

$$L = (\hat{C}_j)_{\mathcal{H}(X)} \cap \xi^{-1}(\overline{co}(\xi(K)))$$

is compact by Lemma 54.9, and contains K. For each point $x \in L \setminus U$ there is a function $f_x \in Psc(X)$ such that $f_x < 0$ on K and $f_x(x) > 0$. Since $L \setminus U$ is compact we can find functions $f_1, \ldots, f_m \in Psc(X)$ such that $f_j < 0$ on K for $j = 1, \ldots, m$ and

$$L \setminus U \subset \bigcup_{j=1}^{m} \{x \in X : f_j(x) > 0\}.$$

If we set $f = sup\{f_1, \ldots, f_m\}$ then $f < 0$ on K and

$$(55.1) \quad (\hat{C}_j)_{\mathcal{H}(X)} \cap \xi^{-1}(\overline{co}(\xi(K))) \cap \{x \in X : f(x) \leq 0\} \subset U.$$

Choose $\varepsilon > 0$ such that $\varepsilon < d_X((\hat{C}_j)_{\mathcal{H}(X)})$ and $\varepsilon < d_{C_{j+1}}(C_j)$. Then it follows from lemma 54.8 that

$$(55.2) \quad (\hat{C}_j)_{\mathcal{H}(X)} + B_E(0;\varepsilon) \subset (\hat{C}_{j+1})_{\mathcal{H}(X)}.$$

We claim there exists δ with $0 < \delta < \varepsilon$ such that

$$(55.3) \quad (\hat{C}_j)_{\mathcal{H}(X)} \cap \xi^{-1}[\overline{co}(\xi(K)) + B_E(0;\delta)] \cap \{x \in X : f(x) \leq 0\} \subset U.$$

Otherwise we can find a sequence (δ_k), with $0 < \delta_k < \varepsilon$ and $\delta_k \to 0$, and a sequence (x_k) such that

$$x_k \in (\hat{C}_j)_{\mathcal{H}(X)} \cap \xi^{-1}[\overline{co}(\xi(K)) + B_E(0;\delta_k)] \cap \{x \in X : f(x) \leq 0\} \setminus U$$

for every $k \in \mathbb{N}$. For each $k \in \mathbb{N}$ there is a point $b_k \in \overline{co}(\xi(K))$ such that $\|b_k - \xi(x_k)\| < \delta_k$. Hence for each $k \in \mathbb{N}$ there is a unique $y_k \in B_X(x_k;\delta_k)$ such that $\xi(y_k) = b_k$. Then it follows from (55.2) that

$$y_k \in (\hat{C}_{j+1})_{\mathcal{H}(X)} \cap \xi^{-1}(\overline{co}(\xi(K)))$$

for every $k \in \mathbb{N}$. By Lemma 54.9 that set is compact. Hence, after replacing (y_k) by a subsequence, if necessary, we may assume that (y_k) converges to a point y. But then (x_k) also converges to y, and whence it follows that

$$y \in (\hat{C}_j)_{\mathcal{H}(X)} \cap \xi^{-1}(\overline{co}(\xi(K))) \cap \{x \in X : f(x) \leq 0\} \setminus U,$$

contradicting (55.1). This shows the existence of $\delta > 0$ satisfying (55.3). Set

$$V = int(\hat{C}_j)_{\mathcal{H}(X)} \cap \xi^{-1}[\overline{co}(\xi(K)) + B_E(0;\delta)] \cap \{x \in X : f(x) < 0\}.$$

Then $K \subset V \subset U$ and to show that V is X-pseudoconvex it only remains to show that the open set $int\,(\hat{C}_j)_{\mathcal{H}(X)}$ is X-pseudoconvex. To prove this let A be a compact subset of $int(\hat{C}_j)_{\mathcal{H}(X)}$. Choose ρ with $0 < \rho < \delta$ such that $A + B_E(0;\rho) \subset (\hat{C}_j)_{\mathcal{H}(X)}$. Then it follows from Lemma 54.8 that

$$\hat{A}_{Ps(X)} + B_E(0;\rho) \subset \hat{A}_{\mathcal{H}(X)} + B_E(0;\rho) \subset (\hat{C}_j)_{\mathcal{H}(X)}.$$

Thus $\hat{A}_{Ps(X)} \subset int(\hat{C}_j)_{\mathcal{H}(X)}$ and the open set $int(\hat{C}_j)_{\mathcal{H}(X)}$ is X-pseudoconvex, as we wanted.

(b) The case in which E is a separable Banach space with the bounded approximation property can be reduced to the preceding case in the usual manner. We leave the details to the reader as an exercise.

From Theorems 55.1 and 55.2 we obtain at once the following result:

55.3. THEOREM. *Let E be a separable Banach space with the bounded* *approximation property. Let (X,ξ) be a pseudoconvex Riemann domain over E. Let K be a compact subset of X such that $\hat{K}_{Psc(X)} = K$. Then for each $f \in \mathcal{H}(K)$ there are an open neighborhood U of K and a sequence (f_j) in $\mathcal{H}(X)$ such that $f \in \mathcal{H}(U)$ and (f_j) converges to f in $(\mathcal{H}(U), \tau_c)$.*

55.4. THEOREM. *Let E be a separable Banach space with the bounded approximation property. Let (X,ξ) be a holomorphically convex Riemann domain over E, and let U be an open subset of X. Then the following conditions are equivalent:*

(a) $\hat{K}_{\mathcal{H}(X)} \cap U$ *is compact for each compact set* $K \subset U$.

(b) $\hat{K}_{\mathcal{H}(X)} \subset U$ *for each compact set* $K \subset U$.

(c) U *is holomorphically convex and* $\mathcal{H}(X)$ *is sequentially dense in* $(\mathcal{H}(U), \tau_c)$.

(d) U *is holomorphically convex and* $\mathcal{H}(X)$ *is dense in* $(\mathcal{H}(U), \tau_c)$.

PROOF. The proof of Theorem 44.5 applies.

55.5. THEOREM. *Let E be a separable Banach space with the bounded approximation property. Let (X,ξ) be a holomorphically convex Riemann domain over E. Then $\hat{K}_{Ps(X)} = \hat{K}_{\mathcal{H}(X)}$ for each compact set $K \subset X$.*

PROOF. We already know that $\hat{K}_{Ps(X)} \subset \hat{K}_{\mathcal{H}(X)}$. Suppose that $\hat{K}_{Ps(X)} \neq \hat{K}_{\mathcal{H}(X)}$ and let $a \in \hat{K}_{\mathcal{H}(X)} \setminus \hat{K}_{Ps(X)}$. Then there is a function $u \in Ps(X)$ such that $u < 0$ on K and $u(a) > 0$. Fix $\varepsilon > 0$ such that $4\varepsilon < d_X(\hat{K}_{\mathcal{H}(X)})$. Since u is upper semicontinuous and K is compact we can easily find δ with $0 < \delta < \varepsilon$ such that $u < 0$ on $K + B_E(0;2\delta)$. Let $T \in \mathcal{L}(E;E)$ be a finite rank operator such that $\|T \circ \xi(x) - \xi(x)\| < \delta$ for every $x \in \hat{K}_{\mathcal{H}(X)}$. Define $\sigma \in \mathcal{H}(\hat{K}_{\mathcal{H}(X)};X)$ by

$$\sigma(x) = \xi_x^{-1}[\xi(x) + (T \circ \xi(x) - \xi(x)) - (T \circ \xi(a) - \xi(a))].$$

Then $\sigma(a) = a$ and $\sigma(K) \subset K + B_E(0;2\delta)$. Hence $u \circ \sigma(a) > 0$ whereas $u \circ \sigma(x) < 0$ for every $x \in K$. Let E_o be the finite dimensional subspace of E generated by the set $T \circ \xi(E) - T \circ \xi(a) + \xi(a)$, and let $X_o = \xi^{-1}(E_o)$. Then $\sigma(\hat{K}_{\mathcal{H}(X)}) \subset X_o$ and if follows from Theorem 53.9 that

$$\sigma(a) \notin [\sigma(K)]\hat{}_{Ps(X_o)} = [\sigma(K)]\hat{}_{\mathcal{H}(X_o)}.$$

Hence there is a function $f \in \mathcal{H}(X_o)$ such that

$$\sup_K |f \circ \sigma| = \sup_{\sigma(K)} |f| < |f \circ \sigma(a)|.$$

Since $f \circ \sigma \in \mathcal{H}(\hat{K}_{\mathcal{H}(X)})$, an application of Theorem 55.3 yields a function $g \in \mathcal{H}(X)$ such that $\sup_K |g| < |g(a)|$, a contradiction. This completes the proof of the theorem.

NOTES AND COMMENTS

Theorem 52.11 is a classical result of H. Cartan and P. Thullen [1]. Its generalization to the case of Banach spaces, Theorem 52.10, is due to M. Schottenloher [1]. The implication (d) ⇒ (e) in Theorem 52.6 is due to A. Hirschowitz [3].

The main result in Section 53, Theorem 53.7, is due to K. Oka [4]. The proof of Theorem 53.7 given here is due to L. Hörmander [3]. Theorem 53.10 is a weak version of results obtained by R. Narasimhan [1] and E. Bishop [1].

The main results in Section 54, Theorems 54.7 and 54.11, are due to Y. Hervier [1]. M. Schottenloher [4] has obtained Fréchet space versions of these theorems.

Theorems 55.1 and 55.2 are due to J. Mujica [5]. Theorems 55.3, 55.4 and 55.5 sharpen results of M. Schottenloher [4]. The idea of the proof of Theorem 55.5 given here is due to M. Schottenloher [5].

CHAPTER XIII

ENVELOPES OF HOLOMORPHY

56. ENVELOPES OF HOLOMORPHY

In this section we introduce the notion of envelope of holomorphy. We show that the envelope of holomorphy of a given Riemann domain always exists.

56.1. DEFINITION. Let (X,ξ) be a Riemann domain over E and let $\mathcal{F} \subset \mathcal{H}(X)$. A morphism $\tau : X \to Y$ is said to be an $\mathcal{F}$-*envelope of holomorphy* of X if:

(a) τ is an $\mathcal{F}$-extension of X.

(b) If $\mu : X \to Z$ is an $\mathcal{F}$-extension of X then there is a morphism $\nu : Z \to Y$ such that $\nu \circ \mu = \tau$.

A morphism $\tau : X \to Y$ is said to be an *envelope of holomorphy* of X if τ is an $\mathcal{H}(X)$-envelope of holomorphy of X.

If $\tau : X \to Y$ and $\tau' : X \to Y'$ are two $\mathcal{F}$-envelopes of holomorphy of X then the Riemann domains Y and Y' are easily seen to be isomorphic. In other words, the $\mathcal{F}$-envelope of holomorphy of X, it it exists, is unique up to isomorphism.

56.2. DEFINITION. Let I be an index set. For each point $a \in E$ consider the collection of all pairs (U,φ) such that U is an open neighborhood of a and $\varphi = (\varphi_i)_{i\in I} \subset \mathcal{H}(U)$. Two such pairs (U,φ) and (V,ψ) are said to be equivalent if there is an open neighborhood W of a with $W \subset U \cap V$ such that $\varphi_i = \psi_i$ on W for every $i \in I$. We shall denote by $\mathcal{H}_a^I$ the collection of all equivalent classes. The members of $\mathcal{H}_a^I$ are called *germs of holomorphic I-families* at the point a. The germ of (U,φ) at a wil be denoted by φ_a. Clearly $\mathcal{H}_a^I$ is an algebra. Next consider the collection

$$\mathcal{H}_E^I = \bigcup_{a\in E} \mathcal{H}_a^I,$$

where the algebras $\mathcal{H}_a^I$ are regarded as disjoint sets. For each $\varphi_a \in \mathcal{H}_E^I$ let $N(\varphi_a)$ denote the collection of all sets of the form

$$N(U,\varphi) = \{\varphi_b : b \in U\},$$

where (U,φ) varies over all representatives of the germ φ_a. The set $\mathcal{H}_E^I$ will be endowed with the unique topology such that $N(\varphi_a)$ is a neighborhood base at φ_a for each $\varphi_a \in \mathcal{H}_E^I$.

56.3. PROPOSITION. *Let* $\pi : \mathcal{H}_E^I \to E$ *be defined by* $\pi(\varphi_a) = a$ *for each* $\varphi_a \in \mathcal{H}_E^I$. *Then* $(\mathcal{H}_E^I, \pi)$ *is a Riemann domain over* E.

PROOF. We first show that $\mathcal{H}_E^I$ is Hausdorff. Let φ_a and ψ_b be two distinct points of $\mathcal{H}_E^I$. If $a \neq b$ then we can find a representative (U,φ) of φ_a and a representative (V,ψ) of ψ_b such that U and V are disjoint. Then the sets $N(U,\varphi)$ and $N(V,\psi)$ are also disjoint. Next suppose $a = b$. Let (U,φ) be a representative of φ_a, let (V,ψ) be a representative of ψ_a and let W be a connected open neighborhood of a such that $W \subset U \cap V$. Then the sets (W,φ) and (W,ψ) are necessarily disjoint, for otherwise the Identity Principle would imply that $\varphi_i = \psi_i$ on W for every $i \in I$, and therefore $\varphi_a = \psi_a$, a contradiction. Thus $\mathcal{H}_E^I$ is a Hausdorff space. Since the mapping π is clearly a local homeomorphism, the proof of the proposition is complete.

56.4. THEOREM. *Let* (X,ξ) *be a Riemann domain over* E *and let* $\mathcal{F} \subset \mathcal{H}(X)$. *Then the* $\mathcal{F}$*-envelope of holomorphy of* X *always exists.*

PROOF. Write $F = (f_i)_{i\in I} \in \mathcal{H}(X)$. Given $x \in X$ let U be a chart in X containing x, let $\varphi_i = f_i \circ (\xi \mid U)^{-1}$ for each $i \in I$, let $\varphi = (\varphi_i)_{i\in I} \subset \mathcal{H}(\xi(U))$ and let $\varphi_{\xi(x)} \in \mathcal{H}^I_{\xi(x)}$ be the germ of $(\xi(U),\varphi)$ at $\xi(x)$. Then the mapping

$$\tau : x \in X \to \varphi_{\xi(x)} \in \mathcal{H}^I_E$$

is clearly well defined and a morphism. Given $\varphi_a \in \mathcal{H}^I_E$ let (V,φ) be a representative of φ_a and define

$$g_i(\varphi_a) = \varphi_i(a) \quad \text{for each} \quad i \in I.$$

Clearly each g_i is well defined. Since $g_i = \varphi_i \circ \pi$ on a neighborhood of φ_a we see that each g_i is holomorphic on $\mathcal{H}^I_E$. For each $i \in I$ and $x \in X$ we have that

$$g_i(\tau(x)) = g_i(\varphi_{\xi(x)}) = \varphi_i(\xi(x)) = f_i(x).$$

If Y is the union of those connected components of $\mathcal{H}^I_E$ which intersect $\tau(X)$ then it is clear that $\tau : X \to Y$ is a F-extension of X. Let (Z,ζ) be another Riemann domain over E and suppose that $\mu : X \to Z$ is an F-extension of X too. Then for each $i \in I$ there is a unique function $h_i \in \mathcal{H}(Z)$ such that $h_i \circ \mu = f_i$. Given $z \in Z$ let W be a chart in Z containing z, let $\psi_i = h_i \circ (\zeta \mid W)^{-1}$ for each $i \in I$, let $\psi = (\psi_i)_{i\in I} \subset \mathcal{H}(\zeta(W))$ and let $\psi_{\zeta(z)} \in \mathcal{H}^I_{\zeta(z)}$ be the germ of $(\zeta(W),\psi)$ at $\zeta(z)$. Then the mapping

$$\nu : z \in Z \to \psi_{\zeta(z)} \in \mathcal{H}^I_E$$

is clearly well defined and a morphism. Given $x \in X$ let U be a chart in X containing x and let W be a chart in Z containing $\mu(x)$ such that $W = \mu(U)$. Then

$$\varphi_i = f_i \circ (\xi \mid U)^{-1} = h_i \circ \mu \circ (\xi \mid U)^{-1} = h_i \circ (\zeta \mid W)^{-1} = \psi_i$$

for every $i \in I$. Hence

$$\nu \circ \mu(x) = \psi_{\zeta \circ \mu(x)} = \varphi_{\xi(x)} = \tau(x)$$

and in particular $\nu(\mu(X)) = \tau(X) \subset Y$. Since each connected component of Z intersects $\mu(X)$ we see that $\nu(Z) \subset Y$. This completes the proof.

The envelope of holomorphy of a Riemann domain X will be denoted by $E(X)$.

EXERCISES

56.A. Let (X,ξ) be a Riemann domain over E, and let $\tau : X \to Y$ be a holomorphic extension of X. Show that $Y = E(X)$ if and only if Y is a domain of holomorphy.

56.B. Let (X,ξ) be a Riemann domain over E. Show that $X = E(X)$ if and only if X is a domain of holomorphy.

57. THE SPECTRUM

In this section we show that if X is a Riemann domain then the spectrum of $(\mathcal{H}(X), \tau_c)$ can be endowed with the structure of a Riemann domain. We also show that in certain cases the envelope of holomorphy of X can be identified with a subset of the spectrum.

57.1. DEFINITION. Let (X,ξ) be a Riemann domain over E. Let $S_c(X)$ denote the spectrum of the topological algebra $(\mathcal{H}(X), \tau_c)$. Let $\pi : S_c(X) \to E$ be defined by the condition

$$(57.1) \qquad \varphi \circ \pi(h) = h(\varphi \circ \xi)$$

for every $h \in S_c(X)$ and $\varphi \in E'$. And let $\varepsilon : x \in X \to \hat{x} \in S_c(X)$ be defined by $\hat{x}(f) = f(x)$ for every $f \in \mathcal{H}(X)$.

It follows from the Mackey-Arens Theorem that the mapping

π is well defined, for each $h \in S_c(X)$ defines a continuous linear functional $\tilde{h}$ on E'_c by $\tilde{h}(\varphi) = h(\varphi \circ \xi)$ for every $\varphi \in E'$. Furthermore, since $\varphi \circ \pi(\hat{x}) = \varphi \circ \xi(x)$ for every $x \in X$ and $\varphi \in E'$, we see that $\pi \circ \varepsilon = \xi$.

57.2. THEOREM. *Let (X, ξ) be a Riemann domain over E. Then there is a topology on $S_c(X)$ such that:*

(a) *$(S_c(X), \pi)$ is a Riemann domain over E.*

(b) *The evaluation mapping $\varepsilon : X \to S_c(X)$ is a morphism.*

PROOF. Let $h \in S_c(X)$. Then there are a compact set $K \subset X$ and a constant $c > 0$ such that $|h(f)| \leq c \sup_K |f|$ for every $f \in \mathcal{H}(X)$. By applying the preceding inequality to f^m, taking mth root, and letting $m \to \infty$ we obtain that

$$(57.2) \qquad |h(f)| \leq \sup_K |f|$$

for every $f \in \mathcal{H}(X)$. In this case we shall write $h \prec K$. Let $0 < r < d_X(K)$. Given $t \in B_E(0;r)$ choose $\rho > 1$ such that $\rho t \in B_E(0;r)$. It follows from the Cauchy inequalities that

$$|h(P_t^m f)| \leq \sup_K |P_t^m f| \leq \rho^{-m} \sup_{K+\bar{\Delta}\rho t} |f|$$

for every $f \in \mathcal{H}(X)$ and $m \in \mathbb{N}_0$. If we define

$$(57.3) \qquad h_t(f) = \sum_{m=0}^{\infty} h(P_t^m f)$$

for every $f \in \mathcal{H}(X)$, then clearly h_t is a continuous linear functional on $(\mathcal{H}(X), \tau_c)$ and

$$(57.4) \qquad |h_t(t)| \leq \sum_{m=0}^{\infty} \rho^{-m} \sup_{K+\bar{\Delta}\rho t} |f|$$

for every $f \in \mathcal{H}(X)$. Using the formula $P_t^m(fg) = \sum_{j=0}^{m} P_t^{m-j} f \cdot P_t^j g$, we see that $h_t(fg) = h_t(f) h_t(g)$ for all $f, g \in \mathcal{H}(X)$. Thus h_t is a continuous complex homomorphism of $(\mathcal{H}(X), \tau_c)$ and it follows from (57.4) that

(57.5) $$|h_t(f)| \leq \sup_{K+\overline{\Delta}t} |f|$$

for every $f \in \mathcal{H}(X)$. Thus $h_t \prec K + \overline{\Delta}t$.

For each $h \in S_c(X)$ let $N(h)$ denote the collection of all sets of the form

$$N_r(h) = \{h_t : t \in B_E(0;r)\},$$

where $0 < r < d_X(K)$ for some compact set $K \subset X$ such that $h \prec K$. We claim that there is a unique topology on $S_c(X)$ for which $N(h)$ is a neighborhood base at h for each $h \in S_c(X)$. It is sufficient to prove that given $h \in S_c(X)$, $N_r(h) \in N(h)$ and $h_a \in N_r(h)$, we can find $N_\rho(h_a) \in N(h_a)$ such that $N_\rho(h_a) \subset N_r(h)$. Let K be a compact subset of X such that $h \prec K$ and $0 < r < d_X(K)$. Choose a compact set $L \subset X$ such that $h_a \prec L$ and choose $\rho > 0$ such that $\rho < d_X(L)$ and $B_E(a;\rho) \subset B_E(0;r)$. Then $N_\rho(h_a) \in N(h_a)$, and if we assume for a moment that

(57.6) $$(h_a)_t = h_{a+t}$$

for every $t \in B_E(0;\rho)$, then it is clear that $N_\rho(h_a) \subset N_r(h)$. To prove (57.6) observe that for each $f \in \mathcal{H}(X)$ we have that

$$\sum_{m=0}^{\infty} P^m_{a+t}(x) = f(x+a+t) = \sum_{k=0}^{\infty} P^k_t f(x+a) = \sum_{k=0}^{\infty} \sum_{j=0}^{\infty} P^j_a(P^k_t f)(x),$$

with uniform convergence for $x \in K$. Since $h \prec K$ it follows that

$$h_{a+t}(f) = \sum_{m=0}^{\infty} h(P^m_{a+t}f) = \sum_{k=0}^{\infty} \sum_{j=0}^{\infty} h(P^j_a(P^k_t f)) = \sum_{k=0}^{\infty} h_a(P^k_t f) = (h_a)_t(f)$$

and (57.6) has been proved.

We next show that $S_c(X)$ is a Hausdorff space. Let h_1 and h_2 be two distinct points of $S_c(X)$. Then there is a function $f \in \mathcal{H}(X)$ such that $|h_1(f) - h_2(f)| = 3\varepsilon > 0$. Choose a compact set $K \subset X$ such that $h_j \prec K$ for $j = 1, 2$, and choose r with $0 < r < d_X(K)$ such that f is bounded, by c say, on $K + B_E(0;r)$. Then for each $t \in B_E(0;r)$, $m \in \mathbb{N}$, $0 < \rho < 1$ and $j = 1, 2$ we have that

$$|h_j(P^m_{\rho t} f)| \leq \sup_K |P^m_{\rho t} f| \leq c\rho^m.$$

If ρ is sufficiently small then

$$|(h_j)_{\rho t}(f) - h_j(f)| \leq \sum_{m=1}^{\infty} |h_j(P^m_{\rho t} f)| \leq \varepsilon$$

and the neighborhoods $N_{\rho r}(h_1)$ and $N_{\rho r}(h_2)$ are disjoint.

We next show that $\pi : S_c(X) \to E$ is a local homeomorphism. More precisely, we show that π maps $N_r(h)$ homeomorphically onto $B_E(\pi(h);r)$ for each $h \in S_c(X)$ and each $N_r(h) \in \mathcal{N}(h)$. Indeed, if $t \in B_E(0;r)$ and $\varphi \in E'$ then using (57.3) we get that

$$\varphi(\pi(h_t)) = h_t(\varphi \circ \xi) = h(\varphi \circ \xi) + h(\varphi(t))$$

$$= \varphi(\pi(h)) + \varphi(t) = \varphi(\pi(h) + t).$$

Thus

$$\pi(h_t) = \pi(h) + t \tag{57.7}$$

for every $h_t \in N_r(h)$, and the desired conclusion follows.

We next show that $\varepsilon : X \to S_c(X)$ is a morphism. Indeed, if $x \in X$, $0 < r < d_X(x)$, $t \in B_E(0;r)$ and $f \in \mathcal{H}(X)$ then

$$(x + t)\hat{}(f) = f(x + t) = \sum_{m=0}^{\infty} P^m_t f(x) = \hat{x}_t(f).$$

Hence ε maps $B_X(x;r)$ into $N_r(\hat{x})$, and ε is therefore continuous. Since we already know that $\pi \circ \varepsilon = \xi$, we conclude that ε is a morphism.

57.3. THEOREM. *Let (X,ξ) be a Riemann domain over E. For each*

$f \in \mathcal{H}(X)$ let $\hat{f} : S_c(X) \to \mathbb{C}$ be defined by $\hat{f}(h) = h(f)$ for every $h \in S_c(X)$. Then:

(a) *$\hat{f} \in \mathcal{H}(S_c(X))$ and $\hat{f} \circ \varepsilon = f$ for each $f \in \mathcal{H}(X)$.*

(b) *For each compact set $L \subset S_c(X)$ there is a compact set $K \subset X$ such that $\sup_L |\hat{f}| \leq \sup_K |f|$ for every $f \in \mathcal{H}(X)$.*

(c) *The extension mapping*

$$G : f \in (\mathcal{H}(X), \tau_c) \to \hat{f} \in (\mathcal{H}(S_c(X)), \tau_c)$$

is a topological isomorphism between $(\mathcal{H}(X), \tau_c)$ and a complemented subalgebra of $(\mathcal{H}(S_c(X)), \tau_c)$.

PROOF. (a) Let $f \in \mathcal{H}(X)$ and let $h \in S_c(X)$. Choose a compact set $K \subset X$ such that $h \prec k$, and choose r with $0 < r < d_X(K)$ such that f is bounded, by c say, on $K + B_E(0;r)$. Then

$$|h(P_t^m f)| \leq \sup_K |P_t^m f| \leq c$$

for each $t \in B_E(0;r)$, and therefore

$$\hat{f}(h_{\lambda t}) = h_{\lambda t}(f) = \sum_{m=0}^{\infty} h(P_{\lambda t}^m f) = \sum_{m=0}^{\infty} \lambda^m h(P_t^m f)$$

for each $t \in B_E(0;r)$ and $|\lambda| < 1$. Hence the function $\lambda \in \Delta \to \hat{f}(h_{\lambda t}) \in \mathbb{C}$ is holomorphic for each $t \in B_E(0;r)$, thus proving that $\hat{f}$ is G-holomorphic. On the other hand for each $t \in B_E(0;r)$ and $0 < \rho < 1$ we can write

$$|\hat{f}(h_{\rho t}) - \hat{f}(h)| \leq c \sum_{m=1}^{\infty} \rho^m,$$

proving that $\hat{f}$ is continuous at h. Thus $\hat{f}$ is holomorphic on $S_c(X)$, and clearly $\hat{f} \circ \varepsilon = f$.

(b) Let L be a compact subset of $S_c(X)$. Then we can find $h_1, \ldots, h_m \in L$, compact sets $K_1, \ldots, K_m$ in X and $r_1, \ldots, r_m > 0$ such that $h_j \prec K_j$ and $r_j < d_X(K_j)$ for $j = 1, \ldots, m$, and

(57.8) $$L \subset \bigcup_{j=1}^{m} N_{r_j}(h_j).$$

For each $j = 1,\dots,m$ consider the compact set

$$C_j = [\pi(L) - \pi(h_j)] \cap \overline{B}_E(0;r_j).$$

We claim that

(57.9) $$L \subset \bigcup_{j=1}^{m} \{(h_j)_t : t \in C_j\}.$$

Indeed, if $h \in L$ then by (57.8) there exists j with $1 \leq j \leq m$ and $t_j \in B_E(0;r_j)$ such that $h = (h_j)_t$. Then it follows from (57.7) that $\pi(h) = \pi(h_j) + t$, and hence $t \in C_j$. Next we consider the set

$$K = \bigcup_{j=1}^{m} (K_j + \overline{\Delta} C_j),$$

which is compact by Lemma 52.5. If $1 \leq j \leq m$, $t \in C_j$ and $f \in \mathcal{H}(X)$ then using (57.5) we get that

$$|(h_j)_t(f)| \leq \sup_{K_j+\overline{\Delta}t} |f| \leq \sup_K |f|.$$

Thus $\sup_L |\hat{f}| \leq \sup_K |f|$ by (57.9).

(c) The restriction mapping

$$\varepsilon^* : g \in (\mathcal{H}(S_c(X)), \tau_c) \to g \circ \varepsilon \in (\mathcal{H}(X), \tau_c)$$

is always continuous. The extension mapping

$$G : f \in (\mathcal{H}(X), \tau_c) \to \hat{f} \in (\mathcal{H}(S_c(X)), \tau_c)$$

is continuous by (b). And $\varepsilon^* \circ G(f) = f$ for every $f \in \mathcal{H}(X)$. This completes the proof.

57.4. PROPOSITION. *Let (X,ξ) be a Riemann domain over E, and let (Y,η) be a Riemann domain over F. Let $\tau : X \to Y$ be a* *morphism, and let* τ^* *be the restriction mapping*

$$\tau^* : g \in (\mathcal{H}(Y), \tau_c) \to g \circ \tau \in (\mathcal{H}(X), \tau_c).$$

Then the mapping

$$\tau^{**} : h \in S_c(X) \to h \circ \tau^* \in S_c(Y)$$

is a morphism and the following diagram is commutative

$$\begin{array}{ccc} S_c(X) & \xrightarrow{\tau^{**}} & S_c(Y) \\ \varepsilon_X \Big\uparrow & & \Big\uparrow \varepsilon_Y \\ X & \xrightarrow[\tau]{} & Y \end{array}$$

PROOF. Let $h \in S_c(X)$ and let K be a compact subset of X such that $h \prec K$. Then

$$|h \circ \tau^*(g)| = |h(g \circ \tau)| \leq \sup_K |g \circ \tau| = \sup_{\tau(K)} |g|$$

for every $g \in \mathcal{H}(Y)$, and therefore $h \circ \tau^* \prec \tau(K)$. Given ρ with $0 < \rho < d_Y(\tau(K))$, choose $r > 0$ with $r < \rho$ and $r < d_X(K)$. We claim that

$$\tau^{**}(N_r(h)) \subset N_r(h \circ \tau^*). \tag{57.10}$$

Indeed, for every $t \in B_E(0;r)$ and $g \in \mathcal{H}(Y)$ we have that

$$h_t \circ \tau^*(g) = h_t(g \circ \tau) = \sum_{m=0}^{\infty} h[P_t^m(g \circ \tau)] = \sum_{m=0}^{\infty} h[(P_t^m g) \circ \tau]$$

$$= \sum_{m=0}^{\infty} h \circ \tau^*(P_t^m g) = (h \circ \tau^*)_t(g) .$$

This shows (57.10) and we conclude that τ^{**} is continuous. Then it follows easily that τ^{**} is a morphism and that $\tau^{**} \circ \varepsilon_X = \varepsilon_X \circ \tau$.

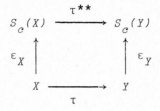

57.5. PROPOSITION. *Let (X,ξ) be a Riemann domain over E. Let $\hat{X}$ denote the union of those connected components of $S_c(X)$ which intersect $\varepsilon(X)$. Then:*

(a) $\varepsilon_X : X \to \hat{X}$ *is a holomorphic extension of* X.

(b) $\varepsilon_X^* : (\mathcal{H}(\hat{X}),\tau_c) \to (\mathcal{H}(X),\tau_c)$ *is a topological isomorphism.*

(c) *If* $\tau : X \to Y$ *is a holomorphic extension of* X *such that* $\tau^* : (\mathcal{H}(Y),\tau_c) \to (\mathcal{H}(X),\tau_c)$ *is a topological isomorphism, then there is a morphism* $\sigma : Y \to \hat{X}$ *such that* $\sigma \circ \tau = \varepsilon_X$.

PROOF. (a) and (b) are direct consequences of Theorem 57.3. To prove (c) we observe that since the mapping $\tau^*: (\mathcal{H}(Y),\tau_c) \to (\mathcal{H}(X),\tau_c)$ is a topological isomorphism, then the morphism $\tau^{**}: S_c(Y) \to S_c(X)$ given by Proposition 57.4 is a bijection, and therefore an isomorphism. Let $\sigma : Y \to S_c(X)$ be defined by $\sigma = (\tau^{**})^{-1} \circ \varepsilon_Y$. Then certainly $\sigma \circ \tau = \varepsilon_X$. Furthermore, since each connected component of Y intersects $\tau(X)$, it follows that $\sigma(Y) \subset \hat{X}$, thus completing the proof.

57.6. COROLLARY. *Let (X,ξ) be a Riemann domain E. Then $S_c(\hat{X}) = S_c(X)$ and $(\hat{X})\hat{} = \hat{X}$.*

57.7. COROLLARY. *Let (X,ξ) be a Riemann domain over $\mathbb{C}^n$. Then $\hat{X} = E(X)$.*

PROOF. Without loss of generality we may assume that X is connected. If $\tau : X \to Y$ is a holomorphic extension of X then Y is also connected and hence $(\mathcal{H}(X),\tau_c)$ and $(\mathcal{H}(Y),\tau_c)$ are Fréchet spaces. Hence $\tau^* : (\mathcal{H}(Y),\tau_c) \to (\mathcal{H}(X),\tau_c)$ is a topological isomorphism by the Open Mapping Theorem. Then the desired conclusion follows from Proposition 57.5.

57.8. PROPOSITION. *Let (X,ξ) be a Riemann domain over E. If we set $A = \{\hat{f} : f \in \mathcal{H}(X)\}$ then $S_c(X)$ is A-separated and metrically A-convex.*

PROOF. Given $h_1 \neq h_2$ in $S_c(X)$ we can find $f \in \mathcal{H}(X)$ such that $h_1(f) \neq h_2(f)$. Thus $\hat{f}(h_1) \neq \hat{f}(h_2)$ and $S_c(X)$ is A-separated.

Given a compact set $L \subset S_c(X)$, Theorem 57.3 yields a compact set $K \subset X$ such that $\sup_L |\hat{f}| \leq \sup_K |f|$ for every $f \in \mathcal{H}(X)$. Choose r such that $0 < r < d_X(K)$. We claim that $d_{S_c(X)}(\hat{L}_A) \geq r$. Indeed, for each $h \in \hat{L}_A$ and $f \in \mathcal{H}(X)$ we have that

$$|h(f)| = |\hat{f}(h)| \leq \sup_L |\hat{f}| \leq \sup_K |f|.$$

Hence $h \prec K$ and therefore $N_r(h) \in N(h)$. Since $\pi(N_r(h)) = B_E(\pi(h);r)$ we conclude that $d_{S_c(X)}(h) \geq r$, as we wanted.

57.9. COROLLARY. *Let E be a separable Banach space with the bounded approximation property. Let (X,ξ) be a Riemann domain over E. Then $\hat{X} = E(X)$.*

PROOF. By Proposition 57.5, $\varepsilon : X \to \hat{X}$ is a holomorphic extension of X. By Proposition 57.8 and Theorem 54.12, $\hat{X}$ is a domain of holomorphy. Then the conclusion follows from Exercise 56.A.

58. ENVELOPES OF HOLOMORPHY AND THE SPECTRUM

In the preceding section we proved that if X is a Riemann domain over a separable Banach space with the bounded approximation property then the envelope of holomorphy $E(X)$ of X can be identified with a subset $\hat{X}$ of the spectrum $S_c(X)$ of $(\mathcal{H}(X),\tau_c)$. In this section we shall prove that actually $E(X) = S_c(X)$.

58.1. DEFINITION. A polynomial $P \in P(\mathbb{C}^n)$ is said to be *normalized* if $P(z) = \sum_\alpha c_\alpha z^\alpha$ and $\max_\alpha |c_\alpha| = 1$.

58.2. LEMMA. *For each polydisc $\Delta^n(0;r)$ with $0 < r < 1$ there* *is a constant* $c = c(n,r) > 0$ *with the following property. If* $P \in P(\mathbb{C}^n)$ *is a normalized polynomial of degree at most* j *in each variable, and*

$$S = \{z \in \Delta^n(0;r) : |P(z)| \leq \theta^j\},$$

where $0 < \theta < 1$, *then* $\lambda(S) \leq -c/\log\theta$.

PROOF. Let $s = (1-r)/2$ and let $M_s = \sup_{\Delta^n(0;s)} |P|$. If $P(z) = \sum_\alpha c_\alpha z^\alpha$ then it follows from the Cauchy integral formulas that $|c_\alpha| \leq M_s s^{-|\alpha|}$ for every α. Since P is normalized there is an α with $\alpha_i \leq j$ for $i = 1,\dots,n$ such that $|c_\alpha| = 1$. Since $0 < s < 1$ it follows that $M_s \geq s^{|\alpha|} \geq s^{nj}$, and hence we can find a point $a \in \Delta^n(0;s)$ such that

$$P(a) > s^{nj}2^{-n}.$$

Let $t = (1+r)/2$ and observe that

$$S \subset \Delta^n(0;r) \subset \Delta^n(a;t) \subset \Delta^n(0;1).$$

Since P is the sum of at most $(j+1)^n$ terms of the form $c_\alpha z^\alpha$, with $|c_\alpha| \leq 1$, we see that $|P| < (j+1)^n$ on $\Delta^n(0;1)$. Thus $\log[(j+1)^{-n}|P|] < 0$ on $\Delta^n(0;1)$, and it follows that

$$\int_{\Delta^n(a;t)} \log[(j+1)^{-n}|P|]\,d\lambda \leq \int_S \log[(j+1)^{-n}|P|]\,d\lambda$$

$$\leq \int_S \log|P|\,d\lambda \leq \lambda(S)\log\theta^j.$$

On the other hand, using the inequality (35.1) we get that

$$\int_{\Delta^n(a;t)} \log[(j+1)^{-n}|P|]\,d\lambda \geq \pi^n t^{2n} \log[(j+1)^{-n}|P(a)|]$$

$$\geq \pi^n t^{2n} \log[(j+1)^{-n} s^{nj} 2^{-n}].$$

Since $\log\theta < 0$ it follows that

$$\lambda(S) \le \frac{\pi^n t^{2n} \log[s^{nj}(2j+2)^{-n}]}{j\log\theta} = \frac{\pi^n t^{2n} n}{\log\theta}\left[\log s - \frac{\log(2j+2)}{j}\right]$$

$$\le \frac{\pi^n t^{2n} n}{\log\theta}(\log s - \log 4) \le -\frac{\pi^n}{\log\theta}\, n\left(\frac{1+r}{2}\right)^{2n} \log\frac{8}{1-r},$$

completing the proof.

58.3. **LEMMA.** *Let (X,ξ) be a Riemann domain over $\mathbb{C}^n$. Let K be a compact subset of X and let $f \in \mathcal{H}(X)$. Then there exist θ with $0 < \theta < 1$ and $k_o \in \mathbb{N}$ with the following property. For each $j, k \in \mathbb{N}$ with $j \ge k \ge k_o$ there exists a nonzero polynomial $P(z_1,\ldots,z_n,w)$ of degree at most j in each of the variables $z_1,\ldots,z_n$, and of degree at most k in w such that*

$$|P(\xi(x), f(x))| \le \theta^{jk^{1/n}}$$

for all $x \in K$. If a is any given point in $\mathbb{C}^n$ then the polynomial P can be taken to be normalized with respect to the variables $z_1 - a_1,\ldots,z_n - a_n, w$.

PROOF. Let $0 < 2\varepsilon < d_X(K)$ and choose $b_1,\ldots,b_m \in K$ such that

$$K \subset \bigcup_{i=1}^{m} \Delta_X(b_i;\varepsilon).$$

Fix $j \ge k$ in $\mathbb{N}$ and let $\mathcal{P}$ be the vector space of all polynomials $P(z_1,\ldots,z_n,w)$ of degree at most j in each of the variables $z_1,\ldots,z_n$ and of degree at most k in w. Then $\mathcal{P}$ has dimension $(j+1)^n(k+1)$. Let p be the largest integer smaller than $m^{-1}(j+1)(k+1)^{1/n}$. Consider all the linear functionals on $\mathcal{P}$ of then form

$$T_{i\alpha}(P) = \frac{\partial^{\alpha_1+\ldots+\alpha_n} P}{\partial z_1^{\alpha_1}\ldots\partial z_n^{\alpha_n}}(\xi(b_i), f(b_i)),$$

with $i = 1,\dots,m$ and $|\alpha| < p$. There are at most mp^n of such linear functionals, and since $mp^n < (j+1)^n(k+1) = \dim \mathcal{P}$, we can find a nonzero $P \in \mathcal{P}$ such that $T_{i\alpha}(P) = 0$ for all $i = 1,\dots,m$ and $|\alpha| < p$. After multiphying P be a constant we may assume that P is normalized with respect to the variables $z_1 - a_1,\dots,z_n - a_n, w$. Let $L = K + \Delta^n(0;2\varepsilon)$ and choose $M > 1$ such that $|f(x)| \le M$ and $|\xi_i(x) - a_i| \le M$ for every $x \in L$ and $i = 1,\dots,n$. Let $Q = P \circ (\xi, f)$. Then $Q(x)$ is the sum of at most $(j+1)^n(k+1)$ terms of the form $c_{\alpha r}(\xi(x) - a)^\alpha f(x)^r$, where $\alpha = (\alpha_1,\dots,\alpha_n)$, with $\alpha_i \le j$, $r \le k$ and $|c_{\alpha r}| \le 1$. Whence it follows that

$$|Q(x)| \le (j+1)^n(k+1)M^{nj+k}$$

for every $x \in L$. By our choice of P, all the homogeneous polynomials of order less than p in the Taylor series expansion of Q at the point b_i vanish. Then it follows from Theorem 7.19 that

$$|Q(x)| \le (j+1)^n(k+1)M^{nj+k}\left(\frac{\|\xi(x) - \xi(b_i)\|}{2\varepsilon}\right)^p$$

for all $x \in \Delta_X(b_i;2\varepsilon)$ and $i = 1,\dots,m$. Thus

$$|Q(x)| \le (j+1)^n(k+1)M^{nj+k}2^{-p} \le (j+1)^{n+1}M^{(n+1)j}2^{-p}$$

for all $x \in K$. Since $(j+1)^{n+1}M^{-(n+1)j} \to 0$ when $j \to \infty$ we can find a constant $c = c(n,M) > 0$ such that

$$|Q(x)| \le c^j 2^{-(p+1)}$$

for all $x \in K$. Since

$$p + 1 \ge \frac{(j+1)(k+1)^{1/n}}{m} > \frac{j\,k^{1/n}}{m}$$

it follows that $c^{j/m(p+1)} \to 1$ when $j, k \to \infty$. Choose θ with $0 < \theta < 1$ such that $\theta 2^{1/m} > 1$, and next choose $k_0 \in \mathbb{N}$ such that $c^{j/m(p+1)} < \theta 2^{1/m}$ for every $j \geq k \geq k_o$. Then

$$|Q(x)| \leq \theta^{m(p+1)} < \theta^{jk^{1/n}}$$

for every $x \in K$ and $j \geq k \geq k_o$. This completes the proof.

In the next lemma $\mathbb{C}^n$ is endowed with the norm $\|z\| = \max|z_j|$, so that the balls are the polydiscs.

58.4. LEMMA. *Let (X,ξ) be a Riemann domain over $\mathbb{C}^n$. Let A be a closed subalgebra of $(\mathcal{H}(X),\tau_c)$ such that*

(a) $P_t^m f \in A$ *for every* $f \in A$, $m \in \mathbb{N}$ *and* $t \in E$.

(b) $\xi_1,\ldots,\xi_n \in A$.

Assume that X is A-separated and metrically A-convex, but X is not A-convex. Then there are a compact set K in X, a polydisc $\Delta^n(a;r)$ in $\mathbb{C}^n$, a function $f \in A$, and a sequence (x_j) of distinct points in X such that

(c) $x_j \in \xi^{-1}(a) \cap \hat{K}_A$ *and* $d_X(x_j) > r$ *for every* $j \in \mathbb{N}$.

(d) $\Delta_X^n(x_j;r) \subset \hat{K}_A$ *for every* $j \in \mathbb{N}$.

(e) *The set*

$$\{z \in \Delta^n(a;r) : f(\xi^{-1}(z) \cap \hat{K}_A) \text{ is finite}\}$$

has Lebesgue measure zero.

PROOF. Since X is not A-convex there is a compact set $L \subset X$ such that $\hat{L}_A$ is not compact. Since X is metrically A-convex we can find $r > 0$ such that $2r < d_X(\hat{L}_A)$. Let (y_j) be a sequence in $\hat{L}_A$ with no cluster point. Since $|\xi_i(y_j)| \leq \sup_L|\xi_i|$ for each $i = 1,\ldots,n$ and $j \in \mathbb{N}$, we see that the sequence $(\xi(y_j))$ is bounded and has therefore a cluster point a in $\mathbb{C}^n$. After replacing (y_j) by a subsequence, if necessary, we may assume that $\xi(y_j) \to a$ and $\|\xi(y_j) - a\| < r$ for every $j \in \mathbb{N}$. Hence for each $j \in \mathbb{N}$ there is a unique point $x_j \in \Delta^n_X(y_j;r)$ such that $\xi(x_j) = a$. If there were a point $x \in X$ such that $x_j = x$ for infinitely many indices, then x would be a cluster point of the sequence (y_j). Hence, after replacing (y_j) and (x_j) by suitable subsequences, if necessary, we may assume that $x_j = x_k$ whenever $j \neq k$. Clearly $d_X(x_j) > r$ and using Lemma 54.8 we get that

$$\Delta^n_X(x_j;r) \subset \Delta^n_X(y_j;2r) \subset [L + \Delta^n(0;2r)]\hat{}_A$$

for every $j \in \mathbb{N}$. If we set $K = L + \overline{\Delta}^n(0;2r)$ then (c) and (d) are verified. Since X is A-separated, the set $\{f \in A : (x_j) \neq f(x_k)\}$ is open and dense in A whenever $j \neq k$. Since A is a Baire space, the set $\bigcap_{j \neq k} \{f \in A : f(x_j) \neq f(x_k)\}$ is also dense in A, and in particular contains a function f. Since the sets $\Delta_X(x_j;r)$ are pairwise disjoint, the function

$$g_{jk} = f \circ (\xi \mid \Delta^n_X(x_j;r))^{-1} - f \circ (\xi \mid \Delta^n_X(x_k;r))^{-1}$$

is well defined and is holomorphic on the polydisc $\Delta^n(a;r)$,and does not vanish at the point a whenever $j \neq k$. Hence the set

$$Z_{jk} = \{z \in \Delta^n(a;r) : g_{jk}(z) = 0\}$$

has Lebesgue measure zero whenever $j \neq k$ by Exercise 35.F. Since one can readily see that

$$\{z \in \Delta^n(a;r) : f(\xi^{-1}(z) \cap \hat{K}_A) \text{ is finite}\}$$

$$\subset \{z \in \Delta^n(a;r) : f(\xi^{-1}(z) \cap \bigcup_{j=1}^{\infty} \Delta^n_X(x_j;r)) \text{ is finite}\} \subset \bigcup_{j \neq k} Z_{jk},$$

(e) follows. The proof of the lemma is now complete.

58.5. **THEOREM.** *Let (X,ξ) be a Riemann domain over $\mathbb{C}^n$. Let* A *be a closed subalgebra of $(\mathcal{H}(X),\tau_c)$ such that*

(a) $P^m_t f \in$ A *for every* $f \in$ A, $m \in \mathbb{N}$ *and* $t \in E$.

(b) $\xi_1, \ldots, \xi_n \in A$.

Suppose that X *is* A*-separated and metrically* A*-convex. Then* X *is* A*-convex.*

PROOF. Suppose that X is not A-convex. Then there are a compact set K in X, a polydisc $\Delta^n(a;r)$ in $\mathbb{C}^n$, a function $f \in$ A, and a sequence (x_j) of distinct points in X with the properties (c), (d) and (e) stated in Lemma 58.4. By Lemma 58.3 there exist θ with $0 < \theta < 1$ and $k_o \in \mathbb{N}$ such that for each j, $k \in \mathbb{N}$ with $j \geq k \geq k_o$ there is a nonzero polynomial $P_{jk}(z_1, \ldots, z_n, w)$ of degree at most j in each of the variables $z_1, \ldots, z_n$, and of degree at most k in w, such that

$$(58.1) \qquad |P_{jk}(\xi(x), f(x))| \leq \theta^{jk^{1/n}}$$

for all $x \in K$, and hence for all $x \in \hat{K}_A$. For $z \in \mathbb{C}^n$ and $w \in \mathbb{C}$ we can write

$$(58.2) \qquad P_{jk}(z,w) = \sum_{s=0}^{k} P_{jks}(z) w^s,$$

where $P_{jks}(z)$ is a polynomial of degree at most j in each of the variables $z_1, \ldots, z_n$. For $j \geq k \geq k_o$ set

$$S_{jk} = \{z \in \Delta^n(a;r) : \max_{s \leq k} |P_{jks}(z)| \leq \theta^{\frac{1}{2}jk^{1/n}}\}.$$

By Lemma 58.3 each of the polynomials P_{jk} can be taken to be normalized with respect to the variables $z_1 - a_1, \ldots, z_n - a_n, w$. Hence for each $j \geq k \geq k_o$ at least one of the polynomials P_{jks} is normalized with respect to the variables $z_1 - a_1, \ldots, z_n - a_n$. Since we may assume that $0 < r < 1$, an application of Lemma 58.2 yields a constant $c = c(n,r) > 0$ such that $\lambda(S_{jk}) \leq - c / \frac{1}{2} k^{1/n} \log \theta$. In particular $\lambda(S_{jk}) \to 0$ when $j, k \to \infty$. For $j \geq k \geq k_o$ set

$$T_{jk} = \Delta^n(a;r) \setminus S_{jk}.$$

Then we can find $\rho > 0$ and $k_1 \geq k_o$ such that $\lambda(T_{jk}) \geq \rho$ for every $j \geq k \geq k_1$. If we set

$$T = \bigcap_{k=k_1}^{\infty} \bigcup_{j=k}^{\infty} T_{jk_1}$$

then $\lambda(T) \geq \rho > 0$. Furthermore, if $z \in T$ then $z \in T_{jk_1}$ for infinitely many values of j, and therefore

(58.3) $$\max_{s \leq k_1} |P_{jk_1 s}(z)| > \theta^{\frac{1}{2} jk^{1/n}},$$

for infinitely many values of j. Fix $z \in T$ and let $w \in f(\xi^{-1}(z) \cap \hat{K}_A)$. For each $j \geq k_1$ and each $s \leq j$ set

(58.4) $$c_{js}(z) = \frac{P_{jk_1 s}(z)}{\max_{t \leq k_1} |P_{jk_1 t}(z)|},$$

and observe that

(58.5) $$\max_{s \leq k_1} |c_{js}(z)| = 1$$

for every $j \geq k_1$. It follows from (58.1), (58.2) (58.3) and (58.4) that

(58.6) $$\left| \sum_{s=0}^{k_1} c_{js}(z) w^s \right| \leq \theta^{\frac{1}{2} jk^{1/n}}$$

for infinitely many values of j. By choosing a sequence of values of j for which (58.6) is true, we can find for each $s \leq k_1$ a cluster point $c_s(z)$ of the sequence $(c_{js})_{j=k_1}^{\infty}$ such that

(58.7) $$\sum_{s=0}^{k_1} c_s(z) w^s = 0.$$

Furthermore, (58.5) guarantees the existence of at least one $s_1 \leq k_1$ such that $|c_{s_1}(z)| = 1$. Thus, for each $z \in T$ the set $f(\xi^{-1}(z) \cap \hat{K}_A)$ has at most k_1 points, the roots of the equation (58.7). Since $\lambda(T) > 0$, this contradicts the conclusion of Lemma 58.4. Thus X most be A-convex and the proof of the theorem is complete.

58.6. COROLLARY. *Let (X,ξ) be a Riemann domain over $\mathbb{C}^n$. Then the set*

$$\tilde{K} = \{h \in S_c(X) : |h(f)| \leq \sup_K |f| \quad \text{for every} \quad f \in \mathcal{H}(X)\}$$

is compact for each compact set $K \subset X$.

PROOF. Let $A = \{\hat{f} : f \in \mathcal{H}(X)\}$. Then $S_c(X)$ is A-separated and metrically A-convex, by Proposition 57.8, and hence $S_c(X)$ is A-convex, by Theorem 58.5. Since $\tilde{K} = (\varepsilon(K))\hat{}_A$ we conclude that $\tilde{K}$ is compact.

58.7. THEOREM. *Let (X,ξ) be a Riemann domain over $\mathbb{C}^n$. Then $S_c(X) = \hat{X} = E(X)$.*

PROOF. Since each $h \in S_c(X)$ belongs to $\tilde{K}$ for some compact set $K \subset X$, it suffices to show that $\tilde{K} \subset \hat{X}$ for each compact set $K \subset X$. Let K be a fixed compact subset of X and let $\mathcal{B}$ be the closure of $\mathcal{H}(X)$ in $C(K)$. Then $\mathcal{B}$ is a commutative Banach algebra whose spectrum is in one-to-one correspondence with $\tilde{K}$. If the spectrum $S(\mathcal{B})$ of $\mathcal{B}$ is endowed with the Gelfand topology then the canonical bijection $\tilde{K} \to S(\mathcal{B})$ is clearly continuous, and is therefore a homeomorphism, since $\tilde{K}$ is compact. Thus $S(\mathcal{B})$ can be canonically identified with $\tilde{K}$. Now, we can write

$$S(\mathcal{B}) = \tilde{K} = (\tilde{K} \cap \hat{X}) \cup (\tilde{K} \setminus \hat{X}),$$

and the sets $\tilde{K} \cap \hat{X}$ and $\tilde{K} \setminus \hat{X}$ are compact and disjoint. By Theorem 31.8 there exists $f \in \mathcal{B}$ such that $h(f) = 0$ for every $h \in \tilde{K} \cap \hat{X}$ and $h(f) = 1$ for every $h \in \tilde{K} \setminus \hat{X}$. Hence f is identically zero on K but $h(f) = 1$ for every $h \in \tilde{K} \setminus \hat{X}$. This is impossible, unless $\tilde{K} \setminus \hat{X}$ is empty. Thus $\tilde{K} \subset \hat{X}$ and the proof is complete.

58.8. COROLLARY. *Let* (X,ξ) *be a pseudoconvex Riemann domain over* $\mathbb{C}^n$. *Then each continuous complex homomorphism of* $(\mathcal{H}(X),\tau_c)$ *is a point evaluation.*

PROOF. X is a domain of holomorphy by Theorem 53.8. Hence $X = E(X)$ by Exercise 56.B. And hence $X = S_c(X)$ by Theorem 58.7.

58.9. THEOREM. *Let* E *be a separable Banach space with the bounded approximation property. Let* (X,ξ) *be a pseudoconvex Riemann domain over* E. *Then each continuous complex homomorphism of* $(\mathcal{H}(X),\tau_c)$ *is a point evalutation.*

PROOF. (a) We first assume that E has a monotone Schauder basis. Given $h \in S_c(X)$ we can find a compact set $K \subset X$ such that $|h(f)| \leq \sup_K |f|$ for every $f \in \mathcal{H}(X)$. Then the compact set K is contained in one of the open sets A_j given by Lemma 54.4. Then A_j is X-pseudoconvex, and hence $\mathcal{H}(X)$ is dense in $(\mathcal{H}(A_j),\tau_c)$ by Theorem 51.1. Hence h can be extended to a continuous complex homomorphism h_j of $(\mathcal{H}(A_j),\tau_c)$. Since $\tau_n(A_j) \subset X_n$ for every $n \geq j$, the mapping

$$f \in \mathcal{H}(X_n) \to h_j(f \circ \tau_n) \in \mathbb{C}$$

defines a continuous complex homomorphism of $(\mathcal{H}(X_n),\tau_c)$ for each $n \geq j$. Then, by Corollary 58.8, for each $n \geq j$ there exists a point $a_n \in X_n$ such that

$$h_j(f \circ \tau_n) = f(a_n)$$

for every $f \in \mathcal{H}(X_n)$, and in particular for each $f \in \mathcal{H}(X)$. If $f \in \mathcal{H}(X)$ then we can readily see that the sequence $(f \circ \tau_n)_{n=j}^{\infty}$ converges to f in $(\mathcal{H}(A_j),\tau_c)$, and whence it follows that

$$(58.8) \qquad h(f) = h_j(f) = \lim_{n\to\infty} h_j(f \circ \tau_n) = \lim_{n\to\infty} f(a_n).$$

In particular the sequence $(f(a_n))$ is bounded for each $f \in \mathcal{H}(X)$, and it follows from Theorem 54.11 that the sequence (a_n) has a cluster point a. Then it follows from (58.8) that $h(f)=f(a)$ for every $f \in \mathcal{H}(X)$.

(b) If E is a separable Banach space with the bounded approximation property then by Theorem 27.4 we may assume that E is a complemented subspace of a Banach space G which has a monotone Schauder basis. Thus there is a closed subspace F of G such that $G = E \times F$. Let $h \in S_c(X)$. If we identify X with $X \times \{0\}$ then the mapping

$$g \in \mathcal{H}(X \times F) \to h(g \mid X) \in \mathbb{C}$$

defines a continuous complex homomorphism of $(\mathcal{H}(X \times F), \tau_c)$. By (a) there is a point $(a,b) \in X \times F$ such that

$$h(g \mid X) = g(a,b)$$

for every $g \in \mathcal{H}(X \times F)$. Let $\pi : X \times F \to X$ denote the canonical projection. If $f \in \mathcal{H}(X)$ then $f \circ \pi \in \mathcal{H}(X \times F)$ and $f \circ \pi \mid X = f$. Whence it follows that $h(f) = f(a)$ for every $f \in \mathcal{H}(X)$, and the proof of the theorem is complete.

58.10. COROLLARY. *Let E be a separable Banach space with the bounded approximation property. Let (X, ξ) be a Riemann domain over E. Then $S_c(X) = \hat{X} = E(X)$.*

PROOF. By Proposition 57.8 $\hat{X}$ is pseudoconvex. Then by Corollary 57.6, Theorem 58.9 and Corollary 57.9 we get that $S_c(X) = S_c(\hat{X}) = \hat{X} = E(X)$.

Theorems 54.12 and 58.9 can be summarized as follows.

58.11. THEOREM. *Let E be a separable Banach space with the bounded approximation property. Let (X, ξ) be a pseudoconvex Riemann domain over E. Then the following conditions are equivalent:*

(a) *X is a domain of existence.*

(b) *X is a domain of holomorphy.*

(c) *X is holomorphically convex.*

(d) *X is metrically holomorphically convex.*

(e) *X is pseudoconvex.*

(f) *Each continuous complex homomorphism of $(\mathcal{H}(X),\tau_c)$ is a point evalutation.*

To conclude this section we prove that if X is a Riemann domain over $\mathbb{C}^n$ then each complex homomorphism of $\mathcal{H}(X)$ is continuous for the compact-open topology.

58.12. THEOREM. *Let (X,ξ) be a pseudoconvex Riemann domain over $\mathbb{C}^n$. Let $f_1,\dots,f_m \in \mathcal{H}(X)$ be m functions without common zeros. Then there are functions $g_1,\dots,g_m \in \mathcal{H}(X)$ such that $f_1(x)g_1(x) + \dots + f_m(x)g_m(x) = 1$ for every $x \in X$.*

PROOF. Without loss of generality we may assume that X is connected. Then X is hemicompact and $(\mathcal{H}(X),\tau_c)$ is a Fréchet algebra. By Corollary 58.8 each continuous complex holomorphism of $(\mathcal{H}(X),\tau_c)$ is a point evaluation. Hence the set $\{f_1,\dots,f_m\}$ is not contained in the kernel of any $h \in S_c(X)$. Then it follows from Proposition 32.13 that the point $(0,\dots,0)$ does not belong to the joint spectrum of $f_1,\dots,f_m$. Hence the functions $f_1,\dots,f_m$ generate the improper ideal.

58.13. THEOREM. *Let (X,ξ) be a pseudoconvex Riemann domain over $\mathbb{C}^n$. Then each complex homomorphism of $\mathcal{H}(X)$ is a point evaluation.*

PROOF. To prove this theorem we apply Proposition 33.6. Observe that condition (a) follows from Theorem 58.12, whereas condition (b) follows from Theorem 53.10.

58.14. COROLLARY. *Let (X,ξ) be a Riemann domain over $\mathbb{C}^n$. Then each complex homomorphism of $\mathcal{H}(X)$ is continuous for the compact-open* *topology.*

PROOF. If suffices to apply Theorem 58.13 to $S_c(X)$.

The preceding results can be summarized as follows.

58.15. THEOREM. *Let (X,ξ) be a Riemann domain over $\mathbb{C}^n$. Then the following conditions are equivalent:*

(a) *X is a domain of existence.*

(b) *X is a domain of holomorphy.*

(c) *X is holomorphically convex.*

(d) *X is metrically holomorphically convex.*

(e) *X is pseudoconvex.*

(f) *Each continuous complex homomorphism of $(\mathcal{H}(X),\tau_c)$ is a point evaluation.*

(g) *Each complex homomorphism of $\mathcal{H}(X)$ is a point evaluation.*

(g) *X is holomorphically separated, and each finitely generated, proper ideal of $\mathcal{H}(X)$ has a common zero.*

NOTES AND COMMENTS

Theorem 56.4 is due to A. Hirschowitz [3].

Theorem 57.2 and 57.3, and Proposition 57.5, are due to H. Alexander [1], who adapted to the case of Banach spaces a construction of H. Rossi [1] in the finite dimensional case.

Theorem 58.5 is due to E. Bishop [2]. Theorem 58.7 has been taken from the book of R. Gunning and H. Rossi [1]. Theorem 58.9 is due to M. Schottenloher [7]. Theorem 58.12 was proved first by H. Cartan [1] for domains of holomorphy in $\mathbb{C}^n$ using sheaf theory.

BIBLIOGRAPHY

H. ALEXANDER

[1] Analytic functions on Banach spaces. Ph. D. thesis. University of California, Berkeley, 1968.

R. ARENS

[1] Dense inverse limit rings. Michigan Math. J. 5 (1958), 169 - 182.

R. ARENS and A. CALDERON

[1] Analytic functions of several Banach algebra elements. Ann. of Math. (2) 62 (1955), 204 - 216.

R. ARON and M. SCHOTTENLOHER

[1] Compact holomorphic mappings on Banach spaces and the approximation property. J. Funct. Anal. 21 (1976), 7 - 30.

S. BANACH

[1] Théorie des Opérations Linéaires. Warsaw, 1932. Republished by Chelsea, New York.

E. BISHOP

[1] Mappings of partially analytic spaces. Amer. J. Math. 83 (1961), 209 - 242.

[2] Holomorphic completion, analytic continuation and the interpolation of seminorms. Ann. of Math. (2) 78 (1963), 468 - 500.

J. BOCHNAK and J. SICIAK

[1] Polynomials and multilinear mappings in topological vector spaces. Studia Math. 39 (1971), 59 - 76.

[2] Analytic functions in topological vector spaces. Studia Math. 39 (1971), 77 - 112.

N. BOURBAKI

[1] Eléments de Mathématique. Topologie Générale, chapitres 1 à 4, nouvelle édition. Hermann, Paris, 1971.

H. BREMERMANN

[1] Über die Äquivalenz der pseudokonvexen Gebiete und der Holomorphiegebiete im Raum von n komplexen Veränderlichen. Math. Ann. 128 (1954), 63 - 91.

[2] Holomorphic functionals and complex convexity in Banach spaces. Pacific J. Math. 7 (1957), 811 - 831.

H. CARTAN

[1] Idéaux et modules de fonctions analytiques de variables complexes. Bull. Soc. Math. France 78 (1950), 29 - 64.

[2] Calcul Differentiel. Hermann, París, 1967.

[3] Formes Différentielles. Hermann, París, 1967.

H. CARTAN and P. THULLEN

[1] Zur Theorie der Singularitäten der Funkkionen mehrerer komplexen Veränderlichen. Regularitäts- und Konvergenzbereiche. Math. Ann. 106 (1932), 617 - 647.

D. CLAYTON

[1] A reduction of the continuous homomorphism problem for F-algebras. Rocky Mountain J. Math. 5 (1975), 337 - 344.

G. COEURE

[1] Analytic Functions and Manifolds in Infinite Dimensional Spaces. North-Holland Mathematics Studies, vol.11. North-Holland, Amsterdam, 1974.

J. F. COLOMBEAU

[1] Differential Calculus and Holomorphy. North-Holland Mathematics Studies, vol. 64. North-Holland, Amsterdam, 1982.

I. CRAW

[1] A condition equivalent to me continuity of characters on a Fréchet algebra. Proc. London Math. Soc. (3) 22 (1971), 452 - 464.

J. DIESTEL

[1] Sequences and Series in Banach Spaces. Graduate Texts in Mathematics, Vol. 92. Springer, New York, 1984.

J. DIEUDONNE

[1] Foundations of Modern Analysis. Academic Press, New York, 1960.

S. DINEEN

[1] The Cartan-Thullen theorem for Banach spaces. Ann. Scuola Norm. Sup. Pisa (3) 24 (1970), 667 - 676.

[2] Bounding subsets of a Banach space. Math. Ann. 192 (1971), 61 - 70.

[3] Runge domains in Banach spaces. Proc. Roy.Irish Acad. 71 (1971), 85 - 89.

[4] Unbounded holomorphic functions on a Banach space. J. London Math. Soc. (2) 4 (1972), 461 - 465.

[5] Complex Analysis in Locally Convex Spaces. North-Holland Mathematics Studies, vol. 57. North-Holland, Amsterdam, 1981.

S. DINEEN and A. HIRSCHOWITZ

[1] Sur le théorème de Levi banachique C.R. Acad. Sci.Paris 272 (1971), 1245 - 1247.

P. DIXON and D. FREMLIN

[1] A remark concerning multiplicative functionals on LMC algebras. J. London Math. Soc. (2) 5 (1972), 231 - 232.

P. DOLBEAULT

[1] Formes différentielles et cohomologie sur une varieté analytique complexe I. Ann. of Math. (2) 64 (1956), 83 - 130.

L. EHRENPREIS

[1] A new proof and an extension of Hartogs' theorem. Bull. Amer. Math. Soc. 67 (1961), 507 - 509.

P. ENFLO

[1] A counterexample to the approximation property in Banach spaces. Acta Math. 130 (1973), 309 - 317.

M. ESTEVEZ and C. HERVES

[1] Application des mesures gaussiennes complexes aux fonctions plurisousharmoniques et analytiques. Bull. Sci.Math. (2) 105 (1981), 73 - 83.

J.P. FERRIER and N. SIBONY

[1] Approximation pondérée sur une sous-variété totalement réelle de $\mathbb{C}^n$. Ann. Inst. Fourier Grenoble 26,2 (1976), 101 - 115.

T. FRANZONI and E. VESENTINI

[1] Holomorphic Maps and Invariant Distances. North-Holland Mathematics Studies, vol. 40. North-Holland, Amsterdam, 1980.

I. GELFAND, D. RAIKOV and G. SHILOV

[1] Commutative Normed Rings. Chelsea, New York, 1964.

A. GROTHENDIECK

[1] Produits Tensoriels Topologiques et Espaces Nucléaires. Memoirs of the American Mathematical Society, number 16. American Mathematical Society, Providence, Rhode Island, 1955.

L. GRUMAN

[1] The Levi problem in certain infinite dimensional vector spaces. Illinois J. Math. 18 (1974), 20 - 26.

L. GRUMAN and C. KISELMAN

[1] Le problème de Levi dans les espaces de Banach à base. C.R. Acad. Sci. Paris 274 (1972), 821 - 824.

A. GUICHARDET

[1] Special Topics in Topological Algebras. Gordon and Breach, New York, 1968.

R. GUNNING and H. ROSSI

[1] Analytic Functions of Several Complex Variables. Prentice-Hall, Englewood Cliffs, New Jersey, 1965.

F. HARTOGS

[1] Zur Theorie der analytischen Funktionen meherer unabhängiger Veränderlichen, insbesondere über die Darstellung derselben durch Reinen, welche nach Potenzen einer Veränderlichen fortschreiten. Math. Ann. 62 (1906), 1 - 88.

M. HERVE

[1] Analytic and Plurisubharmonic Functions in Finite and Infinite Dimensional Spaces. Lecture Notes in Mathematics, vol. 198. Springer, Berlin, 1971.

Y. HERVIER

[1] Sur le problème de Levi pour les espaces étalés banachiques. C.R. Acad. Sci. París 275 (1972), 821 - 824.

E. HILLE and R. PHILLIPS

[1] Functional Analysis and Semi-groups, revised edition. American Mathematical Society Colloquium Publications, vol.31. American Mathematical Society, Providence, Rhode Island, 1957.

A. HIRSCHOWITZ

[1] Sur le non-plongement des varietés analytiques banachiques réelles. C.R. Acad. Sci. París 269 (1969), 844 - 846.

[2] Bornologie des espaces de fonctions analytiques en dimension infinie. Séminaire Pierre Lelong 1969/70, pp. 21- 33. Lecture Notes in Mathematics, vol. 205. Springer, Berlin, 1971.

[3] Prolongement analytique en dimension infinie. Ann. Inst. Forier Grenoble 22, 2 (1972), 255 - 292.

L. HÖRMANDER

[1] Linear Partial Differential Operators. Die Grundlehren der mathematischen wissenschaften, Band 116. Springer, Berlin, 1963.

[2] L^2 estimates and existence theorems for the $\bar{\partial}$ operator.

Acta Math. 113 (1965), 89 - 152.

[3] An Introduction to Complex Analysis in Several Variables, second edition. North-Holland Mathematical Library, vol. 7. North-Holland, Amsterdam, 1973.

J. M. ISIDRO

[1] Characterization of the spectrum of some topological algebras of holomorphic functions. Advances in Holomorphy, pp. 407 - 416. North-Holland Mathematics Studies, vol. 34. North-Holland, Amsterdam, 1979.

B. JOSEFSON

[1] A counterexample in the Levi problem. Proceedings on Infinite Dimensional Holomorphy, pp. 168 - 177. Lecture Notes in Mathematics, vol. 364. Springer, Berlin, 1974.

[2] Weak sequential convergence in the dual of a Banach space does not imply norm convergence. Ark. Mat. 13 (1975), 79 - 89.

P. LELONG

[1] Les fonctions plurisousharmoniques. Ann. Sci. Ecole Norm Sup. (3) 62 (1945), 301 - 338.

[2] Fonctions plurisousharmoniques dans les espaces vectoriels topologiques. Séminaire Pierre Lelong 1967/68, pp. 167-190. Lecture Notes in Mathematics, vol. 71. Springer, Berlin, 1968.

E. LEVI

[1] Sulle ipersuperfici dello spazio a 4 dimensioni che possono essere frontiera del campo di esistenza di una funzioni analitica di due variabili complesse. Ann. Mat. Pura Appl. (3) 18 (1911), 69 - 79.

E. LIGOCKA

[1] A local factorization of analytic functions and its applications. Studia Math. 47 (1973), 239 - 252.

[2] Levi forms, differential forms of type (0,1) and pseudoconvexity in Banach spaces. Ann. Polon. Math. 33 (1976), 63 - 69.

J. LINDENSTRAUSS and L. TZAFRIRI

[1] Classical Banach Spaces I. Ergebnisse der Mathematik and ihrer Grenzgebiete, Band 92. Springer, Berlin, 1977.

M. MATOS

[1] A characterization of holomorphic mappings in Banach spaces with a Schauder basis. Atas do 19º Seminário Brasileiro de Análise. Sociedade Brasileira de Matemática, São Paulo, 1984.

C. MATYSZCZYK

[1] Approximation of analytic operators by polynomials in complex B_0-spaces with bounded approximation property. Bull. Acad. Polon. Sci. 20 (1972), 833-836.

[2] Approximation of analytic and continuous mappings by polynomials in Fréchet spaces. Studia Math. 60 (1977), 223-238.

E. MICHAEL

[1] Locally Multiplicatively - Convex Topological algebras. Memoirs of the American Mathematical Society, number 11. American Mathematical Society, Providence, Rhode Island, 1952.

J. MUJICA

[1] Ideals of holomorphic functions on Fréchet spaces. Advances in Holomorphy, pp. 563-576. North-Holland Mathematics Studies, vol. 34. North-Holland, Amsterdam, 1979.

[2] Complex homomorphisms of the algebras of holomorphic functions on Fréchet spaces. Math. Ann. 241 (1979), 73-82.

[3] The Oka-Weil theorem in locally convex spaces with the approximation property. Séminaire Paul Krée 1977/78, exposé nº 3. Institut Henri Poincaré, París, 1979.

[4] Holomorphic approximation in Fréchet spaces with basis. J. London Math. Soc. (2) 29 (1984), 113-126.

[5] Holomorphic approximation in infinite dimensional Riemann domains. Studia Math. (to appear).

L. NACHBIN

[1] Lectures on the Theory of Distributions. Textos de Matemática, Nº 15. Universidade do Recife, Recife, 1964.

[2] Topology on Spaces of Holomorphic Mappings. Ergebnisse der Mathematik und ihrer Grenzgebiete, Band 47. Springer, New York, 1969.

[3] Holomorphic Functions, Domains of Holomorphy and Local Properties. North-Holland Mathematics Studies, vol. 1. North-Holland, Amsterdam, 1970.

[4] Introdução à Análise Funcional: Espaços de Banach e Cálculo Diferencial. Série de Matemática, monografia nº 17. Organization of American States, Washington, D.C., 1976.

R. NARASIMHAN

[1] Holomorphically complete complex spaces. Amer. J. Math. 82 (1961), 917-934.

[2] Several Complex Variables. The University of Chicago Press, Chicago, 1971.

S. M. NIKOL'SKII

[1] Approximation of Functions of Several Variables and Imbedding Theorems. Die Grundlehren der mathematischen

Wissenschaften, Band 205. Springer, Berlin, 1975.

A. NISSENZWEIG

[1] w* sequential convergence. Israel J. Math. 22 (1975), 266 - 272.

F. NORGUET

[1] Sur les domaines d'holomorphie des fonctions uniformes de plusieurs variables complexes. Bull. Soc. Math. France 82 (1954), 137 - 159.

P. NOVERRAZ

[1] Fonctions plurisousharmoniques et analytiques dans les espaces vectoriels topologiques complexes. Ann. Inst. Fourier Grenoble 19,2 (1969), 419 - 493.

[2] Sur la pseudo-convexité et la convexité polynomiale en dimension infinie. Ann. Inst. Fourier Grenoble 23,1 (1973), 113 - 134.

[3] Pseudo-Convexité, Convexité Polynomiale et Domaines d'Holomorphie en Dimension Infinie. North-Holland Mathematics Studies, vol. 3. North-Holland, Amsterdam, 1973.

[4] Approximation of holomorphic or plurisubharmonic functions in certain Banach spaces. Proceedings on Infinite Dimensional Holomorphy, pp. 178 - 185. Lecture Notes in Mathematics, vol. 364. Springer, Berlin, 1974.

K. OKA

[1] Domaines convexes par rapport aux fonctions rationelles. J. Sci. Hiroshima Univ. 6 (1936), 245 - 255.

[2] Domaines d'holomorphie. J. Sci. Hiroshima Univ. 7 (1937), 115 - 130.

[3] Domaines pseudoconvexos. Tohoku Math. J. 49 (1942), 15 - 52.

[4] Domaines finis sans point critique intérieur. Japan. J. Math. 27 (1953), 97 - 155.

R. OVSEPIAN and A. PELCZYNSKI

[1] The existence in every separable Banach space of a fundamental total and bounded biorthogonal sequence and related constructions of uniformly bounded orthonormal systems in L^2. Studia Math. 54 (1975), 149 - 159.

A. PELCZYNSKI

[1] Any separable Banach space with the bounded approximation property is a complemented subspace of a Banach space with a basis. Studia Math. 40 (1971), 239 - 243.

B. RIESZ and B. SZ.-NAGY

[1] Leçons d' Analyse Fonctionnelle, sixième édition.

Gauthier-Villars, Paris, 1972.

H. ROSSI

[1] On envelops of holomorphy. Comm. Pure Appl. Math. 16 (1963), 9 - 17.

J. SCHAUDER

[1] Zur Theorie stetiger Abbildungen in Funktionalräumen. Math. Z. 26 (1927), 47 - 65.

[2] Eine Eigenschaft der Haarschen Orthogonalsystems. Math. Z. 28 (1928), 317 - 320.

M. SCHOTTENLOHER

[1] Über analytische Fortsetzung in Banachräumen. Math. Ann. 199 (1972), 313 - 336.

[2] Bounding sets in Banach spaces and regular classes of analytic functions. Functional Analysis and Applications, pp. 109 - 122. Lecture Notes in Mathematics, vol. 384. Springer, Berlin, 1974.

[3] Das Leviproblem in unendlichdimensionalen Räumen mit Schauderzerlegung. Habilitationsschrift, Universität München, 1974.

[4] The Levi problem for domains spread over locally convex spaces with a finite dimensional Schauder decomposition. Ann. Inst. Fourier Grenoble 26,4 (1976), 207 - 237.

[5] Polynomial approximation on compact sets. Infinite Dimensional Holomorphy and Applications, pp. 379 - 391. North-Holland Mathematics Studies, vol. 12. North-Holland, Amsterdam, 1977.

[6] Michael problem and algebras of holomorphic functions. Arch. Math. (Basel) 37 (1981), 241 - 247.

[7] Spectrum and envelope of holomorphy for infinite dimensional Riemann domains. Math. Ann. 263 (1983), 213 - 219.

L. SCHWARTZ

[1] Théorie des Distributions, nouvelle édition. Hermann, Paris, 1966.

G. SHILOV

[1] On the decomposition of a commutative normed ring into a direct sum of ideals. Mat. Sb. 32 (1953), 353 - 364 (Russian). Amer. Math. Soc. Transl. (2) 1 (1955), 37 - 48 (English translation).

A. TAYLOR

[1] On the properties of analytic functions in abstract spaces. Math. Ann. 115 (1938), 466 - 484.

L. WAELBROECK

[1] Le calcul symbolique dans les algèbres commutatives. J. Math. Pures Appl. 33 (1954), 147 - 186.

A. WEIL

[1] L' integrale de Cauchy et les fonctions de plusiers variables. Math. Ann. 111 (1935), 178 - 182.

S. WILLARD

[1] General Topology. Addison-Wesley, Reading, Massachusetts, 1970.

W. ZELAZKO

[1] Banach Algebras. Elsevier Publishing Company, Amsterdam, and Polish Scientific Publishers, Warsaw, 1973.

M. ZORN

[1] Characterization of analytic functions in Banach spaces. Ann. of Math. (2) 46 (1945), 585 - 593.

[2] Gateaux differentiability and essential boundedness. Duke Math. J. 12 (1945), 579 - 583.

INDEX